T0139909

Advances in Intelligent Systems and Computing

Volume 1344

The series "Advances in Intelligent Systems and Computing" contains publications on theory, applications, and design methods of Intelligent Systems and Intelligent Computing. Virtually all disciplines such as engineering, natural sciences, computer and information science, ICT, economics, business, e-commerce, environment, healthcare, life science are covered. The list of topics spans all the areas of modern intelligent systems and computing such as: computational intelligence, soft computing including neural networks, fuzzy systems, evolutionary computing and the fusion of these paradigms, social intelligence, ambient intelligence, computational neuroscience, artificial life, virtual worlds and society, cognitive science and systems, Perception and Vision, DNA and immune based systems, self-organizing and adaptive systems, e-Learning and teaching, human-centered and human-centric computing, recommender systems, intelligent control, robotics and mechatronics including human-machine teaming, knowledge-based paradigms, learning paradigms, machine ethics, intelligent data analysis, knowledge management, intelligent agents, intelligent decision making and support, intelligent network security, trust management, interactive entertainment, Web intelligence and multimedia.

The publications within "Advances in Intelligent Systems and Computing" are primarily proceedings of important conferences, symposia and congresses. They cover significant recent developments in the field, both of a foundational and applicable character. An important characteristic feature of the series is the short publication time and world-wide distribution. This permits a rapid and broad dissemination of research results.

Indexed by DBLP, EI Compendex, INSPEC, WTI Frankfurt eG, zbMATH, Japanese Science and Technology Agency (JST).

All books published in the series are submitted for consideration in Web of Science.

More information about this series at http://www.springer.com/series/11156

Noreddine Gherabi · Janusz Kacprzyk
Editors

Intelligent Systems in Big Data, Semantic Web and Machine Learning

Editors
Noreddine Gherabi
ENSA
Sultan Moulay Slimane University
Khouribga, Morocco

Janusz Kacprzyk
Systems Research Institute
Polish Academy of Sciences
Warsaw, Poland

ISSN 2194-5357 ISSN 2194-5365 (electronic)
Advances in Intelligent Systems and Computing
ISBN 978-3-030-72590-7 ISBN 978-3-030-72588-4 (eBook)
https://doi.org/10.1007/978-3-030-72588-4

This Springer imprint is published by the registered company Springer Nature Switzerland AG
The registered company address is: Gewerbestrasse 11, 6330 Cham, Switzerland

Preface

Over the last decade or more there has been an explosion of new and powerful computing and information technology. This explosion has been followed by the emergence of vast amounts of data in a variety of fields such as medicine, biology, finance, and marketing, to just mention a few. The challenge of understanding this data has led to the development of new tools in the field of artificial intelligence and spawned new fields such as data mining, machine learning, and the Semantic Web, all in the broadly perceived context of Big Data.

This book describes important methodologies in the above fields of science and technology. Many approaches are more conceptual and fundamental and are based on mathematical models and formal analyses. Many ideas are given, with generous use of graphics and visualization. Moreover, many engineering solutions, developments, and implementations have been shown.

The book's coverage is broad, ranging from machine learning to the use of data, notably Big Data, on the Semantic Web. Many original topics and issues are covered including various aspects of machine learning, deep learning, Big Data, data processing in medicine, similarity processing in ontologies, semantic image analysis, machine learning techniques for cloud security, artificial intelligence techniques for detecting COVID-19, the Internet of Things, etc.

The book may be a valuable source of new information for graduate and Ph.D. students, researchers, and practitioners in data analysis, data sciences, Semantic Web, Big Data, machine learning, deep learning, computer engineering, and related disciplines.

The contents, issues considered, and their coverage of the consecutive papers will now be briefly summarized to facilitate the use of the volume for the interested user.

Mourade Azrour, Jamal Mabrouki, Yousef Farhaoui, and Azidine Guezzaz ("Experimental evaluation of proposed algorithm for identifying abnormal Messages in SIP network") are concerned with the Abstract Session Initiation Protocol (SIP) which is meant as an application layer protocol developed to initiate, modify, and stop a multi-media session through an Internet network. SIP becomes more and more popular and is used in many applications exemplified by the

Telephony over the Internet (ToIP). However, SIP has some security problems which are addressed in the papers. The authors concentrate on the problem of detection of the SIP malformed messages attacks and propose an algorithm that can determine if the received message is normal or abnormal. The effectiveness and efficiency of the algorithm are confirmed by simulation experiments that show the attack can be detected in minimum time.

Abdelhadi Daoui, Noreddine Gherabi, and Abderrahim Marzouk ("Smart tourism recommender system using semantic matching") deal with a problem that is implied by a widespread diffusion of information technologies and a rapid growth of the use of the social network, and hence an explosion of data volumes in the Web. This calls for automatic filtering of data to customize the data to be displayed to the users according to their needs. Recommendation systems have been designed to facilitate and personalize data access to the users, reflecting their specific needs, and among important application areas, one can mention the broadly perceived e-commerce and tourism. In this paper, the authors focus on designing an intelligent recommendation system for the tourism field, and the system is used to analyze images stored in users' mobile devices to identify the fields of interest of these users yielded as keywords. These fields of interest provide then a basis for the recommendation of tourist services and places to visit suitable to each user according to his or her preferences.

Islam A. Heggo and Nashwa Abdelbaki ("Data-driven information filtering framework for dynamically hybrid job recommendation") are concerned with the recommender systems which become one of the core models in practically all modern Web applications. They are obviously found in numerous applications like video recommendations by Youtube and product recommendations by Amazon. Similarly, the recommendation systems are in great demand for the challenging e-recruitment domain. In the e-recruitment domain, it is mandatory to build an intelligent system that digs deeper into thousands of jobs to eventually filter out relevant jobs for each job-seeker. We explain what is needed to build such a sophisticated and practical hybrid recommendation engine. This engine is employed in practice in a dynamic e-recruitment portal and its efficiency is proven in the real-world market. Moreover, the applied recommendation system approach is used in several other industries to help the field of e-recruitment. The project aims at recommending relevant jobs for job-seekers to solve the problem of disrupting job-seekers with irrelevant jobs to improve the users' satisfaction and loyalty.

M. Oujaoura and B. Minaoui ("Semantic image analysis for automatic image annotation") are concerned with the semantic image analysis which is a process aimed at automatically extracting an accurate semantic concept from image visual content. In the paper, light is shed on a study conducted to boost the efficiency of semantic image analysis for improving automatic image annotation. This research study explores different aspects of semantic analysis to overcome the semantic gap problem. In the first part of this study, the authors evaluate the impact of how different low-level features are related to the image annotation performance. In the second part, an advantage of using two complementary classifiers for object classification is dealt with. The final part is devoted to the assessment of the usefulness of regrouping adjacent regions of segmented images for the discrimination of objects.

Noreddine Gherabi ("New method for data replication and reconstruction in distributed databases") dealt with data replication between servers which is a critical and important step for sharing data between multiple remote sites in a distributed database. Data replication is used to replicate data to remote sites, improve system performance with respect to data access, and reduce the connection time to the database. A new algorithm for dynamic data replication in distributed databases is proposed. The algorithm is adaptive to the changes in the schema objects which then replicate these objects in the central scheme. By using the strengths of communication between mobile agents, such a communication can facilitate the extraction, reconstruction, and replication of schemes from remote sites to the central site. It is shown that the algorithm proposed can be combined with the mechanisms of the distributed database management systems.

Azidine Guezzaz, Ahmed Asimi, Younes Asimi, Mourade Azrour, and Said Benkiranc ("A distributed intrusion detection approach based on machine learning technique for a cloud security") consider problems related to a rapid evolution of computer networks in which a large amount of data generated by systems require an accurate monitoring and high security related to the data and resources calling for the security of personal information. The cloud computing technology makes it possible for the users and organizations to store and process data in a privately owned cloud or on a third-party server and makes data access mechanisms more relevant. It is useful to monitor and protect data housed on the cloud by detecting and stopping intrusive activities and attacks. For this reason, many security tools are available to monitor transactions in the cloud and have been used to prevent networks from intrusions. It is also very useful to choose a cloud provider that considers security and protection to be an essential goal. The intrusion detection and prevention is a recent technology used to secure data in the cloud computing environments. In this paper, there is proposed and designed a new distributed intrusion detection system as a service to improve the cloud security. All solutions that are suggested to validate various parts of the proposed system for detecting the intrusions and defending against attacks in the cloud are presented.

Sajida Mhammedi, Hakim El Massari, and Noreddine Gherabi ("Cb2Onto: OWL ontology learning approach from Couchbase") are concerned with some problems related to big data which are collections of large datasets of both unstructured and structured data characterized by volume, variety, and velocity, implying the heterogeneity and complexity of data which exceeds the capacity of traditional systems to cope with them. Therefore, it is crucial to have a unified conceptual view of this as well as an efficient representation of knowledge for big data management. The ontology may be a tool to understand and provide the meaning to process big data. However, the construction of an ontology by hand is an incredibly challenging and error-prone process. The learning of an ontology from an existing resource may be a reasonable alternative, and that is why in this work an approach is proposed to learn the OWL ontology from data in the Couch base database by the application of six mapping rules using the Ontop reasoner to evaluate the consistency of the extracted ontology.

Bahaa Eddine Elbaghazaoui, Mohamed Amnai, and Abdellatif Semmouri ("Data profiling over Big Data Area—A survey of big data profiling: state-of-the-art, use cases and challenges") deal with an important problem the essence of which is that before using any application we need to understand the dataset at hand and its metadata. The process of discovering metadata, known as data profiling, focuses on the examination of the datasets and the collection of metadata such as statistics or informative summaries about that data. The authors discuss the importance of data profiling and shed light on the area of data profiling in the big data context. In addition, they present data profiling cases and present state-of-the-art of data profiling systems and techniques. Finally, directions and challenges for the future research in the area of data profiling are pointed out.

Atimad Harir, Said Melliani, and L. Saadia Chadli ("Generalization of the fuzzy conformable differentiability with application to fuzzy fractional differential equations") study the concept of generalized conformable differentiability for fuzzy valued functions. The existence of the solutions of the fuzzy fractional differential equations involving the generalized conformable differentiability is studied. Moreover, some applications to ordinary fuzzy fractional differential equations with fuzzy initial values are shown.

Said El Kafhali and Mohamed Lazaar ("Artificial intelligence for healthcare: roles, challenges, and applications") are concerned with opportunities that may emerge due to the use of an intelligent technology-based healthcare for improving the quality and effectiveness of medical care which imply an increase of the patients' wellness. All over the world, the rising healthcare costs and the appearance of many illnesses imply the necessity to focus on the people-centered environment, not just on the hospital-based ones. The future of healthcare may change completely using Artificial Intelligence (AI) that can profoundly change how we prevent, diagnose, and cure illnesses. It is basically a decision-making machine that can exponentially increase the efficiency of healthcare organizations. Recently, many publications have appeared that use the AI technology to monitor and control the spread of the COVID-19 (Coronavirus) pandemic. There are however many obstacles and barriers in the use of artificial intelligence in healthcare exemplified by data integration that is complex, trust issues, time, energy limitations, etc. This chapter provides a survey of the artificial intelligence-driven healthcare and identifies the proposed models. It identifies the existing approaches to the design of models for the artificial intelligence-driven healthcare showing the roles, challenges, applications, and future opportunities of artificial intelligence for healthcare.

Jamal Mabrouki, Mourade Azrour, Amina Boubekraoui, and Souad El Hajjaji ("Intelligent system for the protection of people") are concerned with the protection of people against assault, capturing, kidnapping, dealing with drugs, rape, etc. that are common all over the world. The purpose of this paper is to propose a relatively small and simple device (a "gadget") which, when actuated, gives a protection solution that can help individuals and their families in danger to get protected against their aggressors. The solution comprises of a switch, a Raspberry Pi, an

Arduino UNO, a Raspberry Pi camera module, a Global Positioning System (GPS) module, a heartbeat sensor, a bell, and a vibration engine. The whole design and arrangement are centered around the Arduino and Raspberry Pi. The sensors incorporated in Arduino read information persistently and if there arises an occurrence of a deviation from a predefined limit estimation of the sensor, the data (GPS directions and photograph/video) is sent to the applicable specialists via Raspberry Pi. This solution sidesteps manual impedance at a crucial time that can represent the deciding moment of the situation when the casualty is in an unfortunate circumstance. It is simple, cheap, convenient, and easy to understand.

Sajida Mhammedi and Noreddine Gherabi ("Heterogeneous integration of Big Data using Semantic Web technologies") propose to use the Semantic Web technologies which offer means to deal with large data volumes from a semantic point of view and a meaningful content of unstructured data so that they open new benefits for Big Data research and applications. Big Data is a new tendency related to a wide availability of datasets including structured, semi-structured, and unstructured data collected from different sources. Their integration faces many problems as it is difficult to process this information using traditional database and software tools and techniques. This article attempts to give a comparative study of methods for the integration of Big Data with Semantic Web, describing how Semantic Web makes Big Data smarter, analyzing the difficulties and possibilities of such an integration, and finally summarizing future challenges and solutions.

Youness Madani, Mohammed Erritali, and Belaid Bouikhalene ("Fake news detection approach using parallel predictive models and Spark to avoid misinformation") deal with issues related to a wide proliferation of the Internet and social media which suffer from false messages and false news since the outbreak of the COVID-19 pandemic. The intention of them is often to mislead the readers and/or make them believe in something that is not true. All that increases the need for automatic methods that can detect fake news on the social media. The authors propose a classification model, based on machine learning and deep learning, to classify COVID-19 tweets into two classes using the Apache Spark and the Python API Tweepy. The idea proposed uses the features of tweets to detect the fake news. Experimental results show that the random forest algorithm gives the best results with an accuracy equal to 79% and that the sentiment of tweets plays an important role in the detection of the fake news. By applying the proposed model on our COVID-19 dataset, 67% of tweets are classified as real while 37% are classified as fake.

My Abdelouahed Sabri, Youssef Filali, Fathi Soumia, and Abdellah Aarab ("Detection, analysis and classification of skin lesions: Challenges and opportunities") consider issues related to skin cancer which is one of the deadliest cancers in the world. Computer-Aided Diagnosis (CAD) systems are a widely used solution for the detection and the classification of skin cancers. Such systems can significantly reduce the physicians' effort and time with a high classification accuracy. Several challenges are encountered while setting up such systems. In the literature, two categories of approaches to such systems are proposed which depend directly on the size of the skin lesion images datasets available. Thus, for small datasets, machine learning-based approaches are the most commonly used, starting with the identification of a lesion region. Usually, the analyzed areas contain a lot of noise. A good lesion identification can make it possible to extract relevant features and in the end can yield an excellent classification accuracy. For large datasets, deep learning-based approaches are the most widely used and the most efficient. Very promising hybrid ideas have recently been proposed combining the power of machine learning and deep learning approaches. This chapter presents and analyzes various challenges and opportunities encountered in each skin lesion classification implementations and proposes solutions.

M. El Abassi, Med. Amnai, and A. Choukri ("Deduplication over Big Data integration") propose a new method to solve the deduplication problem using the Scala over Spark framework. The concept of data quality is very important for good data governance in order to improve the interaction between the different collaborators of one or more organizations concerned, and the existence of duplicate or similar data may imply serious data quality concerns. A panorama of methods for the calculation of distance/similarity between the data as well as algorithms for the elimination of similar data is presented and compared.

Alla Hajar, Youssef Balouki, and Moumoun Lahcen ("Flight arrival delay prediction using supervised machine learning algorithms") are concerned with air transport that has rapidly grown over the last years (in general, without taking into account the recent fall due to COVID-19), and the growth of which over the next two decades is expected to double. This will cause the density of traffic to considerably increase which will result in traffic delays that are the most relevant performance indicators in the air transport systems as they hurt passengers, airports, and airlines and imply losses. Pilots, air traffic controllers, and other aviation personnel are questioned in this study, and a survey is performed to identify the importance of flight delay reductions. Flight delay prediction studies have been conducted in different ways. The approach proposed in this work is based on machine learning. The model presented is able to predict whether a scheduled flight will be on time or delayed. Relevant and filtered features are used that, to the best of our knowledge, have not been in full adopted in the previous studies. Holidays, high seasons, specific day of the week, and the importance of the particular airport are used to add new features to enhance the accuracy of the prediction system. The resulting model is deployed and used as a flight delay prediction tool to inform the airport and airline personnel about flight delays in advance to avoid losses and terminal congestions.

Nouhaila Bensalah, Habib Ayad, Abdellah Adib, and Abdelhamid Ibn El Farouk ("Arabic machine translation based on the combination of word embedding techniques") discuss some issues related to the automatic machine translation from one language to another which is gaining more and more importance with an increased volume of user-generated content on the Web, a wide availability of textual information and its gigantic quantity. Hence, it is becoming increasingly common to adopt automated analytic tools for this purpose, notably derived from Machine Learning (ML) to represent such kind of information. The authors propose a new method called the Enhanced Word Vectors (EWVs) generated using the Word2vec and FastText models. These EWVs are then used for training and testing a new Deep Learning (DL) architecture based on the Convolutional Neural Networks (CNNs) and Recurrent Neural Networks (RNNs). Moreover, a special preprocessing of Arabic sentences is carried out. The performance of the proposed scheme is validated and compared with the Word2vec and FastText using a widely employed dataset. From the experimental results, it is found that in most of the cases, the proposed approach attains better results as compared to the Word2vec and FastText models alone.

Mourade Azrour, Jamal Mabrouki, Yousef Farhaoui, and Azidine Guezzaz ("Security analysis of Nikooghadam et al.'s authentication protocol for Cloud-IoT") are concerned with aspects of the IoT which has grown rapidly and has changed the technology and everyday life. It has entered virtually all domains including agriculture, industry, environment, etc. Furthermore, as a natural question one can ask here is how we can secure private data? For this purpose, many authentication protocols have been proposed for controlling the access to private data in the IoT environment. Recently Nikooghadam et al. presented a lightweight authentication and session key agreement protocol for the Cloud-IoT. Afterward, they have proved that their proposed protocol can stand against several attacks. Nevertheless, in this paper, we show that Nikooghadam et al.'s protocol is vulnerable to password guessing attacks.

Jamal Mabrouki, Maria Benbouzid, Driss Dhiba, and Souad El Hajjaji ("Internet of Things for monitoring and detection of agricultural production") are concerned with problems related to the fact that the world population grows exponentially so that it is fundamental to audit the current cultivating practices to satisfy the security of food availability and nourishment needs. An insightful sensor frameworks give more data on water needs and harvests. This data can be utilized to computerize the water supply framework. The data procured in the initial step is moved to the cloud. The Unclassified Exceptional Readiness for Distributed Power Generation is then a significant performance measure that a developer of a cell phone application can utilize. This paper presents and assesses the idea of a remote detection device.

Assia Ennouni, My Abdelouahed Sabri, and Abdellah Aarab ("Plant diseases detection and classification based on image processing and machine learning") are concerned with an important problem that is related to the use of artificial intelligence and machine learning to attain sustainable development and make it possible to solve real-life problems in countries like Marocco who heavily rely on agriculture and food production. All species of plants are subjected to various types of diseases that can cause huge damage. Although the observation of variations in the infected part of the leaf plant is very important, it is not enough because the perception of the human eye is not strong enough. The identification of plant diseases is a very important task in the agriculture area as it can contribute to a huge gain in agricultural productivity, quality, and quantity. To detect plant diseases in an earlier stage, efficient and precise techniques are needed to assist farmers. The authors present, first, an overview of plant diseases from leaves images and different disease classification approaches, followed by an overview of classification techniques that can be used for plant leaf disease detection.

We wish to express our deep gratitude to all the authors for their excellent contributions. Special thanks are due to anonymous peer referees whose insightful comments and constructive remarks and suggestions have greatly helped improve the contributions. We also wish to thank Dr. Tom Ditzinger, Dr. Leontina di Cecco, and Mr. Holger Schaepe from the Engineering Editorial, SpringerNature for their dedication and help to implement and finish this large publication project maintaining the highest publication standards.

Khouribga, Morocco — Noreddine Gherabi
Warsaw, Warsaw — Janusz Kacprzyk
Autumn 2020

Contents

Experimental Evaluation of Proposed Algorithm for Identifying Abnormal Messages in SIP Network

Mourade Azrour, Jamal Mabrouki, Yousef Farhaoui, and Azidine Guezzaz

Abstract Session Initiation Protocol (SIP) can be defined as an application layer protocol that was developed in order to initiate, modify, and stop a multimedia session through internet network. In recent decades, SIP is coming more and more popular as it is adopted by many typical applications such as telephony over IP (ToIP). Due to its popularity, SIP has some security problems that we have to be resolve. In this paper, we concentrate our efforts on SIP malformed messages attack detection. Therefore, we propose our algorithm that can determine whether the received message is normal or abnormal. The simulation results confirm that our algorithm is efficient and can detect the attack in a minimum time.

Keywords Session initiation protocol · Attack · Security · Malformed message · Telephony over IP · VoIP

1 Introduction

Session Initiation Protocol (SIP) [1] is one of the recent invented protocols used to manage a multimedia session among various participants. SIP is a text based protocol created on the base of the HTTP [2] and SMTP [3] protocols. It is also the most popular signaling protocol, thinks to its simplicity and adoption by the 3rd Generation Partnership Project (3GPP) to be the signaling protocol for IP Multimedia Subsystem (IMS). In order to secure SIP communications, Internet Engineering Task Force (IETF) has recommended the authentication security service for SIP. As result, this service has received much importance and it is enhanced in [4, 17].

Unfortunately, SIP is the new target of the attackers. Therefore, it suffers from various problems such as flooding attack [18], SPAM over IP Telephony attack [19]

M. Azrour (✉) · Y. Farhaoui
Department of Computer Science, IDMS Team, Faculty of Sciences and Techniques, Moulay Ismail University, Errachidia, Morocco

J. Mabrouki
Laboratory of Spectroscopy, Molecular Modeling, Materials, Nanomaterial, Water and Environment, CERNE2D, Faculty of Science, Mohammed V University, Rabat, Morocco

A. Guezzaz
Technology High School, M2SC Team Essaouira, Cadi Ayyad University, Marrakesh, Morocco

N. Gherabi and J. Kacprzyk (eds.), *Intelligent Systems in Big Data, Semantic Web and Machine Learning*, Advances in Intelligent Systems and Computing 1344,
https://doi.org/10.1007/978-3-030-72588-4_1

and malformed message attack [20]. This last one is an attack that can be used to discover the system weakness, to have an unauthorized access, or to cause Deny of Service (DoS) attack. The problem was addressed in different studies [20, 24] but the attack still exists.

In this paper, we have proposed a simple algorithm based on our recent survey study [25]. The proposed algorithm is able to detect malformed SIP message. Furthermore, it can identify the type of abnormal received message.

The remainder of this paper is organized as following. The section two details the information about SIP protocol. In the section three, we present related works which are aimed to detect malformed messages. Our proposed algorithm is detailed in the section four. The section five is reserved for detailing the experimental evaluations and the discussion of obtained results. Finally, the last section concludes our proposed work.

2 SIP Overview

2.1 SIP Architecture

The architecture of SIP is composed on a proxy server, redirect server, registrar server, location server, and user agents. The role of each element is labeled in the following:

- User Agent (UA): It can be a SoftPhone (software) or HardPhone (IP phone). Its principal role is generates, sends and receives SIP requests. It can act as User Agent Client (UAC) or as User Agent Server (UAS).
- Registrar Server: It has the responsibility to save the location of each User Agent.
- Proxy Server: It is the principal server that is linked directly to fixed or portable stations. It can act as a server or client.
- Redirect Server: It is a server that receives SIP requests, converts the SIP address of a destination to IP address and sends them back to the client.
- Location server: It offers its services to proxy, redirect, and registrar servers. It allows for them to search or record the position of the user agent.

2.2 SIP Messages

Generally, for establishing the communication in SIP networks, the user and server can exchange two types of SIP message that are the request and response. Accordingly, the request message is sent by a client to a server. Furthermore, the response message is transmitted by a server to a client. On the other hands, each SIP message is constituted on three distinct parts: a start-line, header fields, and an optional message-body. In addition, the headers and body message are separated with an empty line.

```
INVITE sip:192.169.1.8:5060 SIP/2.0
Via: SIP/2.0/UDP
192.168.1.2:16999;rport;branch=z9hG4bKPj820
50b9f03844659a2b24641722ca1a1
Route: <sip:192.169.1.2:5060;lr>
Max-Forwards: 70
From: <sip: azrour@192.169.1.2>;
tag=3ea7b9ce8a9b0db
To: <sip:ouanan@192.169.1.8>
Call-ID: 5b3d61c6c6654bf0989d21b9e4c92300
CSeq: 63287 INVITE
Allow: INFO, PRACK, SUBSCRIBE, NOTIFY,
REFER, INVITE, ACK, BYE, CANCEL,
UPDATE
Contact: <sip: azrour@192.168.1.2:16999>
User-Agent: StarTrinity.SIP 2017-04-05 14.41
UTC
Expires: 3600
```

(a)

```
SIP/2.0 202 Accepted
CSeq: 2 SUBSCRIBE
Call-ID: 959093180754@192.168.1.2
From:
<sip:azrour@192.168.1.3>;tag=z9hG4bK03
540318
To:
<sip:azrour@192.168.1.3>;tag=f96da656
Via: SIP/2.0/UDP 192.168.1.2:48110;
  rport=48110;branch=z9hG4bK10362;
  received=192.168.1.2
Expires: 184000
Contact: "" <sip:192.168.1.3:5060>
Content-Length: 0Content-Length: 0
```

(b)

Fig. 1 **a** Example of SIP request; **b** Example SIP response

The start-line is useful to distinguish the request message from response message. Therefore, the start-line of the request message comprehends three different values that are: a method name, a Request-URI, and the protocol version. These values are disconnected by a single space character. The first of the three values can be one of the six original methods: REGISTER, INVITE, ACK, CANCEL, BYE and OPTIONS, defined in RFC3261. However, there are the other optional methods that are: SUBSCRIBE, REFER, MESSAGE, NOTIFY, UPDATE, INFO, and PRACK, which are labeled in other RFC's. The second value denotes IP address and the port for example sip: 192.169.1.8:5060. The third value denotes the protocol version. The latest version of SIP is 2.0.

On the other sides, the start-line of response message contains the Status-Line that holds the protocol version, a numeric Status-Code, and its associated textual phrase. The Fig. 1(a) and (b) depicts an example of SIP request and the response messages respectively.

2.3 SIP Operation

For establishing the communication between two users, the three phases must be performed: initiation phase, communication phase, and termination phase. The first one is executed to invite the callee to accept the communication. In the case he accepts, the second phase is performed to exchange data. Then, the communication is established. For ending the communication the third phase must be achieved. The example of the communication is detailed in the following.

As depicted in Fig. 2, whenever user AZROUR (caller) wish to establish the communication with user FARHAOUI (callee), he firstly generates and sends a request

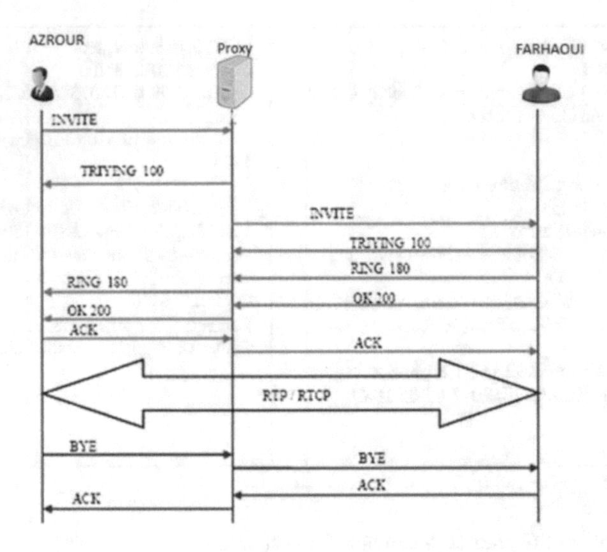

Fig. 2 Invite scenario

message INVITE to the proxy server. After the authentication of the received request message, the proxy forwards it to the user FARHAOUI (callee). If the callee is available and accepts the call, he generates the "200 OK" response message, which is sent back to the caller via the proxy server. Upon receiving the callee response "OK", the caller AZROUR send an "ACK" message to ensure that is ready to talk. Afterword, the caller and the callee can exchange their data (voice, video...) through Real-time Transport Protocol (RTP). Besides, for stopping the communication the callee (or caller) sends the request message BYE to the caller (or callee)

3 Related Works

Since, the discovering of the malformed message attack, the numerous research works in the literature suggested several techniques [20–26] allowed to the SIP server to identify abnormal SIP messages based on the header values. Therefore, In 2007, Geneiatakis et al. [24] proposed a new mechanism to detect SIP malformed messages. The proposed work is based on attack signature detection and it is composed on two parts based on the SIP syntax described on RFC3261 [27]. The First part is applied to detect any type of malformed message, and for any SIP methods. So, it can be named a general signature. In the contrary to the first part, the second part includes additional rules specific to each SIP method. For that reason, it is optional and it is not applicable to all messages. In addition to the two signatures, Geneiatakis et al. defined the third signature that can detect SQL Injection attacks. The last signature analyzes and checks the validity of the Authorization field data. The analysis will determine if the attributes of Authorization field (username, realm) contains the SQL code or not. The Geneiatakis et al.'s framework woks will with the known attack but it has some limitations because the defined rules cannot cover all types of malformed messages.

In 2012, Ferdous et al. [28] proposed a new approach which aim to classify received SIP messages as "good" or "bad" depending on whether their structure and content are acceptable or not. A "bad message" can be malformed, crooked, or malicious. Ferdous et al.'s framework consists of two stages filtering methodologies. The first stage filtering is named the lexical analyzer; it is able to extract the features from the incoming messages. Then, the extracted information are parsed in order to determinate if they are the part of the language by the formal grammar specified the SIP protocol. In the other hand, the second stage is called the structure and the content analyzer, this stage is based on machine learning Support Vector Machines (SVM) which has been previously trained to classify SIP messages by statistically learning from example of normal and abnormal messages. The advantage of Firdous et al.'s proposed method is the detection and classification of the malformed message. Nonetheless, it has some limitations such as: the SVM has to be trained automatically on real time.

In 2013, Seo et al. [28] proposed SIPAD: SIP-VoIP anomaly detection using a stuteful rule tree, the proposed work aims to secure SIP environment against malformed message attacks and flooding attacks. Seo et al. use an anomaly detection approach by defining legitimate cases. The proposed approach can identify unknown variant types of attacks. Furthermore, it doesn't need to maintain a large amount of attack signatures. SIPAD verifies whether the received message matches the predefined rules. In order to apply the RFC3261 rules, the authors translate the RFC3261 Augmented Backus Naur Form (ABNF) rules to regular expressions. Consequently, any incoming message that has unmatched or undefined headers is considered as undesirable message. The rules defined by Seo et al. are based on the relationship between SIP messages, headers, and the states. These rules can familiarize to a new standard by adding or adjusting the existing rules. The Seo et al.'s method can be adapted to new standard and it is also very fast and efficient than the previous methods.

In 2015, Su and Tsai [23] proposed a new system framework which has two roles: the first is to filter malformed SIP messages that conflict with the SIP protocol, this role is achieved by the malformed message detection module. The second role is to determinate whether the SIP server is under flooding attack or not. To detect malformed SIP message attack Su and Tsai have used string comparison. Therefore, any incoming message must firstly pass the first module which applies the RFC3261 SIP standard format as the basis for identifying malformed packet. Once a message is determined to be malformed the system updates the black list in the server database in order to block the future messages coming from the same source.

4 The Proposed Method

In this section, we present our proposed algorithm that can detect and classify the SIP malformed messages based on their forms and contents. In addition to check if the message is normal or abnormal our method can determine the category of malformed message. Our proposed algorithm defines four classes of invalid messages: Invalid Method in First-Line, Invalid syntax, Invalid values, and Existing of SQL code. Each class is detailed as following.

• Invalid Method in First-Line

As described in section II, SIP message is started by First-line. If the message is a request the First-line contains a method name, a Request-URI, and the protocol version separated by a single space character. If it is a response it contains the protocol version, a numeric Status-Code, and its associated textual phrase. Therefore, our algorithm inspects the first line by checking the correctness of method by comparing its values with 13 passible ones (see Subsect. 2.2). As a result, if any message is send with an invalid method or invalid Status-Code it will be considered as abnormal, even if the rest of the message is valid.

• Invalid Syntax

The syntax of SIP message is considered as invalid if one or more obligatory fields are not existed or if an unique field is duplicated. For example if the field "from" is duplicated the intermediate servers will be disturbed and will not know the exact source of the message (generally there is one source of each message).

• Invalid values

When the syntax of the message is correct but, one or more values of the fields is not valid, or it is null, or contains a value that can causes undesirable result in the system, the message will be considered as abnormal. For example, if the CSeq field contains a string value, this value cannot be incremented for each new request within a dialog.

•Existing of SQL code

SQL (structured query language) injection is type of SIP malformed message in which an attacker can execute SQL code by exploiting the Authorization header field. Upon received the user's request which contains the Authorization field the server request its database to extract user's registered parameters. Therefore, if a malicious has injected SQL code in Authorization filed he/she can insert, alter, or delete the database server information without having the authorized access.

5 Experimental Evaluation

5.1 Description of Dataset

The principal role of proposed systems is to detect and classify the malformed messages. For this reasons, we have developed SIP server application that includes the proposed algorithm under JAVA programing language and using Jain SIP API, the developed server is illustrated in Fig. 3.

With the aim to test our proposed algorithm, we have installed the SIP Client SIPDroid in two smart phones. In two laptops, we have installed the SIP Client X-

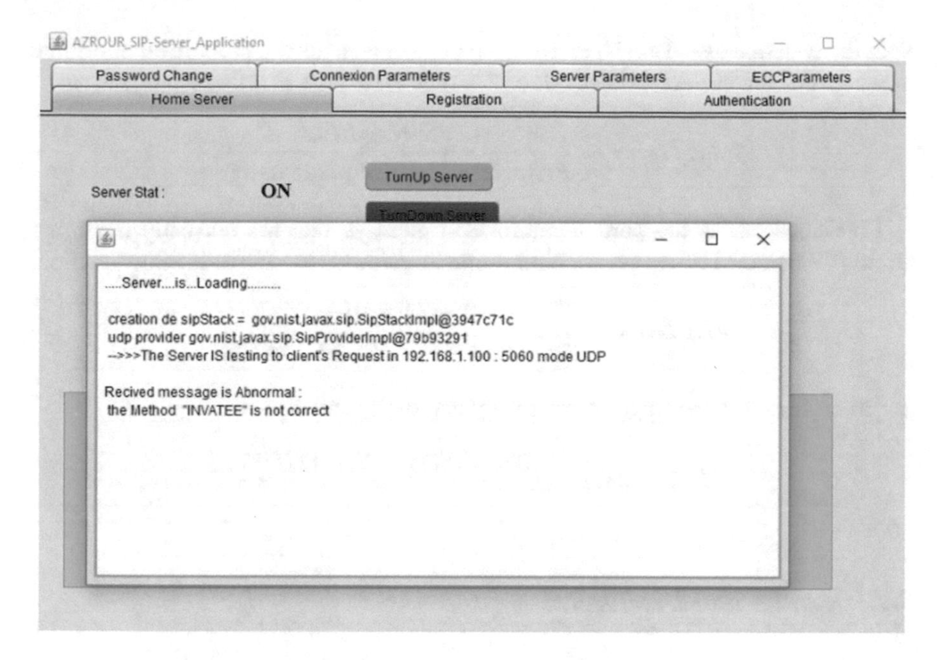

Fig. 3 The developed SIP server

lite, Star Trinity SIP Tester, Sip Scan, and PROTOS. The two latest ones are capable to execute the attacks. In addition, we have developed a SIP client to send malformed messages. In the third laptop, we have installed our developed SIP Server that can receive and analyses the SIP messages (Fig. 3). After setting up our local network, we have tested five various scenarios. In the first scenario, we tested 1000 messages with a 10% that have incorrect method. In the second scenario, the total is 1000 massages, where 10% contains invalid syntax. The same number of messages is tested in the third scenario, but 10% have invalid values. The same rate of message have evaluated in the fourth scenario. However, the application sends 10% that are injected with SQL code.

In order to evaluate our proposed algorithm and for treating the obtained results, we take the advantage of the usage of confusion matrix. Therefore, we evaluate our algorithm, we analyze four principal attributes that are sensitivity, specificity, precision, and accuracy. These attributes are detailed as following:

Sensitivity called also True Positive Rate (TPR): refers to the total number of detected abnormal messages divided by the total number of existed malformed message. It is calculated as following.

$$Sensitivity = \frac{TruePositive}{TruePositive + FalseNegative}$$

Specificity called also True Negative Rate (TNR): denotes the proportion of normal message that are correctly detected. It is computed using the following equation.

$$Specificity = \frac{TrueNegative}{TrueNegative + FalsePositive}$$

Precision: means the part of malformed message that are correctly abnormal among the detected message. For computing its value we use the following equation.

$$Precision = \frac{TruePositive}{TruePositive + FalsePositive}$$

Efficiency: is the average of sensitivity and the specificity

$$Efficiency = \frac{Sensitivity + Specificity}{2}$$

5.2 Results and Discussion

In our experimental test, the installed SIP clients are firstly registered in SIP server which implements our method. The clients generate about 10.000 messages (INVITE, REGISTER, OPTION), 5000 messages are malformed and others are normal. After receiving the incoming message our application executes the analyzer method to check if the message is normal or not.

In order to demonstrate the capability of our proposed algorithm to classify correctly type of received SIP messages. Explicitly, to discover that is normal or abnormal. We have considered the four values for each scenario as depicted in the Fig. 4. From those results, we can show that the effectiveness of our method is between 88% and 100% depending on the type of malformed message.

As sensitivity and specificity are two features that can precise the ability of given method to organize appropriately and to cluster correctly the type of received SIP messages. In our experiment, we have discovered that the sensitivity diverges between 86.71% and 100%, and the specificity balanced between 89% and 100%. Furthermore, the two attributes precision and efficiency, that are computed using the sensitivity and specificity, have values between 89.40% and 100%, and between 87.89% and 100%, respectively. We can remark also that in the first scenario, the all values are 100% because it is simple to verify the correctness of the 13 existed methods. In the contrary, it is difficult to predict all possible values. As a result, the minimum values are recorded in the third scenario.

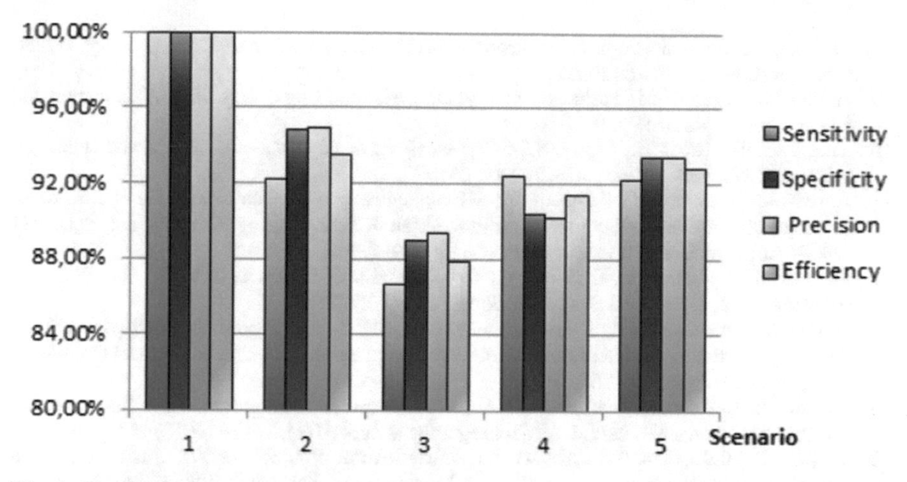

Fig. 4 The obtained results

6 Conclusion

In this paper, we have focus on the SIP malformed message attack. This last one is applied by attackers in order to have various objectives including discover the weakness of SIP servers, to have an authorized access, to cause DoS attack, or to cause unexpected situation for the system. After listing some previous related works, we have proposed our algorithm which can detect the abnormal SIP messages and determine their classes. In order to test our method, we have developed SIP server application, which implements our proposed algorithm. Then, we have done five tests. The obtained results confirm that our algorithm is it is efficient and have low false alarm.

References

1. Johnston, A.B.: SIP: understanding the Session Initiation Protocol, 2nd edn. Artech House, Boston (2004)
2. Fielding, R., et al.: Hypertext transfer protocolHTTP/1.1 (1999)
3. Postel, J.: Simple mail transfer protocol. Inf. Sci. (1982). https://tools.ietf.org/html/rfc821
4. Li, C.-T.: Secure smart card based password authentication scheme with user anonymity. Inf. Technol. Control **40**(2), (2011). https://doi.org/10.5755/j01.itc.40.2.431.
5. Miller, V.S.: Use of elliptic curves in cryptography. In: Conference on the Theory and Application of Cryptographic Techniques, p. 417426 (1985)
6. Sun, D.-Z., Huai, J.-P., Sun, J.-Z., Li, J.-X., Zhang, J.-W., Feng, Z.-Y.: Improvements of Juang's password-authenticated key agreement scheme using smart cards. IEEE Trans. Ind. Electron. **56**(6), 22842291 (2009). https://doi.org/10.1109/TIE.2009.2016508.
7. Zhu, W., Chen, J., He, D.: Enhanced authentication protocol for session initiation protocol using smart card. Int. J. Electron. Secur. Digit. Forensics **7**(4), 330342 (2015)

8. Tsai, J.L.: Efficient nonce-based authentication scheme for session initiation protocol. IJ Network Security **9**(1), 1216 (2009)
9. Azrour, M., Ouanan, M., Farhaoui, Y.: Cryptanalysis of Zhang et al.'s SIP authentication and key agreement protocol, p. 9
10. Thandra, P.K., Rajan, J., Murty, S.S.: Cryptanalysis of an efficient password authentication scheme. IJ Network Secur. **18**(2), 362368 (2016)
11. Azrour, M., Ouanan, M., Farhaoui, Y.: SIP authentication protocols based on elliptic curve cryptography: survey and comparison. Indonesian J. Electric. Eng. Comput. Sci. **4**(1), 231 (2016). https://doi.org/10.11591/ijeecs.v4.i1.pp231-239
12. Azrour, M., Farhaoui, Y., Ouanan, M.: Cryptanalysis of Farash et al.'s SIP authentication protocol. Int. J. Dynamical Syst. Diff. Equations **8**(1/2) (2018)
13. Azrour, M., Ouanan, M., Farhaoui, Y.: A new secure SIP authentication scheme based on elliptic curve cryptography. In: International Conference on Information Technology and Communication Systems, pp. 155–170 (2017)
14. Azrour, M., Ouanan, M., Farhaoui, Y.: A new enhanced and secured password authentication protocol based on smart card. Int. J. Tomography Simul. **31**(1), 14–26 (2018)
15. Toy, N., Senthilnathan, T.: Light weight authentication protocol for WSN using ECC and hexagonal numbers. Indonesian J. Electric. Eng. Comput. Sci. **15**(1), 443450 (2019). https://doi.org/10.11591/ijeecs.v15.i1.pp443-450
16. Adamu, J., Hamzah, R., Rosli, M.M.: Security issues and framework of electronic medical record: a review. Bull. Electric. Eng. Inf. **9**(2), 565572 (2020). https://doi.org/10.11591/eei.v9i2.2064
17. Khalifa, O.O., Roslin, R.J.B., Bhuiyan, S.S.N.: Improved voice quality with the combination of transport layer & audio codec for wireless devices. Bull. Electric. Eng. Inf. **8**(2), 665-673 (2019). https://doi.org/10.11591/eei.v8i2.1490
18. Zargar, R.H.M., Moghaddam, M.H.Y.: SIP flooding attacks detection and prevention using Shannon, Renyi and Tsallis Entropy. Int. J. Hybrid Inf. Tech. **7**(5), 257272 (2014). https://doi.org/10.14257/ijhit.2014.7.5.24
19. Azrour, M., Farhaoui, Y., Ouanan, M., Guezzaz, A.: SPIT detection in telephony over IP using K-means algorithm. Procedia Comput. Sci. **148**, 542551 (2019). https://doi.org/10.1016/j.procs.2019.01.027
20. Tsiatsikas, Z., Kambourakis, G., Geneiatakis, D., Wang, H.: The devil is in the detail: SDP-driven malformed message attacks and mitigation in SIP ecosystems. IEEE Access **7**, 2401–2417 (2018). https://doi.org/10.1109/ACCESS.2018.2886356
21. Tsiatsikas, Z., Kambourakis, G., Geneiatakis, D., Wang, H.: The devil is in the detail: SDP-driven malformed message attacks and mitigation in SIP ecosystems. IEEE Access **7**, 24012417 (2019). https://doi.org/10.1109/ACCESS.2018.2886356
22. Azrour, M., Ouanan, M., Farhaoui, Y.: Malformed message detection in SIP environment. In: The International Symposium on Computer Sciences and Applications (ISCSA2017), Er-Rachidia, Morocco (2017)
23. Su, M.-Y., Tsai, C.-H.: An approach to resisting malformed and flooding attacks on SIP servers. J. Networks **10**(2), (2015). https://doi.org/10.4304/jnw.10.2.77-84
24. Geneiatakis, D., Kambourakis, G., Lambrinoudakis, C., Dagiuklas, T., Gritzalis, S.: A framework for protecting a SIP-based infrastructure against malformed message attacks. Comput. Networks **51**(10), 25802593 (2007). https://doi.org/10.1016/j.comnet.2006.11.014
25. Azrour, M., Ouanan, M., Farhaoui, Y.: Survey of detection SIP malformed messages. Indonesian J. Electric. Eng. Comput. Sci. **7**(2), 457465 (2017). https://doi.org/10.11591/ijeecs.v7.i2.pp457-465
26. Li, H., Lin, H., Yang, X., Liu, F.: A rules-based intrusion detection and prevention framework against SIP malformed messages attacks. In: 2010 3rd IEEE International Conference on Broadband Network and Multimedia Technology (IC-BNMT), p. 700705 (2010)
27. Ferdous, R., Cigno, R.L., Zorat, A.: Classification of SIP messages by a syntax filter and SVMs. In: 2012 IEEE on Global Communications Conference (GLOBECOM), p. 27142719 (2012)
28. Seo, D., Lee, H., Nuwere, E.: SIPAD: SIPVoIP anomaly detection using a stateful rule tree. Comput. Commun. **36**(5), 562574 (2013). https://doi.org/10.1016/j.comcom.2012.12.004

Smart Tourism Recommender System Using Semantic Matching

Abdelhadi Daoui, Noreddine Gherabi, and Abderrahim Marzouk

Abstract Due to the widespread diffusion of information technologies and the number of social network users who are increasing day after day, today there is an explosion of data on the web, which requires a filtering mechanism to customize the data to be displayed to the users according to their needs. For this reason, the recommendation systems have been designed to facilitate and personalize data access, according to the needs of each user. But each system can recommend services linking to a specific area, which makes the system of recommendations point and more efficient, where we find systems for the field of e-commerce, tourism, etc. In this paper, our work focuses on designing an intelligent recommendation system for the tourism field, which is used to analyze images stored in mobile devices of users in order to identify the fields of interest of these users (keywords). Then, we base on these latter to recommend the tourist services and places suitable to each user according to their preferences.

Keywords Profile identification · Recommender system · Semantic web · Ontologies · CC/PP · Semantic matching

1 Introduction

Currently, with the amount of gigantic data on the web, the search for precise information is becoming more and more difficult. To this end, researchers have developed recommendation systems based on new technologies, precisely semantic web [1], the role of these systems is to properly filter the information represented to the user in order to display only what is interesting to him. Recommendation systems are based in their research on a set of criteria that represent the interests and preferences of the user (user profile), they can ask the user to enter his interests and preferences

A. Daoui (✉) · A. Marzouk
Department of Mathematics and Computer Science, FSTS, Hassan 1st University, Settat, Morocco

N. Gherabi
Sultan Moulay Slimane University, ENSAK, Khouribga, Morocco

N. Gherabi and J. Kacprzyk (eds.), *Intelligent Systems in Big Data, Semantic Web and Machine Learning*, Advances in Intelligent Systems and Computing 1344,
https://doi.org/10.1007/978-3-030-72588-4_2

and store them in files of profile or in a database during the first connection [2] or to detect preferences by analyzing his behavior on the Web [3].

Our work in this paper is to offer a set of recommendations to a user based on his profile, where our proposed method aims to show the user of the current profile the tourist sites and places that may interest him according to his preferences. There are some methods in the literature that propose recommendation systems such as: The authors of the article [4] propose a semantic social recommendation system for Tunisian tourism which is based on ontology. This contains all the concepts and relationships of medical tourism in Tunisia. This information is taken from the services of Tunisian medical tourism providers. The paper [3] presents a method that makes it possible to identify users based on their behaviors on the web, the objective is to recommend products, advertising, personalized content, etc.

The authors of the paper [5] present an automatic user interface adaptation technique based on statistical methods that use a set of properties related to user interaction. These properties can be part of user profiles. With the same goal, Huakang Li, Longbin Lai and Xiaofeng Xuin [6] have developed a technique to identify user preferences from visited web pages, This method is based on the nodes of Wikipedia's category network.

Our recommendation system is based on the similarity calculation to identify the elements that will be recommended. This concept of similarity is used in several fields of research. For example, in our latest articles [7] we have proposed two methods to calculate the semantic similarity between the concepts defined in the same ontology. In the paper [10], the authors propose a new method that relies on semantic similarity measurement for defining the compatibility between semantic web services.

2 Proposed Method

The method proposed in this paper aims to present a new and enhanced approach for building a smart e-tourism recommender system. This latter, will be able to identify the user interests in the tourism domain, then provide the suitable services according to their preferences.

This method is composed of three main components: Keyword planner, profile creation, recommendation.

2.1 Keyword Planner

2.1.1 Keywords Identification

This section is consecrated for identifying the keywords by analyzing and processing the images stored in user devices. This latter, will be assured by another team of research and its result will be considered as input in the current paper.

2.1.2 Weight Allocation

Due to its sensibility, the weight allocation plays a crucial role in our proposed method, and the final results (the services which this method will recommend) will depend on their values to define the tourist services which will be recommended or even their sequence.

In the current section, for computing the weight to allocate to each keyword, we use the number of the tourist objects found in the images stored into user devices. This task is insured in the previous section by analyzing and processing the images stored in user devices.

The allocated weight should be strictly greater than 0 and less or equal to 1. For this reason, we have proposed the formula 1 shown below, which define the rate of presence of the tourist object represented by the keyword K_i ($TOP(K_i)$) found into user devices images. This latter, will be used for computing the weight of the keyword K_i as shown in the formula 2.

$$TOP(K_i) = \frac{NI(K_i)}{NTO(K_i)} \quad (1)$$

Where, $NI(K_i)$ and $NTO(K_i)$ represent successively the number of images that contain the tourism object K_i and the number of presence of the tourism concept K_i in these images, because we can find one tourism object more than once in the same image.

The number of images that contain a tourism object help to define the rate of preference of this tourism object to the owner of these images.

$$W(K_i) = \frac{TOP(K_i)}{\sum_{m=1}^{n} TOP(K_m)} \quad (2)$$

Where, n represents the number of keywords defined from the mobile device images owned by the current user (a tourism object can be present more than once in the same image).

2.2 *Profile Creation*

The purpose of this section is to define the semantic matching between the concepts of "DATAtourisme" ontology and the given keywords, then stock the matched concepts in a CC/PP profile. This goal is assured by the following steps:

2.2.1 Building Paths

In the field of semantic matching, the process of building paths plays a crucial role, because we base on these built paths to define the rate of semantic matching.

In the current paper, this process of building paths is assured by WordNet [15]. However, for each keyword or "DATAtourisme" concept name which we need to compute the semantic matching between them, we use the external resource WordNet to find its hypernym that represents its super class (more generic term), then finding the hypernym of the current hypernym and so on until the root term (root node)in the WordNet.

For example, for the keyword "Event" and the "DATAtourisme" concept name "Festival", we have define four means for "Event" and two means for "Festival". Thus, they are four paths for "Event" and two paths for "Festival", where each path represents a mean of this keyword or this "DATAtourisme" concept name as shown below in the Table 1:

Table 1 Event means and their built paths

<table>
<tr><td rowspan="8">Event</td><td>Path 1</td><td>Event → psychological_feature → abstraction → entity.</td></tr>
<tr><td>Meaning 1</td><td>something that happens at a given place and time.</td></tr>
<tr><td>Path 2</td><td>Event → circumstance → condition → state → attribute → abstraction → entity.</td></tr>
<tr><td>Meaning 2</td><td>a special set of circumstances.</td></tr>
<tr><td>Path 3</td><td>Event → physical_phenomenon → natural_phenomenon → phenomenon → process → physical_entity → entity.</td></tr>
<tr><td>Meaning 3</td><td>a phenomenon located at a single point in space-time.</td></tr>
<tr><td>Path 4</td><td>Event → consequence → phenomenon → process → physical_entity → entity.</td></tr>
<tr><td>Meaning 4</td><td>phenomenon that follows and is caused by some previous phenomenon.</td></tr>
<tr><td rowspan="4">Festival</td><td>Path 1</td><td>Festival → time_period → fundamental_quantity → measure → abstraction → entity.</td></tr>
<tr><td>Meaning 1</td><td>a day or period of time set aside for feasting and celebration.</td></tr>
<tr><td>Path 2</td><td>Festival → celebration → diversion → activity → act → event → psychological_feature → abstraction → entity.</td></tr>
<tr><td>Meaning 2</td><td>an organized series of acts and performances.</td></tr>
</table>

2.2.2 Semantic Matching

In the current paper, the process of semantic matching bases on the set of paths built in the previous section. After defining the paths of the keyword and the "DATA-tourisme" concept name, we compare each meaning of this keyword, which is represented per a path with the all built paths of the "DATAtourisme" concepts name (meanings) in order to define the more similar meanings between this keyword and "DATAtourisme" concept name. These latter (more similar meanings) will be the paths (meanings) that have more hypernym in common.

For example, by analyzing Table 1 in order to define the more similar meanings between the keyword "Event" and "DATAtourisme" concept name "Festival", we find the meanings (meaning 1 of Event and meaning 2 of Festival) which are represented by the following paths:

Event : Event → psychological_feature → abstraction → entity.
Festival : Festival → celebration → diversion → activity → act → event → psychological_feature → abstraction → entity.

Where, we can observe that the path of event represents a sub path of festival, and these two paths have the most common hypernyms compared to other paths illustrated in Table 1.

At this step, after defining the more similar meanings, we need to compute the rate of semantic matching between this keyword and "DATAtourisme" concept in order to define the degree of preference of this tourism concept (tourism service) to the owner of this keyword. These selected paths represent the most similar meanings, and they are defined as concepts (C_K, C_{DT}) builts using wordNet.

The formula proposed for computing the semantic matching is defined as:

$$SMat(C_k, C_{DT}) = \frac{1}{NHyp(C_K) + NHyp(C_{DT}) - NCHyp(C_k, C_{DT}) * 2 + 1} \quad (3)$$

SMat (C_K,C_{DT}) represents the semantic matching between the concept C_K built basing on the given keyword K and the concept C_{DT} built basing on "DATAtourisme" concept DT name, NCHyp(C_K, C_{DT}) represents the number of the common hypernyms between the concepts C_K and C_{DT}, and NHyp(C) represents the number of hypernyms of the concept C.

In general, the output of this formula is greater or equal to zero or less or equal to one.

For example, the rate of sematic matching between the following concepts:

Event : Event → psychological_feature → abstraction → entity.
Festival : Festival → celebration → diversion → activity → act → event → psychological_feature → abstraction → entity.

Built basing on the keywords "Event" and "Festival" is equal to 0.166 as illustrated below:

$$SMat(\mathrm{C}_{\mathrm{Event}}, C_{Festival}) = \frac{1}{3+8-6+1} = 0.166$$

The formula 3 is valid only when the name of "DATAtourisme" concept is simple, that is to say, this name is composed of a single word. For example festival, but when this name is composed of more than one word for example "EntertainmentAndEvent". In this case, we need to compute the semantic matching between this keyword and the set of words composing the "DATAtourisme" concept name. for this reason, we have proposed the formula 4 shown below, which bases on the formula 3 and allows defining the semantic matching between a given keyword and a set of words extracted from "DATAtourisme" concept name. For example, for computing the semantic matching between the keyword "Event" and "EntertainmentAndEvent", we use the formula 3 to define the semantic matching between this keyword and each word composing the term "EntertainmentAndEvent", in our case "Entertainment" and "Event", then we base on these results and the formula 4 to compute the rate of semantic matching desired.

The formula 4 is defined as:

$$SMat(\mathrm{C}_{\mathrm{k}}, \mathrm{Set}_{\mathrm{C_{DT}}}) = \frac{\sum_{i=0}^{SCN-1} \mathrm{SMat}(\mathrm{C}_{\mathrm{K}}, \mathrm{C}_{\mathrm{DT_i}})}{\mathrm{SCN}} \tag{4}$$

Where, $\mathrm{SMat}(\mathrm{C}_{\mathrm{K}}, \mathrm{Set}_{\mathrm{CDT}})$ represents the semantic matching between the concept C_{K} built on the keyword K and the set of concepts $\mathrm{Set}_{\mathrm{CDT}}$ which represents the set of concepts built on the words composing "DATAtourisme" concept name, $\mathrm{C}_{\mathrm{DTi}}$ represents the concept i in the set $\mathrm{Set}_{\mathrm{CDT}}$, and SCN represents the number of the concepts in the $\mathrm{Set}_{\mathrm{CDT}}$.

2.2.3 Concept Weight Calculation

In the process of semantic matching, weight allocation plays a crucial role. For this reason, in a previous section, we have computed the weight of each keyword. Therefore, define the rate of preference of each keyword compared to the current user. The recommendation step will base on the concepts of "DATAtourisme" ontology, which will be considered as more similar to the user's keywords. Thus, we need to compute the weight of these "DATAtourisme" concepts based on the results of semantic matching between these concepts and the keywords.

This method will degrade the rate of preference of the user to a concept of "DATAtourisme" ontology compared to the keyword that suits him. This is quite true, because there is a relation between firstly, the value of semantic matching computed between the concepts of "DATAtourisme" ontology and the keywords, on the other hand, the weight of these latter. Increasing of the semantic matching value implies increasing of the weights of the "DATAtourisme" concepts in a way that does not exceed the value calculated for the keyword. The latter (the weight calculated for the keyword) represents the maximum value for the weight of the concept "DATAtourisme" and only in a single case where they can have the same value, if these two concepts are completely identical.

For computing this weight, we have proposed the formula 5:

$$W(C) = W(K) * SMat(C_K, C) \tag{5}$$

Where, W(C) represents the weight of the concept C which defined into "DATAtourisme" ontology, W(K) represents the weight computed for the keyword K and SMat(C_K, C) represents the semantic matching computed between the keyword K and the concept C.

2.2.4 Keywords Directory Building

This section is devoted to record the semantic matching between the given keywords and the concepts of "DATAtourisme" ontology already computed in order to avoid calculate them more than once. For this reason, we have proposed a keywords directory which aims to store firstly the keyword and the concept of "DATAtourisme" ontology, on the other hand the value of semantic matching between these latter and the computed weight of the "DATAtourisme" concept for the potential use in the future with another user. This keywords directory is represented by an XML file that respects the structure shown below in the Fig. 1.

For validating this XML file in order to control its structure, we have proposed the following DTD file (Fig. 2):

```
<?xml version="1.0" encoding="utf-8" ?>
<Keywords> //the list of keywords
<Keyword ID="Keyword_name">  //one given keyword
<dt_concept> // represents "DATAtourisme" concept.
<SMatching></SMatching> // represents the value of semantic matching between this
        //"DATAtourisme" concept and the keyword which contains it.
<Weight></Weight> // represents the weight computed of this "DATAtourisme"
//concept
</dt_concept>
<dt_concept> </dt_concept>
.
.
.
</ Keyword >
.
.
</ Keywords >
```

Fig. 1 XML File for storing the keywords

```
<!ELEMENT Keywords (Keyword*)>
<!ELEMENT Keyword (dt_concept*)>
<!ELEMENT dt_concept (SMatching, Weight)>
<!ELEMENT SMatching (#PCDATA)>
<!ELEMENT Weight (#PCDATA)>
<!ATTLIST Keyword id ID #REQUIRED>
```

Fig. 2 The corresponding DTD file

2.2.5 CC/PP Ontology

After defining the preferences and interests of each user according to the tourism domain, we store them into an ontology respecting the CC/PP standard.

To create and store the profiles we have several methods such as:

- Using a set of keywords [12].
- Using a weighted keyword matrix [12].
- Use of taxonomies [13].
- Use of ontologies [14].

In our CC/PP file, each component contains three attributes:

1. The keyword.
2. The type of the keyword or of the most similar "DATAtourisme" concept.

3. The weight of the keyword or of the most similar “DATAtourisme” concept.

An example of a CC/PP ontology generated by our method (Fig. 3):

```
  <?xml version="1.0"?>
  <rdf:RDF    xmlns:rdf="http://www.w3.org/1999/02/22-rdf-
syntax-ns#"
xmlns:ccpp=http://www.w3.org/2002/11/08-ccpp-schema#
xmlns:dt="http://www.datatourisme.fr/ontology/core/1.0#">
  xmlns:schema="http:// topbraid.org/schema/">

  <rdf:Description
    rdf:about="http://www.domain.org/ontology/profiles#pr
ofil1">
  <ccpp:Component>
  <rdf:Description
    rdf:about="
http://www.domain.org/ontology/profiles#Party">
  <rdf:type
    rdf:resource="http://www.datatourisme.fr/ontology/cor
e/1.0#EntertainmentAndEvent" />
  <schema:weight>0.2</schema:weight>
  </rdf:Description>
  </ccpp:Component>

  <ccpp:Component>
  <rdf:Description
    rdf:about="
http://www.domain.org/ontology/profiles#Festival">
  <rdf:type
    rdf:resource="http://www.datatourisme.fr/ontology/cor
e/1.0#Festival" />
  <schema:weight>0.3</schema:weight>
  </rdf:Description>
  </ccpp:Component>

    <ccpp:Component>
  <rdf:Description
    rdf:about="
http://www.domain.org/ontology/profiles#culturalsite">
  <rdf:type
    rdf:resource="http://www.datatourisme.fr/ontology/cor
e/1.0#Culturalsite" />
  <schema:weight>0.5</schema:weight>
  </rdf:Description>
  </ccpp:Component>
  </rdf:Description>
  </rdf:RDF>
```

Fig. 3 An example of CC/PP ontology generated by our proposed method

2.3 Recommendation

The recommendation step aims to suggest to the system users the suitable places and services in the tourist domain according to their preferences and interests defined in the first section (keywords identification). In this method, we have based firstly, on the "DATAtourisme" ontology which represents a platform containing data accessible by all the world proposing tourist datasets, It aims at gathering tourist information data produced by Tourist Offices, Departmental Agencies, and Regional Tourism Committees, on the other hand, on the user preferences and interests represented by the keywords defined in the section of the keywords identification. Therefore, the process of semantic matching performed between the concepts of "DATAtourisme" ontology and the user keywords allows defining the rate of preference of the users according to the concepts of "DATAtourisme" ontology. consequently, the suitable places and services which will be recommended represent the instances of the classes defined into "DATAtourisme" ontology and which considered the area of the user preferences basing on the semantic matching computed in the previous section.

3 Test and Validation

To evaluate the performance of our proposed method for identifying the users preferences and interests in order to recommend suitable tourist places and services, we use a computer with an i5 3.4 GHZ in the processor and 4 GO in the RAM. The experiments are performed on files containing different numbers of keywords: 1 keyword, 5 keywords, 10 keywords, 20 keywords, 50 keywords and, 100 keywords.

By analyzing the obtained results figured in Table 2 and Fig. 4, we can see that the proposed method allows minimizing the average time per keyword required to provide suitable tourist places and services according to this keyword, in other words, minimizing the time of the recommendation.

Table 2 Cost time of recommendation process

Number of keywords	Recommendation time (Msec)	The average time per keyword (Msec)
1	3194	3194
5	4180	836
10	5201	520
20	8093	404
50	19343	387
100	21174	212

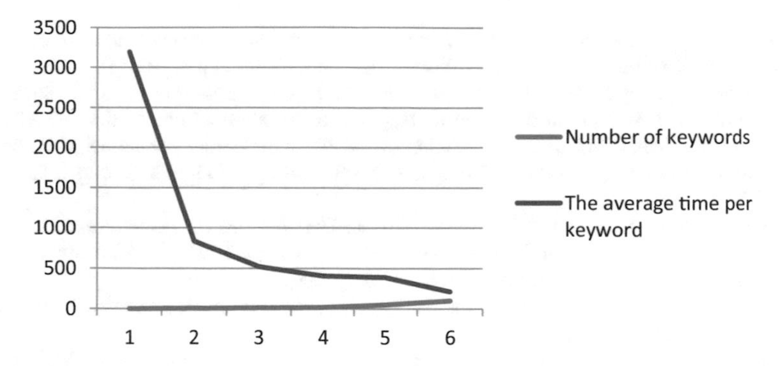

Fig. 4 Overhead of recommendation time

4 Conclusion

The method presented in the current paper aims to identify the user preferences and interests in the tourism domain and stock them into an ontology basing on the CC/PP standard in order to use this one in the process of the recommendation that allows suggesting the suitable tourist places and services for each user.

In the future work, we have interested to adapt this method in order to support the identification of the user preferences basing on its behavior on the social networks.

References

1. Lee, T.B., Hendler, J., Lassila, O.: The semantic web. Sci. Am. **284**, 1–18 (2001)
2. Frikha, M., Mhiri, M., Gargouri, F.: A semantic social recommender system using ontologies based approach for Tunisian tourism. Adv. Distrib. Comput. Artif. Intell. J. **4**(1), 90–106 (2015)
3. Yinghui, Y.: Web user behavioral profiling for user identification. Decis. Support Syst. **49**, 261–271 (2010)
4. Frikha, M., Mhiri, M.,.Gargouri, F., Zarai, M.: Using TMT ontology in trust-based medical tourism recommender system. IEEE (2016)
5. Wassermann, B., Zimmermann, G.: User profile matching: a statistical approach. In: The Fourth International Conference on Advances in Human-oriented and Personalized Mechanisms, Technologies, and Services, CENTRIC (2011)
6. Li, H., Lai, L., Xu, X., Shen, Y., Xia, C.: User interest profile identification using Wikipedia knowledge database. In: International Conference on High-Performance Computing and Communications & 2013 IEEE International Conference on Embedded and Ubiquitous Computing, pp. 2362–2367. IEEE (2013)
7. Daoui, A., Gherabi, N., Marzouk, A.: A new approach for measuring semantic similarity of ontology concepts using dynamic programming. J. Theoret. Appl. Inf. Technol. **95**(17), 4132–4139 (2017)
8. Daoui, A., Gherabi, N., Marzouk, A.: An enhanced method to compute the similarity between concepts of the ontology. In: Noreddine, G., Kacprzyk, J. (eds.) International Conference on Information Technology and Communication Systems, Advances in Intelligent Systems and Computing, vol. 640, pp. 95–107. Springer (2018)

9. Hau, J., Lee, W., Darlington, J.: A semantic similarity measure for semantic web services. In: the 14th international conference on World Wide Web, Chiba, Japan (2005)
10. Burstein, M., Bussler, C., Zaremba, M., Finin, T., Huhns, M.N., Paolucci, M., Sheth, A.P., Williams, S.: A semantic web services architecture. IEEE Internet Comput. **5**(9), 52–61 (2005)
11. Gherabi, N., Bahaj, M.: Outline matching of the 2D shapes using extracting XML data. In: Elmoataz, A., et al. (eds.) ICISP 2012, LNCS, vol. 7340, pp. 502–512. Springer, Heidelberg (2012)
12. Susan G., Mirco S., Aravind C., Alessandro M.: User profiles for personalized information access the adaptive web. In: Brusilovsky, P., Kobsa, A., Nejdl, W. (eds.) The adaptive Web, LNCS, vol. 4321, pp. 54–89. Springer, Heidelberg (2007)
13. Sulieman, D.: Towards semantic-social recommender systems, Doctoral thesis, Cergy Pontoise University France (2014)
14. Frikha, M., Mhiri, M., Gargouri, F.: Toward a user interest ontology to improve social network-based recommender system. In: Sobecki, J., Boonjing, V., Chittayasothorn, S. (eds.) Advanced Approaches to Intelligent Information and Database Systems, Studies in Computational Intelligence, vol. 551, pp. 255–264. Springer, Switzerland (2014)
15. Fellbaum, C. (ed.): WordNet: An Electronic Lexical Database. MIT Press, Cambridge (1998)

Data-Driven Information Filtering Framework for Dynamically Hybrid Job Recommendation

Islam A. Heggo and Nashwa Abdelbaki

Abstract Recommender system is a mature concept that periodically evolves. It is relating to more than one branch of knowledge such as data mining, machine learning, information retrieval, natural language processing and human-computer interaction. Nowadays recommender systems impose themselves to be a core model in all the present modern web applications. They are obviously found in numerous industries such as videos recommendation by Youtube and products recommendation by Amazon. Similarly, recommendation systems are incredibly required in the challenging e-recruitment domain. In e-recruitment domain, it is mandatory to build an intelligent system that digs deeper into thousands of jobs to eventually filter out the relevant jobs for each job-seeker. We explain what it needs to build such sophisticated and practical hybrid engine. The engine is practically employed in a dynamic e-recruitment portal and it has already proven its efficiency in the real-world market. We utilized many applied recommendation approaches in other several industries to empower the field of e-recruitment. Our project aims to recommend relevant jobs for job-seekers to solve the problem of disrupting job-seekers with irrelevant jobs to improve users' satisfaction and loyalty.

Keywords Information retrieval · Hybrid recommender system · Machine learning · Search engine · Ranking · Context-aware · Personalization

1 Introduction

Recommendation Systems can be non-personalized or personalized. Non-personalized recommendation is static for all users. It is not adapting for each individual needs or preferences. It is that kind of recommendations which mostly depends on the popularity of items (the number of viewers, average ratings…etc.). It is obvious

I. A. Heggo (✉) · N. Abdelbaki
School of Communication and Information Technology, Nile University, Giza, Egypt
e-mail: i.heggo@nu.edu.eg

N. Abdelbaki
e-mail: nabdelbaki@nu.edu.eg

N. Gherabi and J. Kacprzyk (eds.), *Intelligent Systems in Big Data, Semantic Web and Machine Learning*, Advances in Intelligent Systems and Computing 1344,
https://doi.org/10.1007/978-3-030-72588-4_3

that this technique is not appropriate in recruitment applications. What if a "Software Engineer at Microsoft" is the most popular job in the site based on the average user rating while a logged-in pharmacist is looking for a suitable job? And vice versa, it should recommend applicants to recruiters personally according to their posted job details not according to the popularity of these applicants in other job application.

Traditionally recommender system was used to be employed in e-commerce applications. For example, Amazon finds for users the most possible interesting products to increase the cross-selling and achieve more sales. Nowadays recommender systems become a core model in different web applications. They are obviously found in numerous industries like videos recommendation by Youtube and products recommendation by Amazon, feeds ranking and friend suggesting by Facebook, results ranking by Google Search and news recommendation by Google News [1].

Our challenge is filtering the massive amount of data to recommend the most relevant jobs for each job-seeker. This challenge is one of the emerging crucial problems all over the industries and business around the world where we start to hear about some terminologies like big data, Hadoop…, etc. These challenges of information overload are supposedly solved by the fields of information filtering, information retrieval and recommendation engines. World moves towards the online services and Internet. Nowadays online recruitment portals become a primary channel for all users whether employer or employee. We aim to ultimately match job-seeker with job. Solving this issue of an online recruitment platform is substantial to achieve the highest level of satisfaction at job-seeker side and recruiter side.

On the job-seeker side, job-seekers need to find an intelligent online system that understands their preference. They want to find easily what they are looking for. It is important to avoid annoying job-seekers with irrelevant jobs, emails and notifications. On the recruiter side, recruiters want to find only the suitable calibers who can fit into the vacancy they offer instead of screening thousands of irrelevant applicants [2].

Rest of this paper explains the details of building the required components for a job recommendation engine. We start by describing the popular techniques of recommending items and the challenges of each technique. Then we explain the actual applied hybrid algorithm for job recommendation and the influential data to produce relevant jobs. We illustrate also the hybrid ranking algorithm that is employed in the recommendation engine and finally we discuss the different offline and online evaluation metrics to evaluate the proposed engine.

2 Popular Recommendation Techniques

Recommender systems are divided into many subcategories. Mainly content-based recommendation and collaborative filtering recommendation are the most researched and discussed methodologies. However, we researched many additional various recommendation methodologies to eventually build a robust hybrid job recommender system that relies on the advantages of each recommendation approach.

Content-Based Recommendation (CBR) is finding the similarity between the content of two profiles, as shown in Fig. 1. It is called cognitive filtering as well. It could be considered as an information retrieval problem. It mainly depends on analyzing the keywords and terms found in item profile and another item profile to calculate how they are similar to each other. CBR relies on the user individual behavior by tracking the user interesting items to recommend items that have similar content. Figure 1 illustrates the basic idea of CBR. Content-based recommendation has a ramp-up problem for new users also called user cold start problem where new users have no enough interactions to get suitable recommendations. Another problem is the data richness and data homogeneity. CBR performance relies on the features associated with the items. So if the content of items is not homogeneous and poor then the recommendation will not produce good results. However, this is not the case in e-recruitment where items here refer to jobs which have rich description and profile. Another problem is stability-plasticity problem. It is the converse of the user cold start problem. Stability-plasticity happens when user extensively use the system and rate a lot of items. Their profiles of preferences are established and difficult to change by minor interactions in the future [1–11]. But in real world users' taste evolves, so this becomes a problem. The solution for this is to gradually discount older ratings to have less influence. At the same time that can lead the recommendation engines to risk and loose information about the long-term users' interests.

Collaborative Filtering Recommendation (CFR) is a well-known technique which has proven its novelty. The fundamental argument of collaborative filtering is that if users X and Y rate n items similarly or have similar behaviors (e.g. buying, watching, listening), hence they will rate or act on other items similarly [3–8, 10, 11]. Figure 2 illustrates the idea of CFR. CFR relies on the crowd behavior of all users and each individual behavior. It tracks the interesting items for the currently processed user

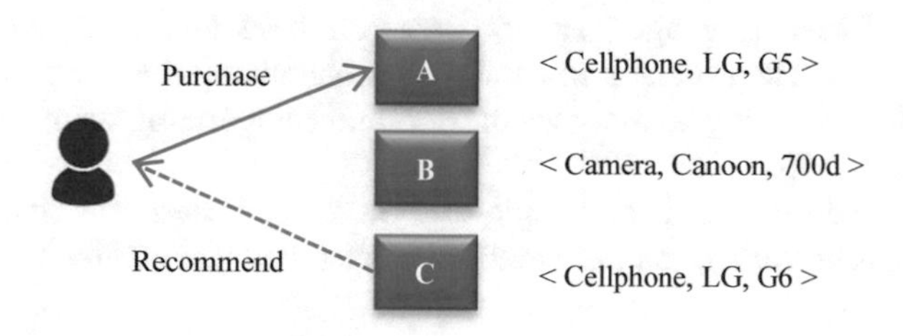

Fig. 1 Content-based recommendations

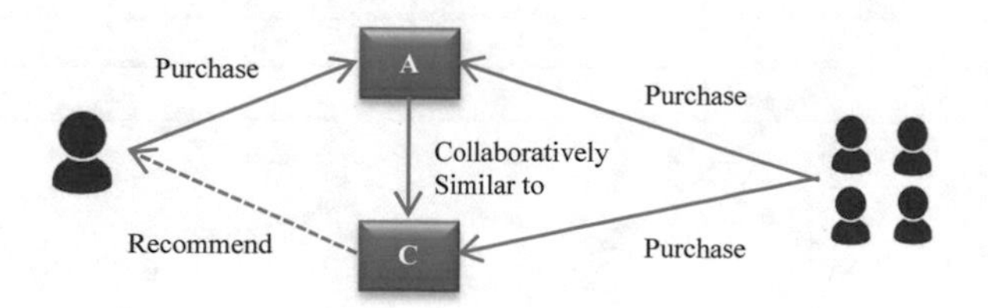

Fig. 2 Collaborative-filtering recommendations

and the crowd interesting items. Then it finds similar users and recommend what they like or finds similar potential items and recommend them.

CFR has two main techniques among the memory-based collaborative filtering. They both work on the user-item matrix that contains a row for each user and column for each item and the corresponding cell tells if the corresponding user likes the corresponding item or not. CFR uses this user-item matrix to find the similar users (user-based CFR) or the similar items (item-based CFR).

CFR is better than CBR in the diversity of recommendation. CBR will always recommend item similar to the items seen, liked or purchased by the user. However, CFR uses the latent relevancy relations between items by investigating the crowd behavior. CFR can recommend new item that user has not ever seen before. Nevertheless, CFR still faces some challenges as well. User-based CFR faces inaccurate recommendation for new users who have little interaction with the system. The same problem happens for the new items which have few ratings and few interactions from the users. This problem is a popular ramp-up problem in the collaborative-filtering approach. It is also known as the cold-start problem. CFR suffers from the sparse data problem (sparsity). This problem happens when the system has an extremely large set of items, but the active users are fewer. They like or view only small subset of the large pool of items. Thus, even the most popular items will have few ratings. It also will face scalability problems as the performance will decrease as long as data is increasing significantly, and the matrix of user-item become large and huge rapidly. As CFR keep recommending only the items that users interacted with and rated them highly, there will be a chunk of items that banned from recommendations because they do not have a lot of rating. That leads to the long tail problems, where a group of items get recommended a lot and another long tail of items does get recommended to users. CFR suffers from the stability-plasticity problem as well like CBR

Demographic-Based Recommendation (DBR) depends mainly on the demographic data of users (e.g. age, gender). Based on these data it works on clustering users into groups. Then it starts to handle all users among the same group equally [5, 6]. It is about creating stereotypes of users and categorizing them based on their demographic data [6, 11].

Following is an illustrative example to show how clustering is done. In Fig. 3, the clustering is performed on just small set of user data like gender, age, education,

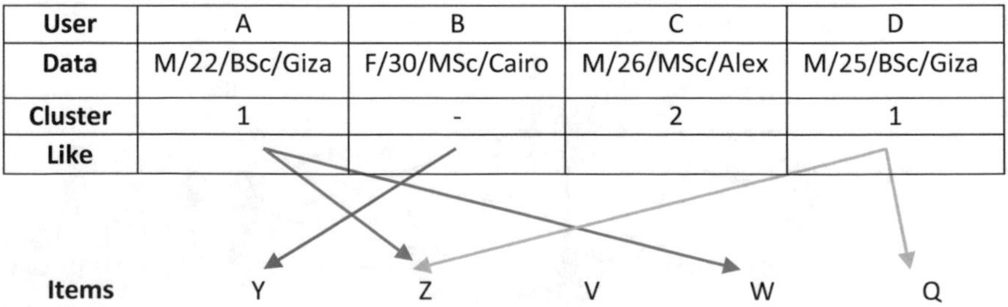

User	A	B	C	D
Data	M/22/BSc/Giza	F/30/MSc/Cairo	M/26/MSc/Alex	M/25/BSc/Giza
Cluster	1	-	2	1
Like				

Items Y Z V W Q

Fig. 3 Visualized example for demographic-based recommendation

Table 1 Example of different clusters used by demographic-based filtering technique

	Gender	Age	Education		City
Cluster 1	Male	22–26 years old	Bachelor's degree	...	Giza
Cluster 2	Male	26–34 years old	MSc degree	...	Alex
Cluster 3	Female	22–25 years old	Bachelor's degree	...	Cairo

location. There is a set of items {Y, Z, V, W, Q}. According to clusters in Table 1, both of users A and D are grouped to the same cluster (Cluster 1). It means that, based on the demographic's recommendation, interesting items for user D are going to be recommended to user A and vise-versa. Based on this methodology, item W will be recommended to user D and by the same way item Q will be recommended to user A.

DBR is not accurate in many cases. Generally, demographically classification is so primitive for quite personalized content. For example, not all 20 years old males who live in Cairo and engineering graduates will have the same tastes. It will not be a highly personalized recommendation in some applications especially where demographics do not play a high influence to decide users' preferences, this is the generalization problems. It is also difficult to change the cluster of users if their tastes are changed because they still have the same demographics. It is the stability-plasticity challenge. Another problem is the gray sheep problem, users that have a different taste than other users, these users have unusual taste. However, it is very important to be used as a supportive model to the core models in some cases.

Knowledge-based Recommendation (KBR) depends on patterns and rules extracted from exploiting deep knowledge of the interesting items for users. That can be done through deep understanding of the market and application context, or through capturing users' preferences via dialogs or questions in some applications, and then the recommendation system considers these preferences to build its own discrimination tree of item attributes [3–8, 14]. KBR is designed basically for complex domains where customers need to specify their preferences explicitly. For example, users need to specify explicitly the price of the apartment that they can afford or the job type (part-time, remotely, full-time..., etc.) that they want to occupy.

KBR will overcome the user cold start problem, but it is not easy to be used as the main recommender technique. Because it is hard to gain the proper deep knowledge, all needed patterns and rules to fit all user needs. It needs the users to state explicitly their preferences every time they visit the site or they have to fill a profile of preferences to get accurate recommendations. It also will not learn from user preferences and behaviors. User may have incomplete, outdated or inaccurate profile. In these cases, this approach will not learn from user behavior and will eventually produce inaccurate results.

We reviewed the various techniques of recommendation concepts and the applied recommendations. We summarized the challenges of each concept in Table 2. We can easily find that each technique has a different type of disadvantages and shortcomings that could be solved by another technique. Definitely there is no individual technique

Table 2 Summarization of recommendation technique shortcomings

Collaborative filtering	• User Cold-Start • Item Cold-Start • Scalability • Stability-Plasticity • Sparsity • Long Tail
Content-based	• User Cold-Start • Stability-Plasticity • Data Homogeneity • Data Richness • Overspecialization
Knowledge-based	• Knowledge Acquisition • Explicit Preferences
Demographic-based	• Generalization & Stereotyping • Gray Sheep • Stability-Plasticity

Table 3 Different approaches of building hybrid recommendation

Weighted	Scores of different recommendation approaches are combined together through predefined weights to produce one recommendation
Cascade	The result of one recommendation approach is refined by another approach
Switching	Switch between different recommendation approaches based on the current context
Mixed	The results of different recommendation approaches are presented together at the same time
Feature Combination	Outputs from different recommendations approached are used as input feature to one recommender
Feature Augmentation	One recommendation approach uses the output of another recommender's output as input feature
Meta-level	Learned model by one recommendation approach is used by another approach

that fits all or able to produce the most efficient recommendations without the support of other techniques.

Each technique has its limitations, so the optimal solution is to combine two or more of those approaches to overcome all the stated shortcomings.

Actually, in hybrid approach there are many of techniques to combine the different techniques. Burke [12, 13] presented different techniques to integrate between the different approaches of recommendations. Some of them are illustrated in Table 3.

The aforementioned approaches are presented in Fig. 4 to show the whole picture of different researched core personalized recommendation techniques.

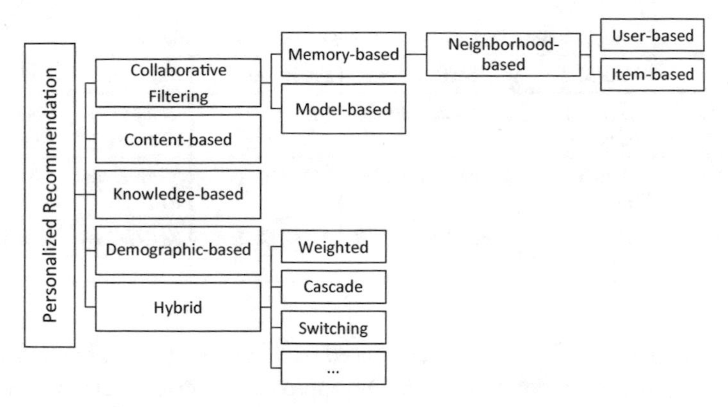

Fig. 4 Full hierarchy of personalized recommendations techniques

3 E-Recruitment Portals Adoption

We witnessed the usage of the aforementioned recommendation techniques in the real practical e-recruitment platforms. For example, LinkedIn and CareerBuilder employed successfully all of the practical CBR, CFR and KBR models in their e-recruitment portals.

Five well known online job recommender systems are investigated for a comparison purpose. The five investigated portals are CareerBuilder, LinkedIn from United Stated, Proactive from England, eRecruiter from Austria and Wuzzuf from the Egyptian market. CareerBuilder is a job search engine and online recruitment platform (https://www.carrerbuilder). LinkedIn is the biggest professional social network (https://linkedin.com) that has been acquired by Microsoft. Proactive is an England online recruitment platform that has applied recommendation modules to its website (https://www.proactiverecruitment.co.uk), eRecruiter is designed for enhancing the functionality and the accurateness of the absolventen.at (https://www.absolventen.at). Wuzzuf is the most popular, first and dominant e-recruitment platform in Egypt (https://wuzzuf.net).

Proactive uses the individual information as job-seeker profile content, including working experience, education, and skills. Not only individual information but also historical behaviors such as job applications and job searches are considered in LinkedIn, CareerBuilder, eRecruiter and Wuzzuf.

Proactive and Wuzzuf capture the user preferences and experiences from job-seeker manually. Not only filling profiles manually but also resume miner (RM) is available in LinkedIn, Careerbuilder and eRecruiter to inspect the job-seekers resume, extract the important keywords and generate the user profile.

Proactive, Wuzzuf and eRecruiter use a cascaded hybrid approach which uses knowledge-based (KBR) and content-based (CBR) recommendation [15, 16]. LinkedIn and CarerrBuilder use successfully a switching hybrid approach which

Table 4 Comparison of recommendations in online recruitment platforms

	LinkedIn	CareerBuilder	Proactive	eRecruiter	Wuzzuf
User profile	Info and behavior	Info and behavior	Info	Info and behavior	Info and behavior
Info capture	Manual and RM	Manual and RM	Manual	Manual and RM	Manual
Approach	KBR, CFR, CBR	KBR, CFR, CBR	KBR, CBR	KBR, CBR	KBR, CBR

uses knowledge-based (KBR), content-based (CBR) recommendation and collaborative filtering (CFR) techniques. Switching approach in LinkedIn is clearly obtained in the context of "who viewed this job also viewed" which is called browsemap by Linkedin [17, 18]. The comprehensive comparison of the five platforms is shown in Table 4.

4 Proposed Job Recommendation Engine

Our hybrid algorithm is a hybrid job recommendation model that consists of more than one approach to overcome the shortcomings of each individual recommendation approach. It is based on the aforementioned methodologies, content-based job recommendation (CBJR), collaborative filtering job recommendation (CFJR), demographics-based recommendation (DBJR), knowledge-based job recommendation (KBJR), reciprocal-based job recommendation (RBJR) and context-based job recommendation (CXJR) in addition to our developed modules which are behavioral-based job recommendation (BBJR), concept-based job recommendation (CPBJR) and ontology-based job recommendation (OBJR). The final output of these modules is interpreted eventually to a group of textual search quires and search cut-off filters to hit our job search engine and retrieve matching jobs from our job index. Figure 5 shows our hybrid engine modules through a simplified flowchart. We can clearly find each recommendation module which receives its specific kind of inputs to operate on and propagates the results and its own input to the next module to continue the rest of operations. For example, the behavioral-based recommendations module works on the applying and searching activities of each job-seeker to retrieve some recommended jobs. Then it passes both of applying and searching activities to the collaborative-filtering recommendation module as well. Now the collaborative-filtering module uses this data as well to find other job-seekers who applied or searched similarly to the current processed job-seeker. By tracking these similar job-seekers on other different jobs, the module will collaboratively find the similar jobs that could be interested for the current processed job-seeker and so on. Finally, the models combiner will combine the produced search queries and filters from these various modules to form a final search query. This final search query is sent to hit our job index and retrieve a set of recommended results.

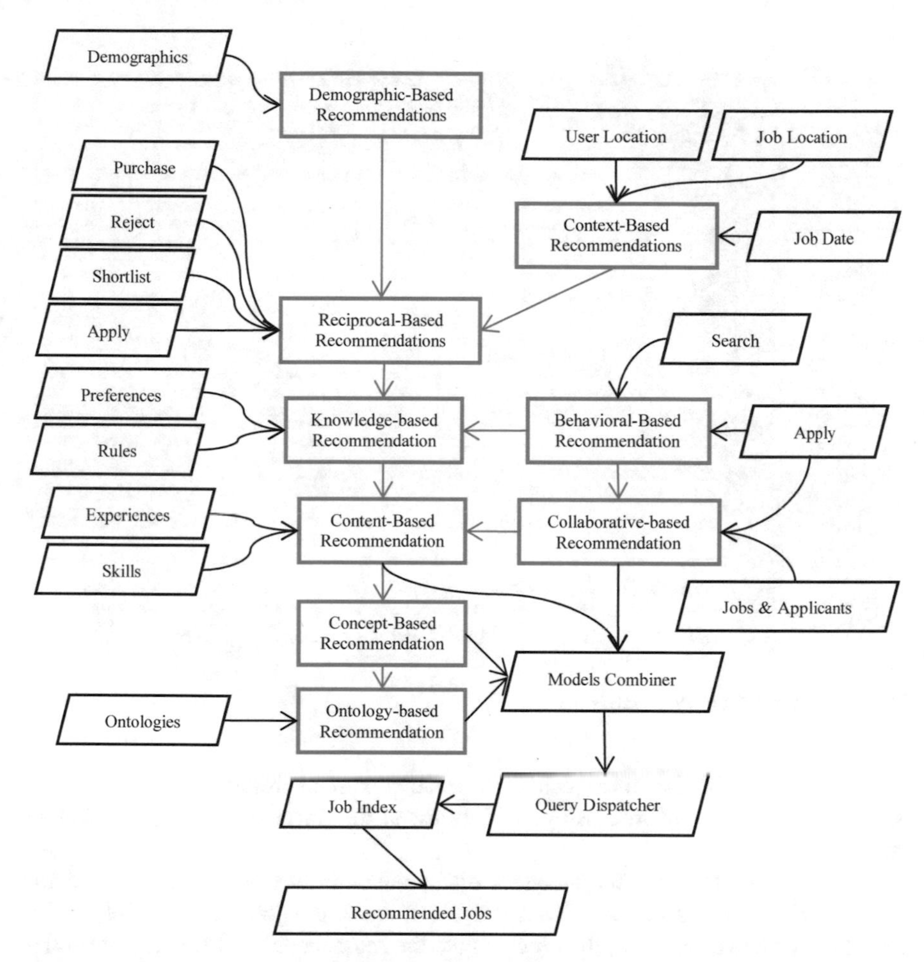

Fig. 5 Modules of our job hybrid recommendation engine

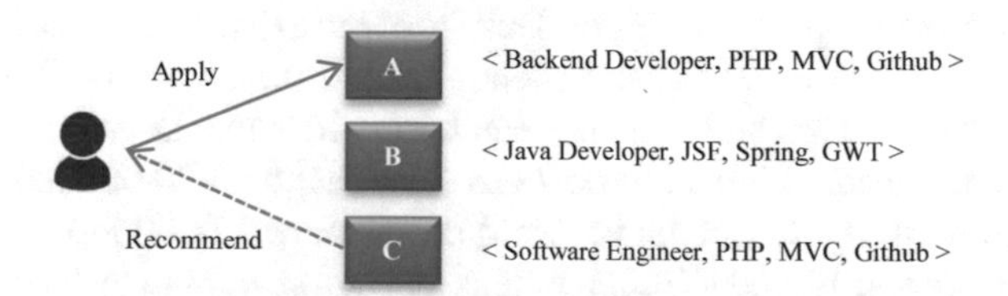

Fig. 6 Content-based job recommendation

4.1 Hybrid Personalized Job Recommendation Algorithm

Content-based job recommender (Fig. 6) and knowledge-based job recommender are the core modules which works on retrieving the jobs that are textually similar to the content of the job-seeker profile and the previous job applications based on the

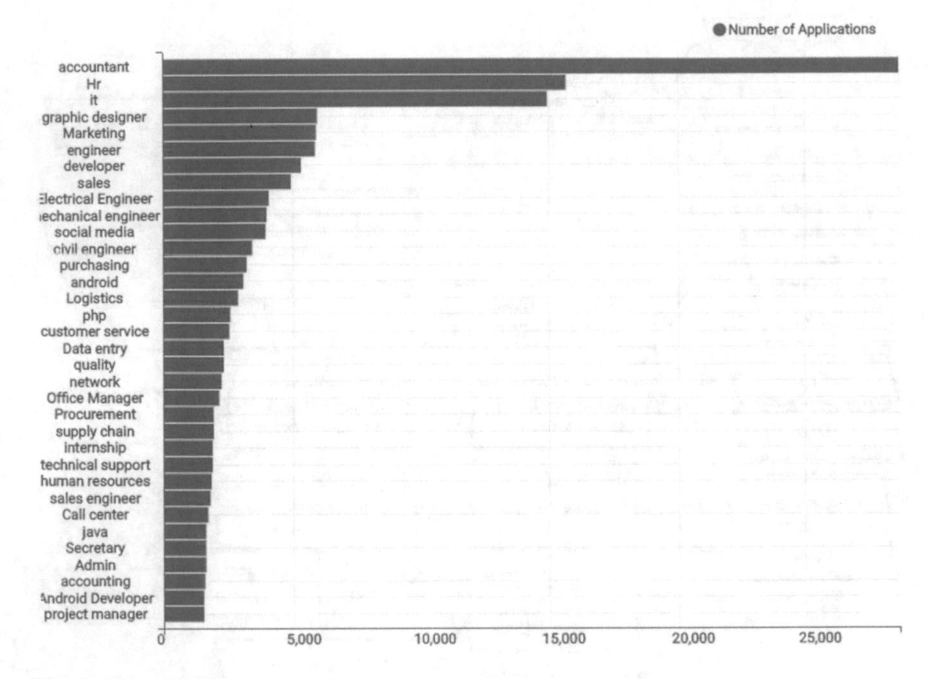

Fig. 7 Job applications via search

concepts of TF-IDF (term frequency—inverse document frequency), VSM (vector space model) and text processing. We discussed this layer particularly in another publication.

Behavioral-based job recommender tracks the activities of job-seeker, it learns from user's previous actions. For instance, it track the jobs that the job-seeker searched for and applied to, then it calculate the most recent and frequent job-titles of these jobs to utilize them in the recommendations even if these job-titles were not present in the job-seeker's profile. Figure 7 shows the real behavior of job-seekers regarding search and apply interactions. Despite of the existence of recommendations and navigations in this e-recruitment portal, but job-seeker did not find what they want in recommendations and uses the search to write some keywords and apply to one of the search resulted jobs. Figure 7 clarifies that about 30,000 job applications are conducted by using the search keyword *accountant,* 17,000 job applications by using keyword *hr* and 15,000 job applications through searching by keyword *graphic designer*

An example to clarify the importance of this module, the job-seeker's profile includes only *human resources executive*, but he searched frequently for *hr* and applied to job from the search results (e.g. personnel officer). Then the behavioral-based job recommender will consider that the *hr* jobs are interesting for that job-seeker. The example is clarified in Fig. 8.

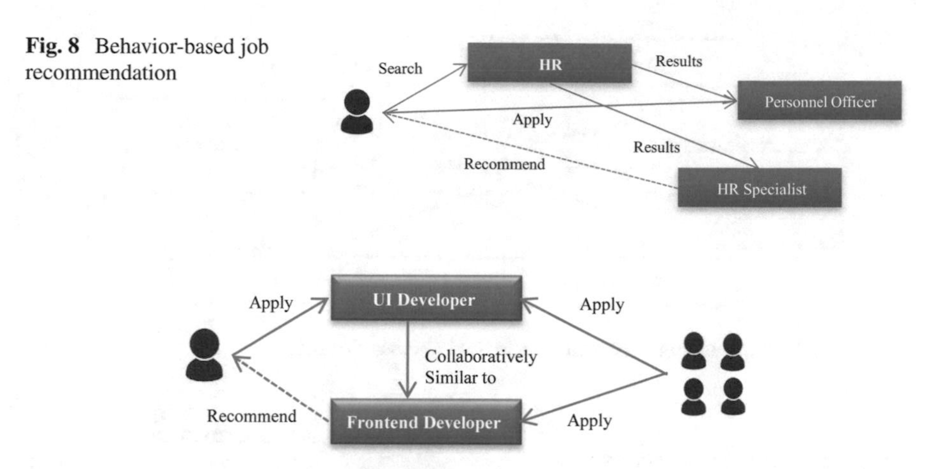

Fig. 8 Behavior-based job recommendation

Fig. 9 Collaborative-filtering job recommendation

Concept-based job recommender tries to augment our search queries by using the job-seeker's profile content (content-based recommendations) to retrieve the recommended jobs. Then it extracts the most frequent keywords found in these content-based recommended jobs to augment the recommendation query and retrieve the conceptually similar jobs. As example, most of the recommended jobs for *javaScript developer* are titled *frontend Engineer* or contain the keyword *jQuery* frequently. It means these keywords will be helpful for further recommendation [19].

Collaborative-filtering job recommender finds what the job-seeker applied to, Thereafter it tries to find the most similar jobs to this job x. It assumes if many applicants of who applied to job x applied also to job y, that means that job y is somehow similar to job x and will be interesting to the job-seeker who applied before to x. A practical example is illustrated in Fig. 9.

CFJR, BBJR and CPBJR modules were also significantly helpful for generating a relational ontology for job-titles and keywords. For example, detecting instances like دليفيرى, دلفري, ديليفيري, ديلفري. They all are different used spelling variations for the same job *delivery boy* jobs in Arabic language which job-seekers collaboratively applied to. Another example is مسئول موارد بشرية, شئون موظفين, موظف شئون عاملين. They are detected as synonyms like *human resources officer*, *personnel specialist*, *hr*. Figure 10 shows an example of automatically detected and manually annotated frequent vehicle driving job titles in Arabic e-recruitment platform

Demographic-based job recommender is based on clustering job-seekers based on demographic-data and extracting the specific rules of each segment behaviors with regards to the demographic-data required by the employer. For example, DBJR recommend job that requires males for males only (Another case-study of education-based clustering is explained in the data analysis section). Figure 11 shows how DBJR works.

Knowledge-based job recommender solved the cold-start problem by retrieving explicitly preferred results and the tolerated results of DBJR, the results that cannot

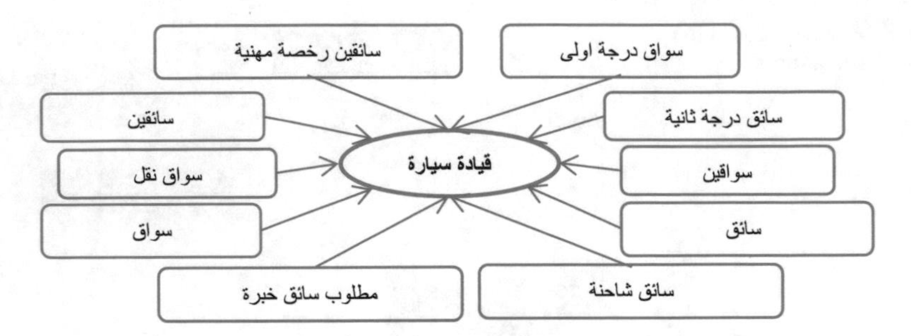

Fig. 10 Relational ontology between vehicle driving jobs in Arabic

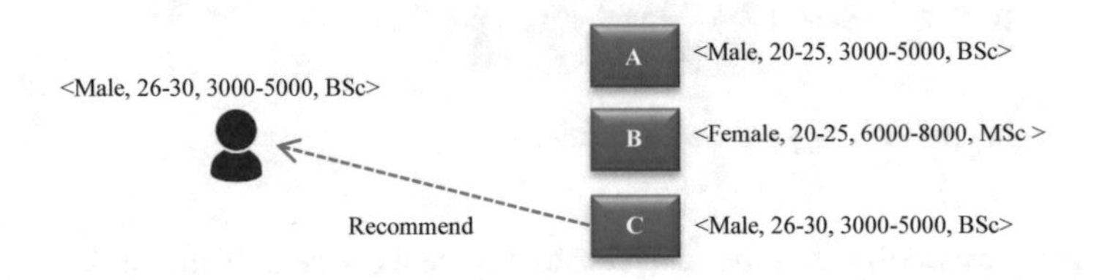

Fig. 11 Demographics-based job recommendation

retrieved through content-based recommendation neither demographic-based recommendation. We analyzed our data to extract the generic rules of our users' behaviors. As example, we figured out that employers tolerate their jobs' age requirements but they are rarely tolerating regarding their jobs' gender requirements. It means that 22 years old job-seeker can be interested in another job which requires 24 minimum age and the employer will probably tolerate and accept this applicant (Reciprocal-based job recommender system).

Reciprocal-based job recommender system is about considering the bi-directional interest of both job-seekers and recruiters. We are inspired by the dating applications where both of requester and recipient should be interested in each other. As analyzing the behavior, interactions and interest of job-seekers (one-way analysis), we realized that analyzing also the behavior and reactions of recruiters is an extremely influential factor that should be considered (two-ways analysis) [20]. The same example we used in KBJR is the age difference between job-seeker age and the required age by employer. Job-seekers do not care about this difference and keep applying but this interest should be verified by the employer interest as well. The approach is clarified in Fig. 12. Another example about job recency is illustrated by graphs in the data analysis section.

Ontology-based job recommender is a simple recommendation version which tries to fetch some other jobs which are considered among the same hierarchy/specialty of the previous recommended jobs [21], like SEO, social media, e-marketing, market research are previously grouped and labeled as marketing jobs. A practical example of ontological related marketing job titles in the recruitment market is shown in Fig. 13.

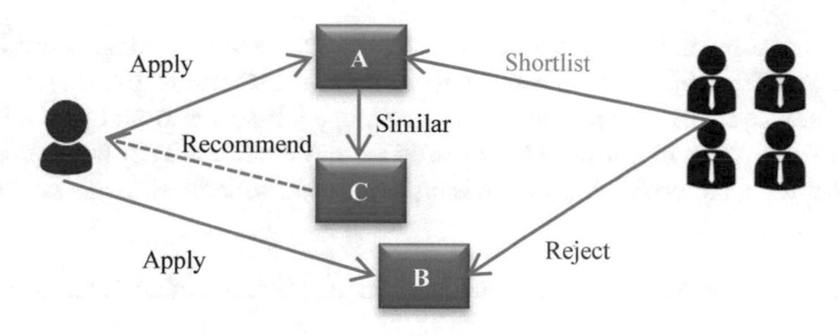

Fig. 12 Reciprocal-based job recommendation

Fig. 13 Relational ontology between marketing jobs

Context-based or context-aware job recommender system takes various types of contextual information into consideration during recommendations. Such contextual information includes posted date of jobs, location of job-seeker and location of jobs. The use of such contextual information incredibly improves the quality of the recommendations. Context-based recommender systems are highly powerful because the underlying domain-knowledge is relevant to a wide variety of domain specific preferences [10]. Our context-aware job recommender system is a time-sensitive recommender system. The recent ratings of job-seekers have higher weights than old ratings. Job-seekers may apply to some jobs long time ago then become interested in another type of jobs or a higher level of seniority. The old ratings of the crowd also have no meaning for the current collaborative filtering recommendations. For example, the recommendations for a job may be very different at the time of publish from the recommendations generated several years later for job-seekers. We analyzed deeper the factor of job posted date and obtained the results in the data analysis section. Our context-aware job recommender system is location-based recommender system as well. It considers the location of job-seeker and the location of jobs. Job-seekers are often interested in location-based recommendations. As we explained in our studies in the data analysis section, job-seekers wish to determine the closest work venue based on their previous history of preferences for other jobs. For example, software engineers from Cairo wish to land a software engineering job in Cairo not Alexandria. This location-based context-aware recommender is user-specific locality which means that the geographical location of job-seekers has an influential role in their preferences. For example, a job-seeker from Cairo might not have the same job preferences as a job-seeker from Alexandria. They both can be software engineers but each job-seeker wishes the closest suitable software engineering job.

Algorithm. Generally, the algorithm of hybrid job recommendation flows as follow

```
For each user u
  Process KBJR refined by DBJR, BBJR, CXBJR and RBJR
  Run CBJR refined by all of KBJR, DBJR and BBJR
  Run CFJR by using crowd behavior and individual BBJR
  If (recommendations are few)
    Generate CPBJR and OBJR in conjunction with CBJR
Return top ranked jobs J
```

We illustrated visually the integration of variant components of the hybrid job recommendation algorithm that we explained above. Figure 14 shows the supposed recommended entities (green jobs) based on the job-seeker individual interactions (user A) on various entities (e.g. blue jobs, profile), the job-seekers crowd interactions and the reciprocal reactions of employers. For example, user A the currently processed job-seeker will have the recommended job F because it is similar to his/her prefilled profile. Job G will get recommended to user A because it is collaboratively similar to job Z which user A has applied to.

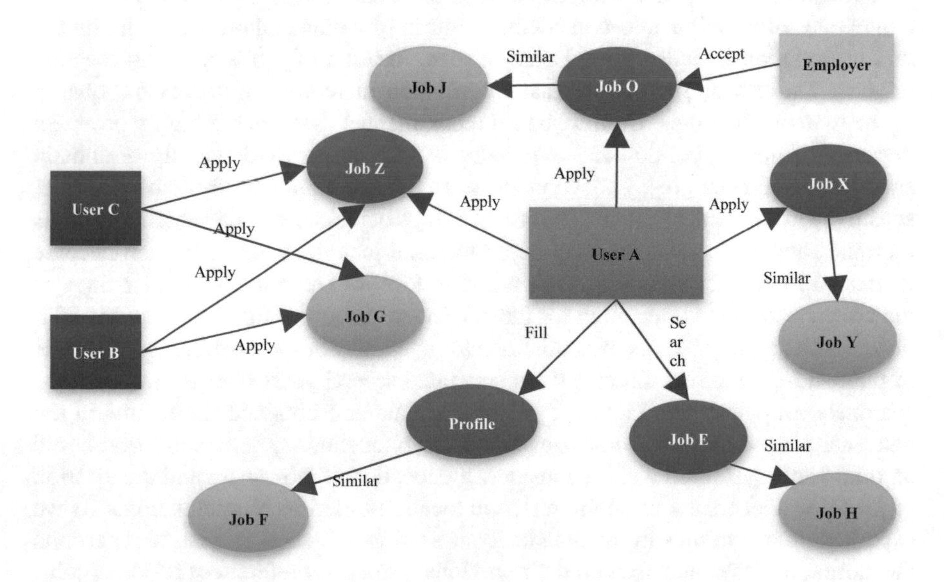

Fig. 14 Results of combining some modules of the hybrid job recommendation algorithm

4.2 *Recommendation Retrieval Queries (Textual Data)*

The algorithm uses the generated keywords that are generated by the hybrid personalized job recommendation algorithm to transform them eventual to the recommendation queries. These extracted keywords from each job-seeker's profile and previous behavior are used to search in our job index. As example, keywords and job titles of the jobs that job-seeker recently applied to, previous experiences, skills, previously conducted search queries. This information helps the recommender system to understand the preferences of each job-seeker. This combined data generated from different recommendation modules is manipulated to be fed into our search engine to retrieve the textually matched jobs. We discussed the details of this textual similarity engine in another publication.

4.3 *Recommendation Retrieval Cutoff Filters (Attributes-Based Data)*

It is not just about matching the textual content to produce meaningful recommendations. There is the attribute-based matching which is responsible for cutoff irrelevant jobs from recommendations. It manipulates the matching of the important attribute-based data that extracted from job-seeker and job profile (illustrated in the data section) like age, gender, educational level, salary…, etc. All these attributes are further manipulated also by our ranking model to rank the most relative jobs first as discussed in the ranking section.

5 Data Analysis

Our data sample is collected from a real online robust e-recruitment platform. We divided the data into two main divisions as shown in Fig. 15. The first part is the profile-based data, and the other one is the behavioral-based data.

Profile-based data is the profile data that statically filled by the user manually during the process of registering to the e-recruitment platform. For job-seeker, these filled data could be attribute-based or textual-based. Attribute-based data refers to the structured data such as gender, country, seniority, and educational level as illustrated in Fig. 16. But the textual-based fields are the unstructured data where users can fill them freely with any possible data, such as previous occupations, activities, skills, and trainings. The same structure of the job profile data. The job profile is divided into attribute-based data and textual-based data. Attribute-based data contains fields such as required min/max age, gender, and offered salary. Textual-based fields of the job are job title, job description…, etc.

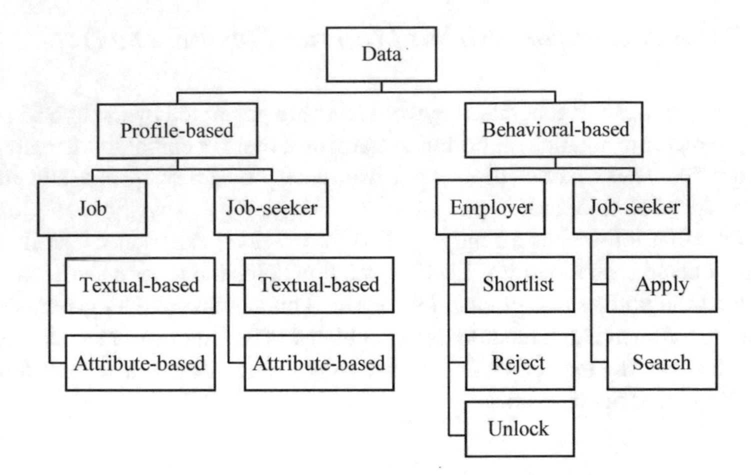

Fig. 15 Dataset hierarchy

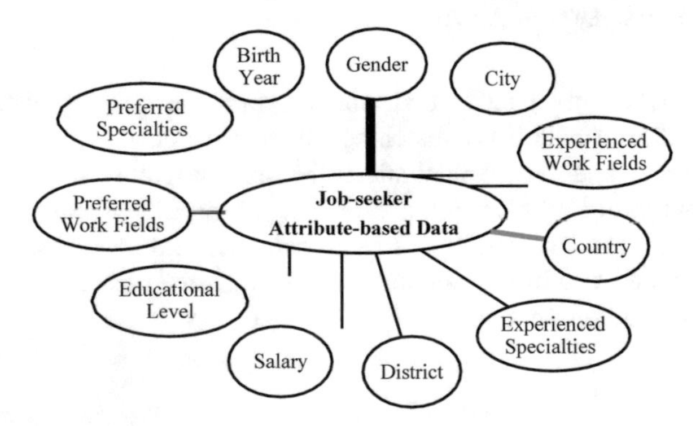

Fig. 16 Job-seeker attribute-based data

Behavioral-based data reflects the behavior of the user while interacting with the platform. Certainly every user type has different actions to perform. Job-seeker can search for specified jobs in addition to apply to the suitable job. The employer receives the applications of job-seekers, screens job-seekers' profiles to shortlist or reject to finally accept only who fit into that position and unlock the contact information of the job-seeker as a final step to be able of contacting him/her.

5.1 *Feature Selection of Job-Seeker and Job*

Feature selection is the phase of deciding which job-seeker features and job features should be considered to be used in the recommendation algorithm. This phase is another data mining task which should be analyzed at the early stages of building a job recommendation engine. This process objective is to find the most influential features on the recommendation quality. The most influential features/attributes are defined based on four pillars: user surveys, job-seekers' online behavior, employers' online behavior and our domain-knowledge. Briefly this pure data mining task involves anomaly detections to ignore outliers, and association rules as support, confidence, lift and conviction to assert which job attributes will attract the job-seekers to apply, and which job-seeker attributes will bias the employer to shortlist.

Following is an example of analyzing job-seekers behavior and employer interactions with respect to job recency (posted date). The graph in Fig. 17 demonstrates the interest of job-seekers mainly is in recent jobs. Their applying trend reaches the peak on jobs that are posted only one day ago, but the applying demand starts in inclining. The applying demand for jobs older than 7 days forms 20% only of the total demand. This generic rule means that jobs posted within 7 days ago gain 80% of the job-seekers interest.

The employer activities are similar also to job-seekers activities frequency. As it is illustrated in the Fig. 18 that employers first-screening phase whether shortlisting or rejecting has an increase in the first day after posting the job then a little decrease

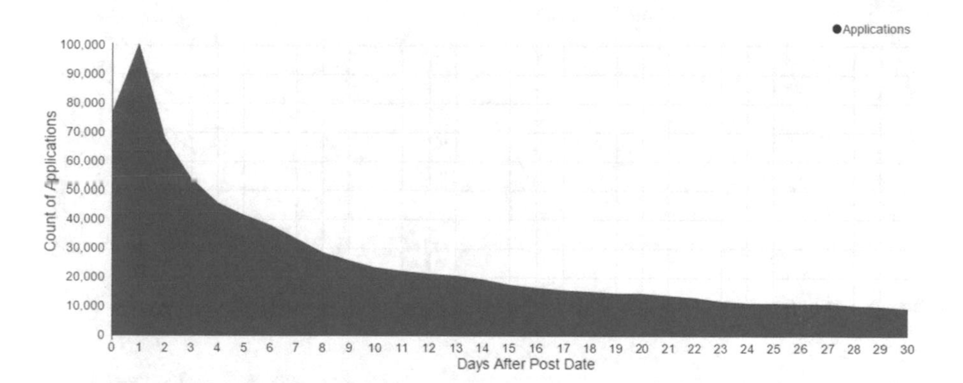

Fig. 17 Job applications frequency along time after job posting

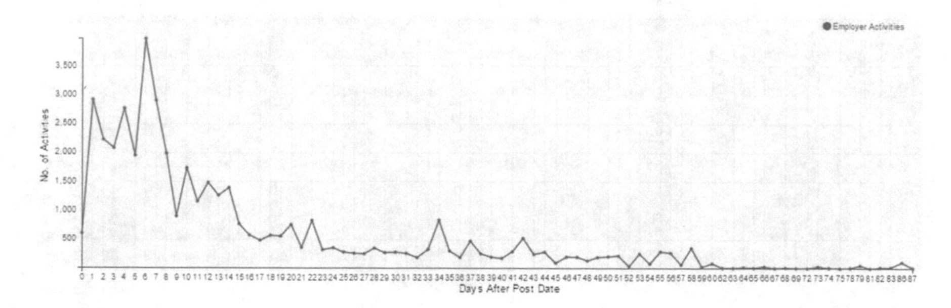

Fig. 18 Employer first-phase reactions on job applications along time after job posting

occurs before having a peak at the sixth day. Finally activities frequency has a sharp incline after the sixth day.

Employers behave similarly as well in the second-screening phase which is unlocking job-seeker contact information and purchasing this profile. But the different was that activities peak comes at the first day after the job publish date but eventually the same as above activities, they gradually decline as shown in Fig. 19.

The aforementioned recency analysis shows us that time is an important factor in e-recruitment domain. This makes it mandatory to build a context-aware job recommender system that recommend recent jobs for job-seekers and guarantee fresh instant applications for recruiters as we previously mentioned in the hybrid algorithm section.

By similar analysis to user surveys, job-seekers' online behavior and employers' online behavior and the usage of association rules of data mining beside to our domain knowledge, we generated Fig. 20 which is a specific cluster sample of education-based clustering for blue-collars job-seekers.

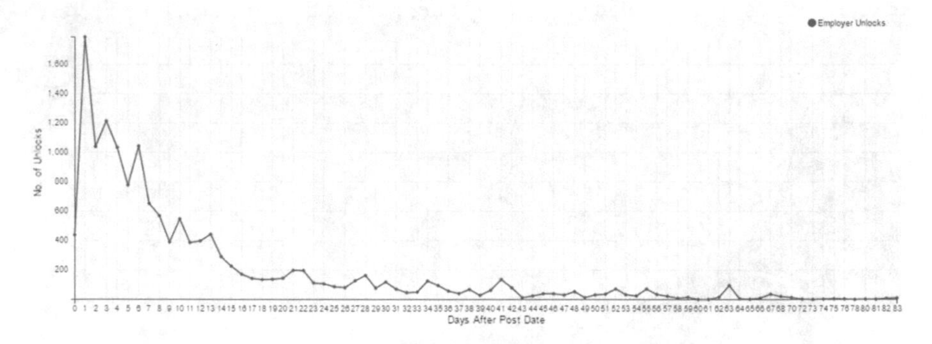

Fig. 19 Employer second-phase reactions on job applications along time after job posting

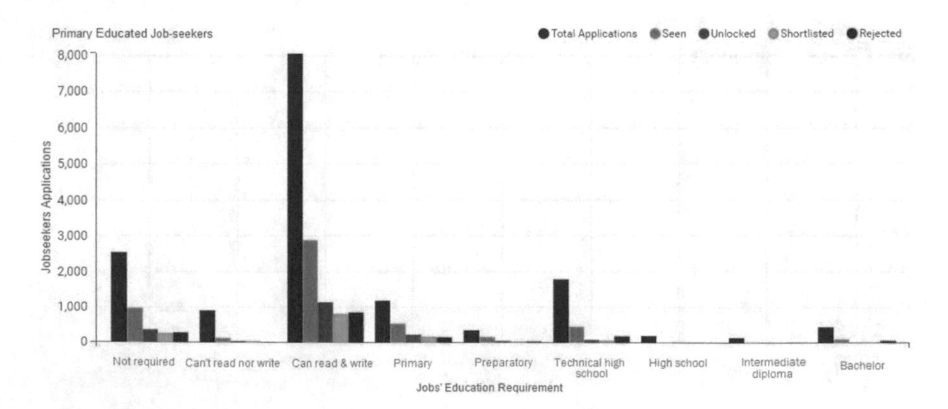

Fig. 20 Behavior of employers and job-seekers based on education clustering

Gray bar is the total job applications, orange bar is the seen applicants by the employer/recruiter, blue bar is the count of unlocked applicants by the employer/recruiter, green bar is the count of shortlisted applicants, red bar is the count rejected applicants and finally values on the y-axe are the different education levels required by the job.

Above chart shows clearly the interest of primary education job-seekers through the gray bar; on the other hand, it clarifies the employers' reactions through the other bars (orange, blue, green and red). It is easy to obtain that there is a high job-seekers interest in applying to jobs which require the following education: "not required", "can read & write" and "technical high school". Because of few jobs requiring primary education, the applications rate of primary education job-seekers on this segment is lower than others, therefore the count of jobs in each segment should be considered as well. Equation 1 shows exactly how to calculate a normalized applying interest for each cluster of job-seekers (e.g. primary educated job-seekers). It calculates the maximum percentage of total job applications to total jobs per each education degree

$$Applying\ Interest_{primary\ educated\ JS} = \max_{edu} \frac{Count\ of\ Apps_{edu}}{Count\ of\ Jobs_{edu}} * 100 \quad (1)$$

On the other side; employers who posted jobs with "technical high school" as a required education are not interested in those primary education job-seekers. But the other employers who posted "not required", "can read & write" and "primary" as education requirements are interested in this cluster of job-seekers, this observation is deducted through analyzing seen, shortlist and unlock activities. Equation 2 shows exactly how to calculate a normalized profile purchasing interest for employers on each cluster of job-seekers (e.g. primary educated job-seekers). It calculates the maximum percentage of total profile purchases to total viewed profile per each education degree

$$Purchase\ Interest_{primary\ educated\ JS} = \max_{edu} \frac{Count\ of\ Purchases_{edu}}{Count\ of\ Viewed\ Apps_{edu}} * 100 \quad (2)$$

Actually, we need to maximize both of the job-seekers interest (applying rate) and employers' interest (purchasing rate). Therefore, the equation to calculate the overall reciprocal interest for each job-seekers cluster (e.g. primary educated job-seekers) across all educational levels will be defined as follow

$$Overall\ Reciprocal\ Interest_{primary\ educated\ JS} = \max_{edu}(Purchase\ Interest_{edu} + Applying\ Interest_{edu}) \quad (3)$$

Similar analysis concept is applied on the salary feature whether the expected salary of job-seekers or the offered salary by employers. It aims to determine how users behave towards the differences in this factor. The idea of this analysis is to find the most efficient formula that can optimally increase the acceptance rate of job-seekers to the offered salary and acceptance rate of employers to the applicant expected salary.

Figure 21 illustrates the behaviors of job-seekers and employers in different contexts. The graph is divided into five groups, each group declares different situation where the gray bar is the total cleansed job applications, orange bar is the seen applicants by the employer, blue bar is the count of unlocked applicants/ purchased profiles by the employer, green bar is the count shortlisted applicants and red bar is the count of rejected applicant.

The first bar group visualizes the job applications where job-seekers' expected salaries were among the job offered salaries ranges (in range). It shows that roughly 47% only of the job applications were in range (job-seeker salary is higher than or equal to the job minimum salary and lower than or equal to the job maximum salary).

The second group visualizes the job applications where job-seekers' expected salaries were lower or higher than the job offered salaries ranges (out of range). It shows that roughly 53% of the job applications were out of range (job-seeker salary is less than the job minimum salary or higher than the job maximum salary).

Although the job-seekers applying rate is higher in the out of range jobs but the seen rate, shortlisting rate and unlocking rate (purchase CV) were higher for the in range job applications. Not only but also the rejection rate was higher in the out of range job applications. The seen rate, shortlisting rate, purchase CV rate and rejection rate percentages out of the corresponding totals were 51%, 53% 56% and 45% respectively in the in range job applications while the percentages in the out of range job applications were 48%, 46%, 42% and 54% respectively

The third group is the corresponding totals of the cleansed job applications where noisy data and salary outliers from job-seekers and jobs are omitted.

Thereafter the ideal scenario is defining the customized formula that maximize the jobs-seekers interest (total applications the gray bar) and employers reciprocal

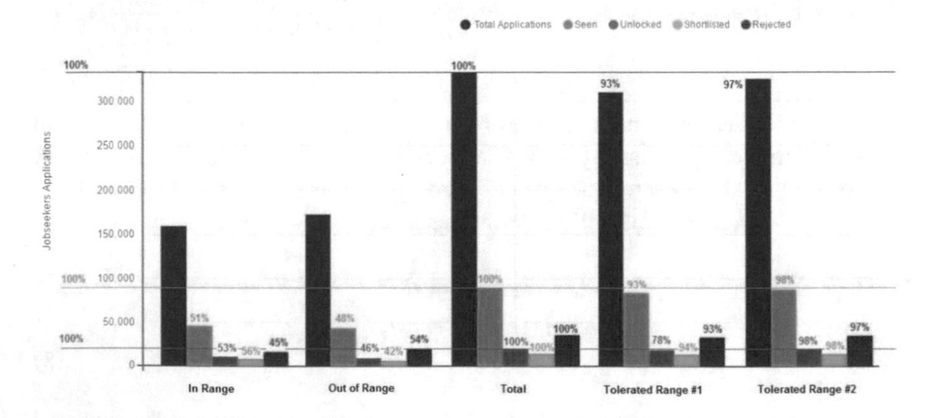

Fig. 21 Behavior of employers and job-seekers based on salary variances

interest (seen, shortlisted and purchased profiles). The fourth and fifth ones visualize the job applications where job-seekers' expected salaries were within the job offered salaries ranges but plus/minus different margins (μ, φ) according to Eq. 7.

Similar analyses are applied on all attributes to summarize the important attributes from job-seeker profile and job profile. Extracted attributes from job-seeker profile are gender, age, city, district, education, salary, experienced main work fields, experienced specialties, preferred main work fields, job-seeker keywords and preferred specialties. Extracted attributes from job profile are gender, minimum age, maximum age, city, district, minimum and maximum salary, education, job main work field, job specialty, job keywords, and job posted date.

6 Recommendations Ranking Algorithm

One of the important components in the recommendation engine is the results ranking methodology. The ranking algorithm attempts to order most related jobs in an ascending order by combining three major factors which are relevancy, proximity and recency. Therefore, the first recommended job should be the newest (recency) and nearest (proximity) most relevant (relevancy) job to each job-seeker.

6.1 *Recommendations Relevancy Ranking (RRR)*

Recommendation Relevancy Ranking (RRR) discusses some of the formulas that contribute in calculating the relevancy score. RRR retrieves a set of recommended jobs J which is similar to the constructed job-seeker profile P. That job-seeker profile P is generated by the aforementioned hybrid recommender engine. For example, the job-seeker profile P contains the filled data (knowledge), individual actions (behavior), crowd actions (collaborative)...etc. For example, job-seeker wrote *Angular.js* among his/her skills (knowledge), searched and applied to *Frontend Developer* (behavior), and job-seekers who applied to *Frontend Developer* also applied to *UI Developer* (collaborative). All of these keywords *Angular.js, Frontend Developer* and *UI Developer* will form the job-seeker profile P to find the relevant jobs for these constructed terms. As shown in Eq. 1, Jobs J that contains more of these terms will have higher relevancy score

$$J < P_1^{job\ title}, \ldots P_k^{job\ title}, P_1^{Skill}, \ldots P_y^{Skill}, \ldots P_1^{Keyword}, \ldots P_z^{Keyword} > \quad (4)$$

As we explained in the algorithm section, demographics data is taken into considerations for demographics-based job recommendations (DBJR). For example, gender matching is processed to filter out and rank jobs. Jobs that require the same job-seeker's gender U^{gender} will have higher relevancy score than the non-matching jobs. The same concept is applied on other fields such as educational level (educational level case-study is explained in the data analysis section). Another relevancy factor is the numerical attributes relevancy like age and salary (salary case-study is explained in the data analysis section). Age matching formula is defined as follow to retrieve only the jobs requiring the same age range of job-seeker's age U^{age}

$$J^{minAge} \leq U^{ge} \leq J^{maxAge} \tag{5}$$

However according to our conducted data analysis, there is some tolerance can be added to the age. So, the correct formula to recommend jobs regarding the age is

$$J^{minAge} - \delta \leq U^{age} \leq J^{maxAge} + \alpha \tag{6}$$

These two tolerance parameters δ and α are not constant. They are variables adjusted and increased along the recommendations generation. They are important influencer of the recommendation ranking formula. Therefore, the relevant jobs with lower values of δ and α are most likely to have a higher relevancy score than similar jobs with higher values of δ and α. For example, the job that requires 25 minimum age ($\delta = 0$) will get a higher relevancy score than any job requires 26 minimum age ($\delta = 1$) when recommending job for a 25-year-old job-seekers.

The same concept is applied for salary matching. The most suitable job to job-seeker is the job that matches the exact desired job-seeker's salary. Actually job-seekers can accept job if the offered salary is a bit lower than the expected salary. But certainly job-seeker will not mind if the recommended job offers higher salary than the expected one.

$$J^{minSalary} - \varphi * U^{salary} \leq U^{Salary} \leq J^{maxSalary} + \mu * U^{salary} \tag{7}$$

6.2 Recommendations Proximity Ranking

Recommendations Proximity Ranking (RPR) works on the job geographical locations and the job-seeker geographical location. It retrieves jobs within a dynamically changed diameter around the job-seeker location. RPR returns only jobs that are around the job-seeker's residence $U^{residence}$ by a variable radius β

$$J^{location} \leq U^{residence} + \beta \tag{8}$$

This β variable radius is adjusted and expanded along the recommendations generation. Radius β is used among an equation to define the proximity score of job. It means that recommended jobs with lower values of β have higher proximity score than other recommended jobs with β values. Because lower value of β means that this recommended job is closer to the job-seeker's residence $U^{residence}$ but higher

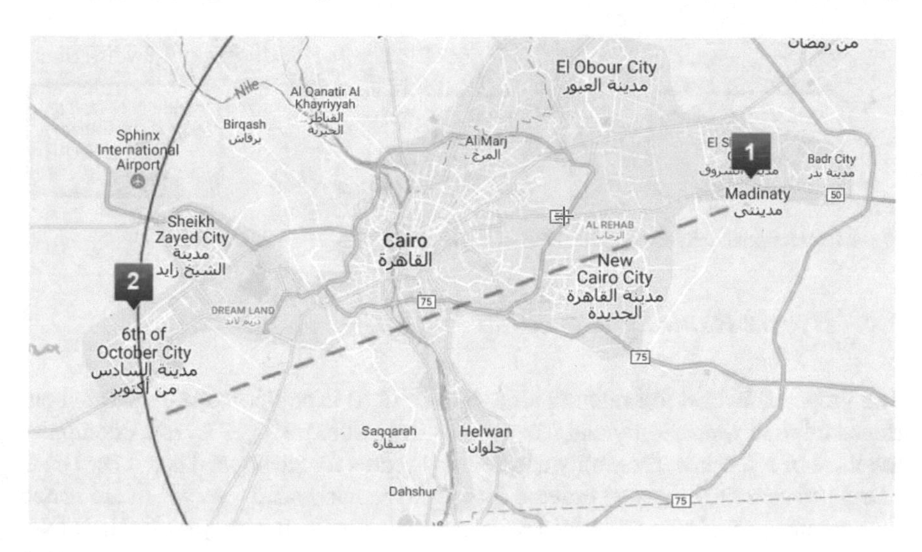

Fig. 22 Illustration of proximity recommendation

value of β means that this recommended job is farther to the job-seeker's residence $U^{residence}$.

Figure 22 illustrates a case study based on Egypt map. It shows that Madinaty residents will get nearby jobs first before farther jobs, but they will not get jobs farther than the upper limit of β. According to this case study example, the upper limit of β is at 6th of October City. Location labeled with number 1 is Madinty while location labeled with number 2 is 6th of October City and the red dotted line refers to β. This β is not a constant value. Actually, it is a personalized variable based on some personal parameters like the job-seeker ability to relocate and number of recommendations generated to this job-seeker. So if the currently processed job-seeker tends to relocate by stating this explicitly or by applying to far work premises then this β radius will increase and farther jobs will be recommended. Another parameter is the number of recommendations, if this job-seeker got a very low number of recommendations then the value of β increases a little bit to get a reasonable number of recommended jobs.

6.3 *Recency Ranking*

Job recency is another important factor that is extracted based on our data analysis that discussed previously. Therefore, it is a vital factor to consider when thinking in results ranking. Jobs with older date have lower recency score than other recent jobs.

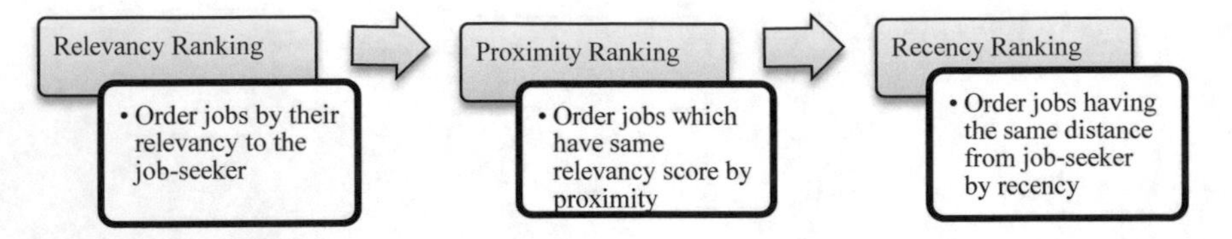

Fig. 23 Recommended jobs ranking phases

6.4 Hybrid Ranking Algorithm

The proposed hybrid job ranking algorithm (HJRA) is composed of the aforementioned relevancy, proximity and recency ranking methods (Fig. 23). HRA computes the three of relevancy, proximity and recency scores for each job. Then it uses each calculated score to return an ordered list of the recommended jobs which are sorted by relevancy, proximity and recency scores respectively. It sorts jobs by their relevancy score. If a relevancy tie occurred (two jobs have an equal relevancy score), it sorts the tied jobs by proximity score. If another proximity tie occurred (two jobs have an equal relevancy and proximity score), it sorts the tied jobs by recency score. The pseudo-code of HRA is illustrated as follow.

```
For each recommended job j to user u
  Compute a relevancy score j.rs between u and j
  Compute a proximity score j.ps between u and j
  Compute a recency score j.rcs between j and time.now()
  Insertion_sort(j,J) by inserting j into sorted J
Return top ranked jobs J, ordered by rs then ps then rcs
```

7 Accuracy Evaluation

It is important when developing a new system to set some kind of criteria for evaluating that proposed system. There are numerous metrics whether offline or online metrics. Offline metrics are precision, recall, F1 measure and normalized discounted cumulative gain (nDCG), however the results of offline metrics may be misleading. Therefore online and real-world evaluations represent probably better methods to evaluate such platforms and assist precisely the users' satisfaction [22].

Online and real-world business metrics are user conversion rate, click-through rate (CTR), time to first click, first click rank. Job application conversion rate is one of the accuracy and business key metrics [22–24], it is about how many actions (apply) are conducted on recommended jobs relative to the total actions (apply) which we will adopt soon. But we evaluated this engine via first click rank and click rank metric which refer to the rank of the clicked jobs. We evaluated by using Google Analytics

Fig. 24 Clicks rank on recommended jobs list

	Event Label	Total Events
		6,610 % of Total: 2.24% (294,861)
1.	1	1,736 (26.26%)
2.	2	994 (15.04%)
3.	3	771 (11.66%)
4.	4	580 (8.77%)
5.	5	476 (7.20%)
6.	10	445 (6.73%)
7.	6	436 (6.60%)
8.	7	397 (6.01%)
9.	8	393 (5.95%)
10.	9	382 (5.78%)

to count the clicks on each job in the list. The result is that the first recommended job got the highest click rate then the second then the third job, the full list of clicks rank on the top 10 results is shown in Fig. 24.

8 Conclusion

We presented our developed data-driven information filtering framework for dynamically hybrid job recommendation. Our framework purpose is improving the domain of e-recruitment by providing online recruitment platforms with a highly personalized content to both of job-seekers and employers to reach the highest level of users' fidelity. Our purpose is designing a hybrid job recommendation algorithm which is based on content-based recommendation, collaborative-filtering recommendation, demographic-based recommendation, behavioral-based recommendation, reciprocal-based recommendation, knowledge-based recommendation, ontology-based recommendations and finally the context-based time sensitive and location-based recommendation system. We also illustrated the data analysis methodologies and the most influential recommendation data that should be fed into the hybrid job recommendation algorithm. We also clarified how these data-attributes should be used as input to our hybrid algorithm in addition to the context-aware ranking criteria for establishing a powerful job recommendation model. Eventually we illustrated the needed metrics to evaluate such engine in the e-recruitment industry. The evaluation of IRE-DJRS proved its efficiency by the click-through rank of real online users who used this functionality extensively.

References

1. Liu, J., Dolan, P., Pedersen, E.R.: Personalized news recommendation based on click behavior. In: Proceedings of the 15th International Conference on Intelligent User Interfaces, Hong Kong, China, pp. 31–40 (2009)
2. Zhu, C., Zhu, H., Xiong, H., Ding, P., Xie, F.: Recruitment market trend analysis with sequential latent variable models. In: Proceedings of the 22nd ACM SIGKDD International Conference on Knowledge Discovery and Data Mining, California, USA, pp. 383–392 (2016)
3. Zheng, S.T., Hong, W.X., Zhang, N., Yang, F.: Job recommender systems: a survey. In: Proceedings of the 7th International Conference on Computer Science & Education (ICCSE 2012), Australia, pp. 920–924 (2012)
4. Lu, Y., El Helou, S., Gillet, D.: A recommender system for job seeking and recruiting website. In: Proceedings of the 22nd International Conference on World Wide Web Companion, pp.963–966 (2013)
5. Owen, S., Anil, R., Dunning, T., Friedman, E.: Mahout in Action (2012)
6. Ricci, F., Rokach, L., Shapira, B., Kantor, P.: Recommender Systems Handbook (2011)
7. Al-Otaibi, S.T., Ykhlef, M.: A survey of job recommender systems. Int. J. Phys. Sci. **7**(29), 5127–5142 (2012)
8. Pandya, S., Shah, J., Joshi, N., Ghayvat, H., Mukhopadhyay, C., Yap, M.H.: A novel hybrid based recommendation system based on clustering and association mining. In: Proceeding of the 10th International Conference on Sensing Technology, China (2016)
9. Hong, W., Zheng, S., Wang, H., Shi, J.: A job recommender system based on user clustering. J. Comput. **8**(8), 1960–1967 (2013)
10. Aggarwal, C.C.: Recommender Systems Text Book (2016)
11. Owen, S., Anil, R., Dunning, T., Friedman, E.: Part 1: recommendations: making recommendations. In: Mahout in Action (2012)
12. Burke, R.: Hybrid web recommender systems. In: The Adaptive Web: Methods and Strategies of Web Personalization, vol. 4321, pp. 377–407 (2007)
13. Burke, R.: Hybrid recommender systems: survey and experiments. User Model User-Adap. Interact. **12**(4), 331–370 (2002)
14. Burke, R.: Knowledge-based recommender systems. Encycl. Libr. Inf. Sci. **69**, 175–186 (2000)
15. Lee, D.H., Brusilovsky, P.: Fighting information overflow with personalized comprehensive information access: a proactive job recommender. In: Proceedings of the Third International Conference on Autonomic and Autonomous Systems, USA (2007)
16. Hutterer, M.: Enhancing a job recommender with implicit user feedback. Fakultät für Informatik, Technischen Universität, Wien (2011)
17. Yuan, J., Shalaby, W., Korayem, M., Lin, D., AlJadda, K., Luo, J.: Solving cold-start problem in large-scale recommendation engines: a deep learning approach. In: Proceeding of IEEE International Conference on Big Data, DC, USA (2016)
18. Wu, L., Shah, S., Choi, S., Tiwari, M., Posse, C.: The browsemaps: collaborative filtering at LinkedIn. In: Proceedings of the 6th Workshop on Recommender Systems and the Social Web, CA, USA (2014)
19. AlJadda, K., Korayem, M., Ortiz, C., Russell, C., Bernal, D., Payson, L., Brown, S., Grainger, T.: Augmenting recommendation systems using a model of semantically-related terms extracted from user behavior. In: Proceeding of the Second CrowdRec Workshop, Austria, pp. 1409–1417. ACM RecSys (2014)
20. Yu, H., Liu, C., Zhang, F.: Reciprocal recommendation algorithm for the field of recruitment. J. Inf. Comput. Sci. **8**(16), 4061–4068 (2011)
21. AlJadda, K., Korayem, M., Grainger, T., Russell, C.: Crowdsourced query augmentation through semantic discovery of domain-specific jargon. In: Proceeding of IEEE International Conference on Big Data (Big Data), USA (2014)
22. Beel, J., Langer, S.: A comparison of offline evaluations, online evaluations, and user studies in the context of research-paper recommender systems. In: Proceeding of the 19th International Conference on Theory and Practice of Digital Libraries, Poland (2015)

23. Liu, R., Ouyang, Y., Rong, W., Song, X., Tang, C., Xiong, Z.: Rating prediction based job recommendation service for college students. In: Proceeding of International Conference on Computational Science and Its Applications (ICCSA), China (2016)
24. Paraschakis, D., Nilsson, B.J., Hollande, J.: Comparative evaluation of top-n recommenders in e-commerce: an industrial perspective. In: Proceeding of IEEE 14th International Conference on Machine Learning and Applications, USA (2015)

Semantic Image Analysis for Automatic Image Annotation

Brahim Minaoui and Mustapha Oujaoura

Abstract The semantic image analysis is a process aimed to extract automatically the accurate semantic concept from image visual content. In this chapter, we shed light on a study conducted to boost the efficiency of semantic image analysis for improving automatic image annotation. We have explored, in this research study, different aspects of semantic analysis to overcome the semantic gap problem. In the first part of this research study, we evaluated the impact of the way in which the different low level features were associated on image annotation performance. In the second part, the benefit of combining two complementary classifiers for objects classification was investigated. The final part of the research study was devoted to the assessment of the usefulness of regrouping adjacent regions of segmented image for objects discrimination.

Keywords Image · Semantic analysis · Annotation · Semantic gap · Generative · Discriminative · Classifier · Segmentation

1 Introduction

Over the last years, semantic image retrieval has become an active research activity [1–6]. The objective of this research study is to develop un efficient image retrieval method by keywords describing semantic content of the image. To achieve this goal, images would to be correctly annotated. This is why multiple research efforts have focused on elaborating automatic techniques to perform image annotation [7–10].

B. Minaoui (✉)
Laboratory of Information Processing and Decision Making, Department of Physics, Faculty of Science and Technology, Sultan Moulay Slimane University, Mghila, PO Box. 523, Beni Mellal, Morocco

M. Oujaoura
Laboratory of Informatics, Mathematics and Communication Systems, Departement of Computer Science, Networks and Telecoms, National School of Applied Sciences, Cadi Ayyad University, Road Sidi Bouzid, PO Box. 63, Safi, Morocco

N. Gherabi and J. Kacprzyk (eds.), *Intelligent Systems in Big Data, Semantic Web and Machine Learning*, Advances in Intelligent Systems and Computing 1344,
https://doi.org/10.1007/978-3-030-72588-4_4

Despite the progress that has been made in this context, the automatic image annotation still suffers from the semantic gap problem [3, 11–14]. To solve this most challenging problem in computer vision, scientist researchers rushed to develop perfect semantic Image analysis techniques to efficiently mapping low level features to high level semantic concepts [15–19]. This research activities lead to propose various semantic image analysis approaches for many sciences and industry applications such as object detection, pattern recognition, semantic segmentation, indexing and retrieval of images, etc. In this chapter, we present a study aimed at highlighting exploration and exploitation of different opportunities offered by some semantic analysis tools for improving automatic image annotation accuracy. This research study includes three parts:

- Comparative study of associating low level features;
- Investigation of combining generative and discriminative classifiers;
- Exploration of regrouping adjacent regions of segmented image.

2 Comparative Study of Associating *Low Level Features*

2.1 Introduction

Recently, in order to improve the quality of image annotation, a number of research efforts [20–29] has focused on the problem of how to extract, from low-level features, the semantic concepts that interpret well the contents of image (objects, themes, events). They have shown that image annotation techniques should integrate different types of low-level visual features of visual data. None of the feature descriptor is sufficient to tackle intra-class diversity and inter-class correlation in an effective or efficient way. Therefore, an effective features combination for image annotation and retrieval has become a desirable and promising perspective from which the semantic gap can be further bridged.

Motivated by this fact, we are interested to search a way to efficiently combine the strength of diverse and complementary features (including color, texture and shape descriptors) for semantic image annotation.

In this research work, in an attempt to achieve this goal, we conducted a comparative study of two approaches for associating three different kinds of visual features: RGB color histogram [30], Haralick texture [31] and Legendre moments descriptors [32].

2.2 Experiments and Results

In this section, we analyze and compare the performance of two approaches for combining low-level visual features for semantic image annotation.

In order to achieve this goal, we conduct two experiments on three image databases ETH-80 [33], COL-100 [34] and NATURE [35]. Figure 1 shows some examples of image objects from these three image databases used in our experiments.

In the phase of learning and classification, we use a training set of 40 images and a test set of 40 images for each image databases.

In all experiments, the features are extracted after image segmentation by region growing [36]. For each region that represents an object, 10 Legendre moments (L00, L01, L02, L03, L10, L11, L12, L20, L21, L30) and 16 elements for RGB color histograms are extracted from each color plane namely R,G and B. The number of

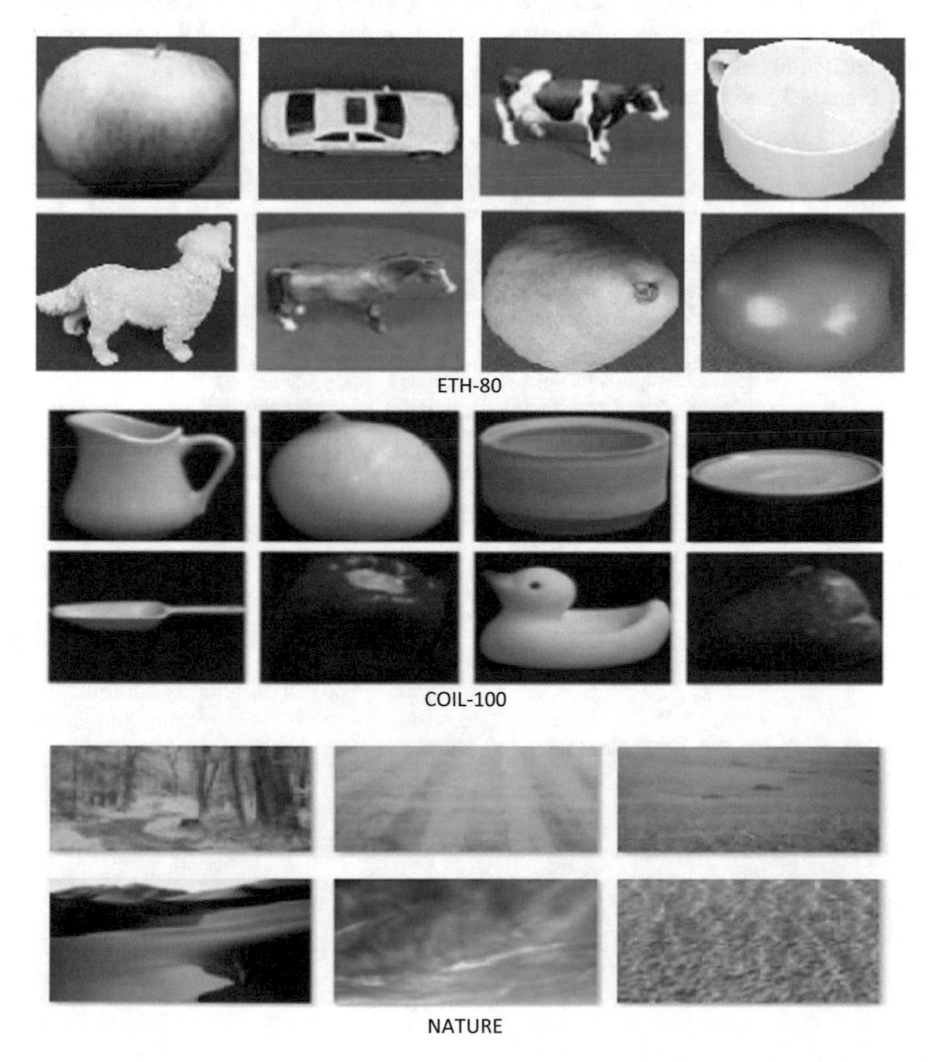

Fig. 1 Some examples of images for objects from ETH-80, COL-100 and NATURE databases

input features extracted using Texture extraction method is 14 Haralick indices for each one of the six co-occurrence matrices leading to 84 textures attributes.

2.2.1 Experiment 1.1

In this experience, we realize a semantic image annotation using an approach that groups all types of low-level visual features, extracted from a region image, in a single vector used as the input of the Bayesian network classifier [37–41].

The experimental approach adopted in this experiment is represented by the Fig. 2.

Table 1 summarizes the global average annotation rate obtained in this experiment for each image database.

Figures 3 and 4 show the confusion matrix.

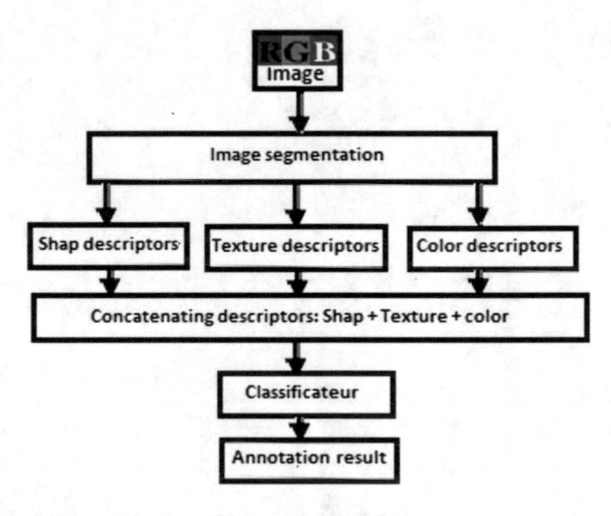

Fig. 2 Experimental approach adopted in experiment 1.1

Table 1 Global average annotation rate and error rate

Database	Global average Annotation rate	Error rate
ETH-80	87.50%	12.50%
COIL-100	82.50%	17.50%
NATURE	90.00%	10%

Fig. 3 Confusion matrix for images of database ETH-80

True \ Predicted	apple	car	cow	cup	dog	horse	pear	tomato
apple	5	0	0	0	0	0	0	0
car	0	4	0	0	1	0	0	0
cow	0	1	4	0	0	0	0	0
cup	1	0	0	4	0	0	0	0
dog	0	1	0	0	4	0	0	0
horse	0	0	0	0	0	5	0	0
pear	0	0	0	0	0	0	5	0
tomato	1	0	0	0	0	0	0	4

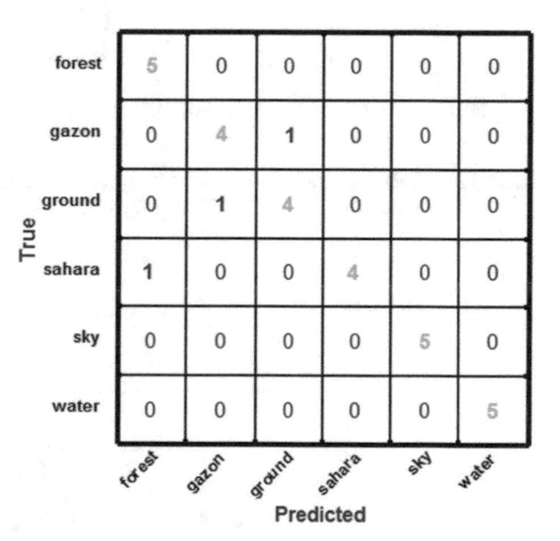

Fig. 4 Confusion matrix for images of database NATURE

2.2.2 Experiment 1.2

In this experiment, we realize a semantic image annotation using an approach that combines all different types of visual features separately in parallel scheme. Each type of descriptors is used as the input of a Bayesian networks classifier. The annotation decision is realized by the vote of combined classifiers corresponding to the different types of descriptors. The outputs of the classifiers are combined in the decision level to produce a final score for semantic image annotation.

The experimental approach adopted in this experiment is represented by Fig. 5.

Table 2 summarizes the global average annotation rate obtained in this experiment for each image database.

Figures 6 and 7 show the confusion matrix.

RGB
Image
Image segmentation
Shap descriptors
Texture descriptors
Color descriptors
Classificateur
Classificateur
Classificateur
Vote and decision
Annotation result

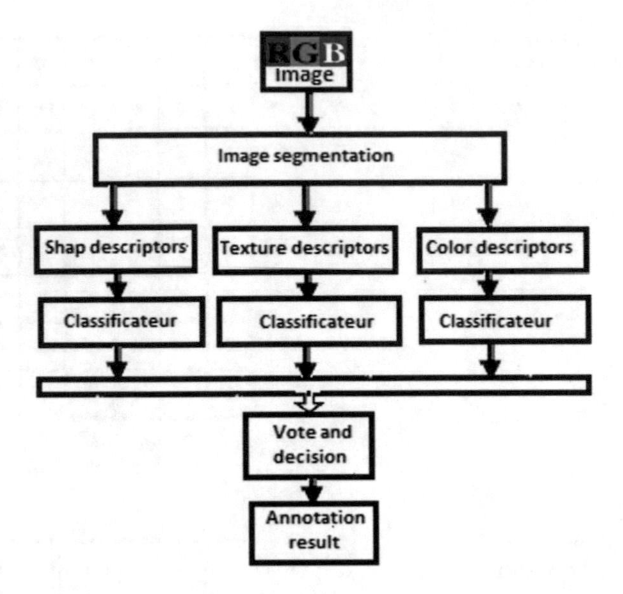

Fig. 5 Experimental approach adopted in experiment 1.2

Table 2 Global average annotation rate and error rate

Database	Global average Annotation rate	Error rate
ETH-80	90.00%	10.00%
COIL-100	85.00%	15.00%
NATURE	93.33%	6.77%

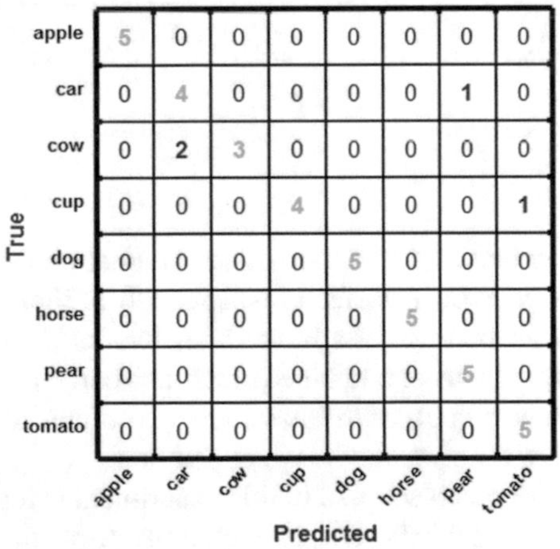

True
apple
car
cow
cup
dog
horse
pear
tomato
5 0 0 0 0 0 0 0
0 4 0 0 0 0 1 0
0 2 3 0 0 0 0 0
0 0 0 4 0 0 0 1
0 0 0 0 5 0 0 0
0 0 0 0 0 5 0 0
0 0 0 0 0 0 5 0
0 0 0 0 0 0 0 5
apple
car
cow
cup
dog
horse
pear
tomato
Predicted

Fig. 6 Confusion matrix for images of database ETH-80

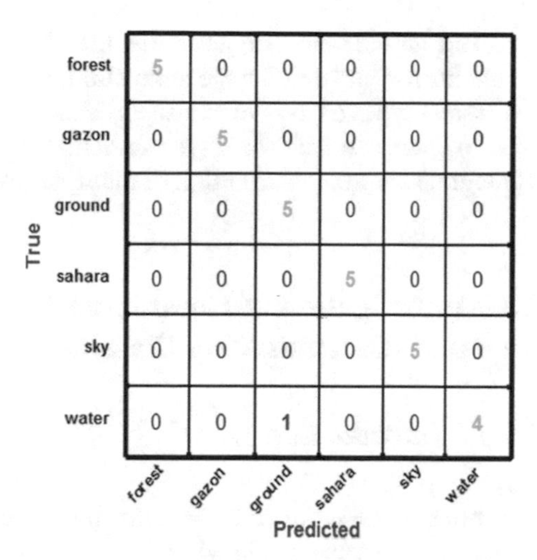

Fig. 7 Confusion matrix for images of database NATURE

2.3 *Analysis of Results*

As observed from Tables 1 and 2, the second approach for combining low-level visual features including color, texture and shape descriptors, produces better global average annotation rates for all the three images databases. Also, analysis of confusion matrix presented by the Figs. 3, 4, 6 and 7, shows that the individual annotation rates obtained for some objects (dog, tomato, and ground) with this approach are better than those obtained with the first combining approach. So, it appears from these remarks that the combination of different types of low level visual features in parallel scheme will improve the semantic image annotation rates. This rate can be further improved by increasing the number of combined descriptors. In the case of descriptor fusion, it is not obvious because the size of the vector of the merged descriptors becomes very large.

2.4 *Conclusion*

In this research work, we have studied two different approaches for combining low-level visual features including color, texture and shape descriptors for semantic image annotation via Bayesian networks classifier.

Analysis and comparative results, obtained from experiments realized on three image databases, have shown that combining the different types of visual features in parallel scheme gives better annotation accuracy than merging these features together in one vector.

Our investigation suggests that the most fruitful approaches, bringing practical benefits for semantic image annotation, will involve an appropriate combination of different types of low-level visual features.

In future research work, we would like to develop other combination schemes that integrate the joint distribution of multiple features.

3 Investigation of Combining Generative and Discriminative Classifiers

3.1 Introduction

In order to overcome the semantic gap, a number of current research efforts focus on robust classifiers achieving automatically multi-level image annotation [42–47]. These classifiers can be characterized as generative and discriminative according to whether or not the distribution of the image and labels is modeled. It was observed that generatively-trained classifiers perform better with very few training examples and provide a principled way of treating missing information, whereas discriminatively-trained classifiers perform better with sufficient training data and provide flexible decision boundaries [48]. Motivated by these observations, several researchers have proposed a variety of techniques that combine the strengths of these two types of classifiers. These hybrid methods, which have delivered promising results in the domains of object recognition [49–51], scene classification [52–56] and automatic image annotation [57, 58], have been explored in different ways. Authors in [50] and [52] propose a classifier switching algorithm to select the best classifier (generative or discriminative) for a given dataset and availability of label. In [51, 55] and [56], authors propose a technique for combining the two classifiers based on a continuous class of cost functions that interpolate smoothly between the generative strategy and the discriminative one. [49, 53, 54] and [57] propose a hybrid generative-discriminative approach in which the features extracted from a generative model are analyzed by a followed discriminative classifier. [58] devise a hybrid generative-discriminative learning approach that includes a Bayesian Hierarchical model (generative model) trained discriminatively.

It appears from this study that the object image annotation problem is basically a classification problem and there are many different modeling approaches for the solution. These approaches can be classified into two main categories such as generative and discriminative. An ideal classifier should combine these two complementary approaches.

In this chapter, in an attempt to highlight the benefit of combining generative and discriminative classifiers, we assess an approach which combines in a parallel scheme the Bayesian networks [37–41] for the generative model and the neural networks [59–61] for the discriminative classifier to accomplish the task of automatic image annotation. The annotation decision is achieved by the vote of combined classifiers.

Each classifier votes for a given keyword. The keyword that has the maximum of votes will be considered as the proper keyword for the annotation of an object in a query image.

3.2 *Experiments and Results*

In this section, we study and compare the performance of discriminative and generative classifiers for automatic image annotation, using in a first time each classifier alone and in a second time the combination of the two different classifiers. In order to achieve this goal, we conduct two experiments on three image databases ETH-80 [33], COIL-100 [34] and NATURE [35]. Figure 3 above shows some examples of image objects from these three image databases used in our experiments.

In the phase of learning and classification, we used a training set of 40 images and a test set of 40 images for each image databases. In all experiments, the features are extracted after image segmentation by region growing. For each region that represents an object, 10 components of Legendre moments (L00, L01, L02, L03, L10, L11, L12, L20, L21, L30) and 16 elements for RGB color histograms are extracted from each color plane namely R, G and B. The number of input features extracted using Texture extraction method is 14 Haralick indices multiplied by 6 co-occurrence matrices. This gives 84 textures attributes.

3.2.1 Experiment 2.1

In this experiment, we provide comparative results of image annotation between the two classifiers: discriminative (neural networks) and generative (Bayesian networks). The experimental method adopted in this experiment is represented by Fig. 5. In first time, we have used three neural networks classifiers to annotate images of all databases. Each neural networks, receiving as input one of the three extracted descriptors, votes for a given keyword. The keyword that has the maximum of votes is considered as the proper keyword for the annotation of an object in a query image.

In second time, we repeated the same operation with Bayesian networks classifier as shown in Fig. 5.

Table 3 summarizes the results of automatic image annotation for each type of classifier, and Figs. 8 and 9 show the confusion matrix.

3.2.2 Analysis of Results

As observed from Table 3, Bayesian networks produce better average annotation rates for all the tree images databases. However, analysis of confusion matrix presented by Figs. 8 and 9 shows that the individual annotation rate obtained for some objects (cow, cup, Sahara and Gazon) with neural networks can be better than those obtained

Table 3 Average annotation rate and error rate

Database	Classifier	Global average Annotation rate	Error rate
ETH-80	neural networks	87.50%	12.50%
	Bayesian networks	90.00%	10.00%
COIL-100	neural networks	82.50%	17.50%
	Bayesian networks	85.00%	15.00%
NATURE	neural networks	90.00%	10.00%
	Bayesian networks	93.33%	6.77%

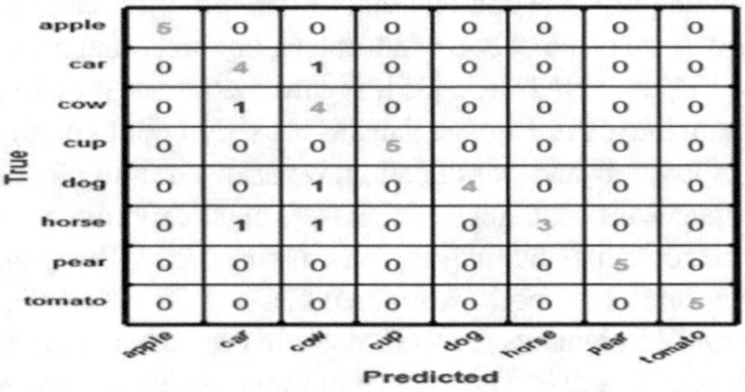

Fig. 8 Confusion matrix for images of database ETH-80 by using neural networks

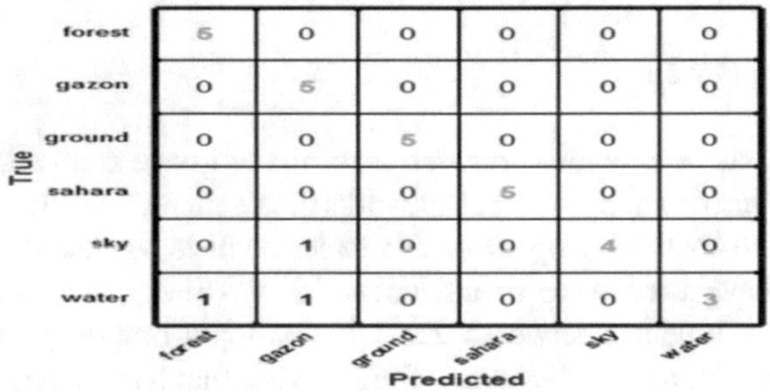

Fig. 9 Confusion matrix for images of database NATURE by using neural networks

with Bayesian networks. So, it appears from these remarks that the combination of these two classifiers will improve the average annotation rates. This constitutes the aim of the next experiment (2.2).

3.3 Experiment 2.2

Based on the remarks released in the previous two experiments, we combined in this experiment, in addition to descriptors, neural networks and Bayesian networks in order to gain the benefit of the complementarity of these two approaches of classification (discriminative and generative). The block diagram shown in Fig. 10 illustrates the principle of this combination. Thus, with the combination of the three types of descriptors described above and the 2 considered types of classifiers, there will be a maximum of votes equal to $3 \times 2 = 6$. Each classifier with each descriptor votes for a given keyword. The keyword with maximum votes will be deemed as the proper keyword for the annotation of an object contained in a query image.

Table 4 shows the average image annotation rate obtained by combining neural networks and Bayesian network classifiers and Figs. 11 and 12 shows the confusion matrix.

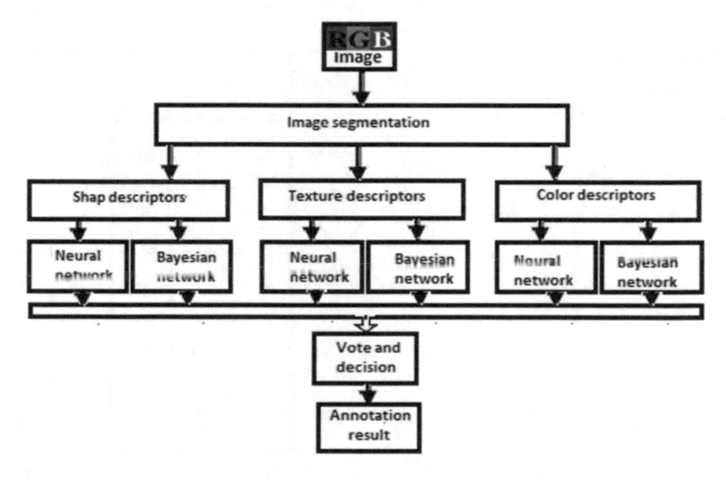

Fig. 10 Block diagram that illustrates the principle of combining discriminative and generative classifiers for automatic image annotation

Table 4 Average annotation rate and error rate

Database	Global average Annotation rate	Error rate
ETH-80	92.50%	7.50%
COIL-100	87.50%	12.50%
NATURE	96.67%	3.33%

apple 5 0 0 0 0 0 0 0
car 0 4 1 0 0 0 0 0
cow 0 2 3 0 0 0 0 0
cup 0 0 0 5 0 0 0 0
dog 0 0 0 0 5 0 0 0
horse 0 0 0 0 0 5 0 0
pear 0 0 0 0 0 0 5 0
tomato 0 0 0 0 0 0 0 5
True
apple car cow cup dog horse pear tomato
Predicted

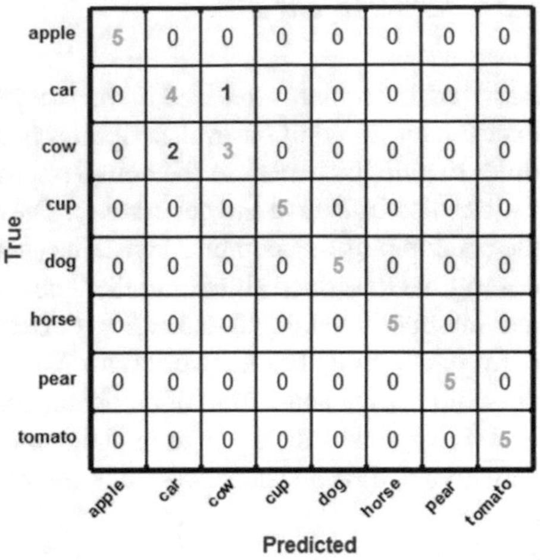

Fig. 11 Confusion matrix for images of database ETH-80

forest 5 0 0 0 0 0
gazon 0 5 0 0 0 0
ground 0 0 5 0 0 0
sahara 0 0 0 5 0 0
sky 0 0 0 0 5 0
water 0 0 1 0 0 4
True
forest gazon ground sahara sky water
Predicted

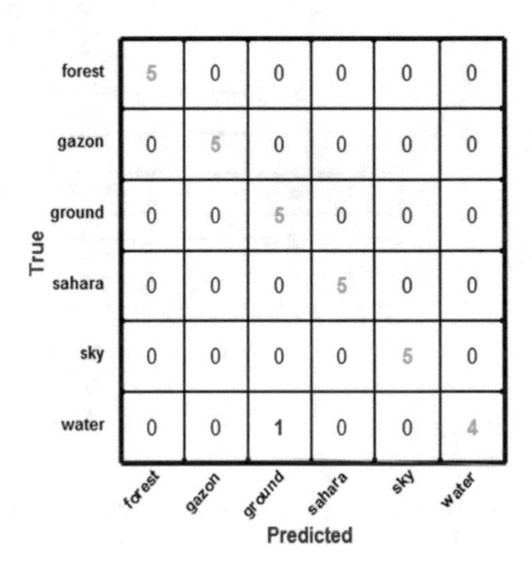

Fig. 12 Confusion matrix for images of database NATURE

3.3.1 Analysis of Results

Comparison between the results obtained in this experiment and the results obtained in experiment 1.2 and 2.1, allows us to notice that the combination of neural networks with Bayesian networks in a parallel scheme, has significantly improved the quality of image annotation. This result is also illustrated by the examples of annotated images presented by Figs. 13 and 14, which shows that the exploitation of complementarities of generative and discriminative classifiers can contribute to the improvement of image annotation. So, it would be interesting to investigate other ways to combine these two different classification approaches to possibly increase the image annotation accuracy.

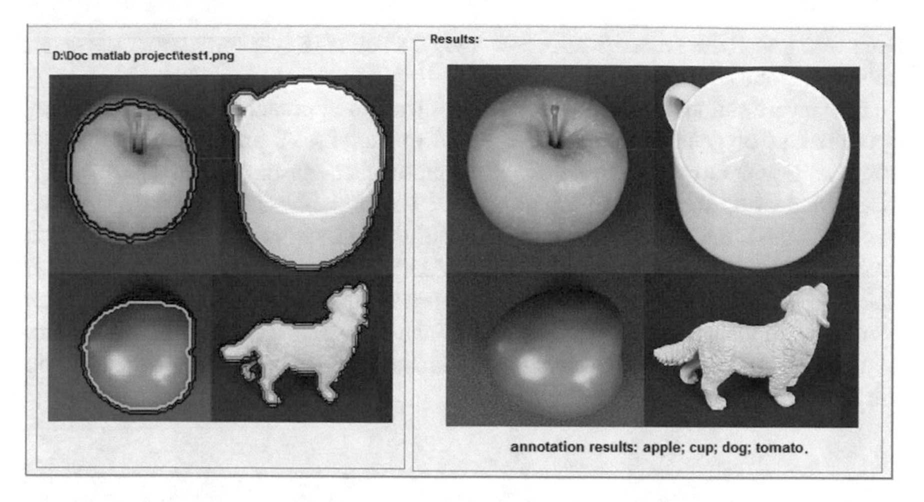

Fig. 13 Example of annotated image

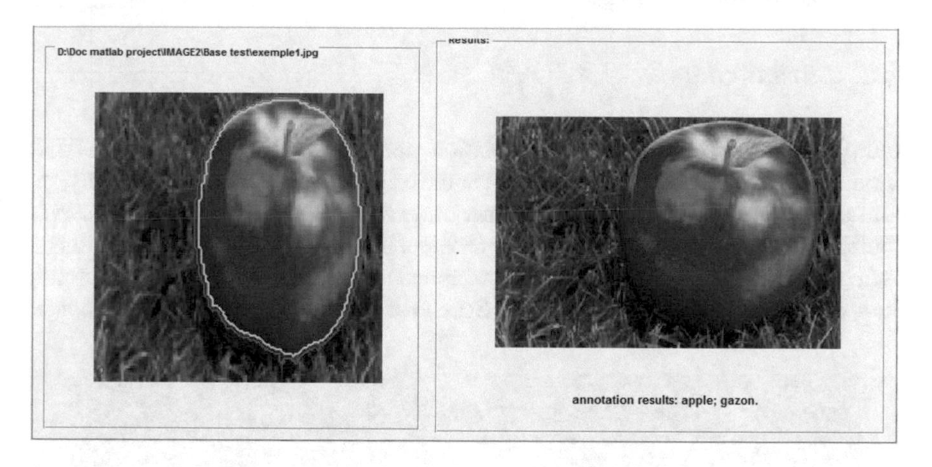

Fig. 14 Example of annotated image

3.4 Conclusion and Future Work

In this research work, we have proposed to build an efficient classifier for automatic image annotation via combining generative and discriminative classifiers which are respectively Bayesian networks and neural networks.

Starting with comparing these classifiers by conducting experiments on three image dataset, we have observed no classifier alone will be sufficient for semantic image annotation. Therefore, we have combined the generative and discriminative classifiers in a parallel scheme in order to join and exploit their strengths. Experimental results show that this approach is promising for automatic image annotation because it gives better classification accuracy than either Bayesian networks or neural networks alone.

Our investigations suggest that the most fruitful approaches will involve some combination of generative and discriminative models. A principled approach to combining generative and discriminative approaches not only gives a more satisfying foundation for the development of new models, but it also brings practical benefits, addresses the extreme data-ambiguity and overfits vulnerability issues in tasks such as automatic image annotation (AIA). In future research work, we would like to develop other hybrid schemes to integrate the intra-class information from generative models and the complementary inter-class information from discriminative models, and to research alternative optimization techniques utilizing ideas from the multi-criteria optimization of literature.

4 Exploration of Regrouping Adjacent Regions of Segmented Image

4.1 Introduction

Although, the automatic image annotation performance has been improved in previous studies, we have observed that some errors are still persistent, namely, the confusion between Car and Cow in some cases as shown in Fig. 15. The observed confusion errors are due to the similarity problem between some segmented regions of both objects. This problem is one of the common problems encountered in extracting the semantic visual content from image. So, one of the main challenges is to recognize

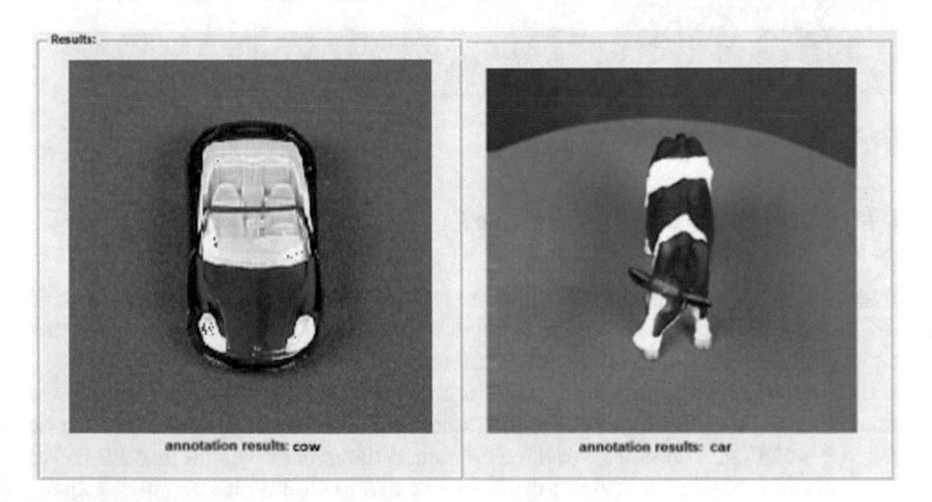

Fig. 15 Confusion between Car and Cow in some cases

homogeneous regions within an image as distinct and belonging to different objects. To overcome this challenge, several segmentation techniques have been developed in order to precisely segment the image without under or over segmentation [62–66]. But, no general segmentation method was proposed to work for all images. These segmentation techniques often have to be combined with domain knowledge in order to effectively solve an image segmentation problem for a problem domain.

In this section, we are interested in evaluating a segmentation process focused on regrouping and finding semantic compact region to correct the confusion errors mentioned above and consequently to increase the annotation rate. To achieve this goal, we conduct the following experiment.

4.2 *Experiment and Results*

The principle of regrouping adjacent region is illustrated in Fig. 16. The image is segmented in different regions using region growing, then the adjacent regions are regrouped to obtain compact region.

In order to study the impact of this segmentation process on the performance of automatic annotation, we conduct an experiment similar to the previous experiment (experiment 22) but with the addition of the regrouping region as illustrated in Fig. 17.

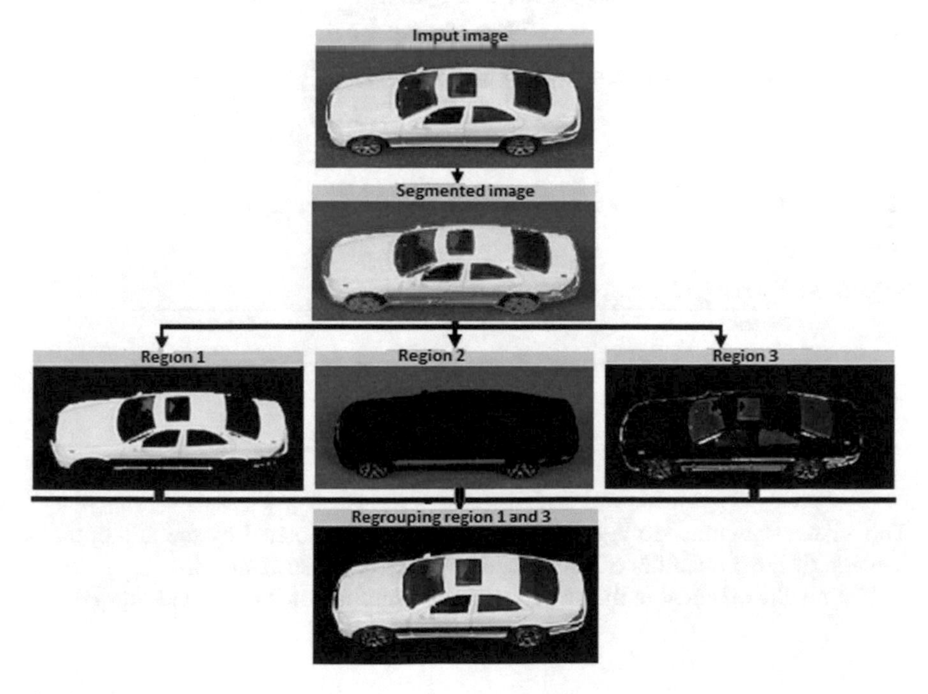

Fig. 16 Principle of regrouping adjacent region

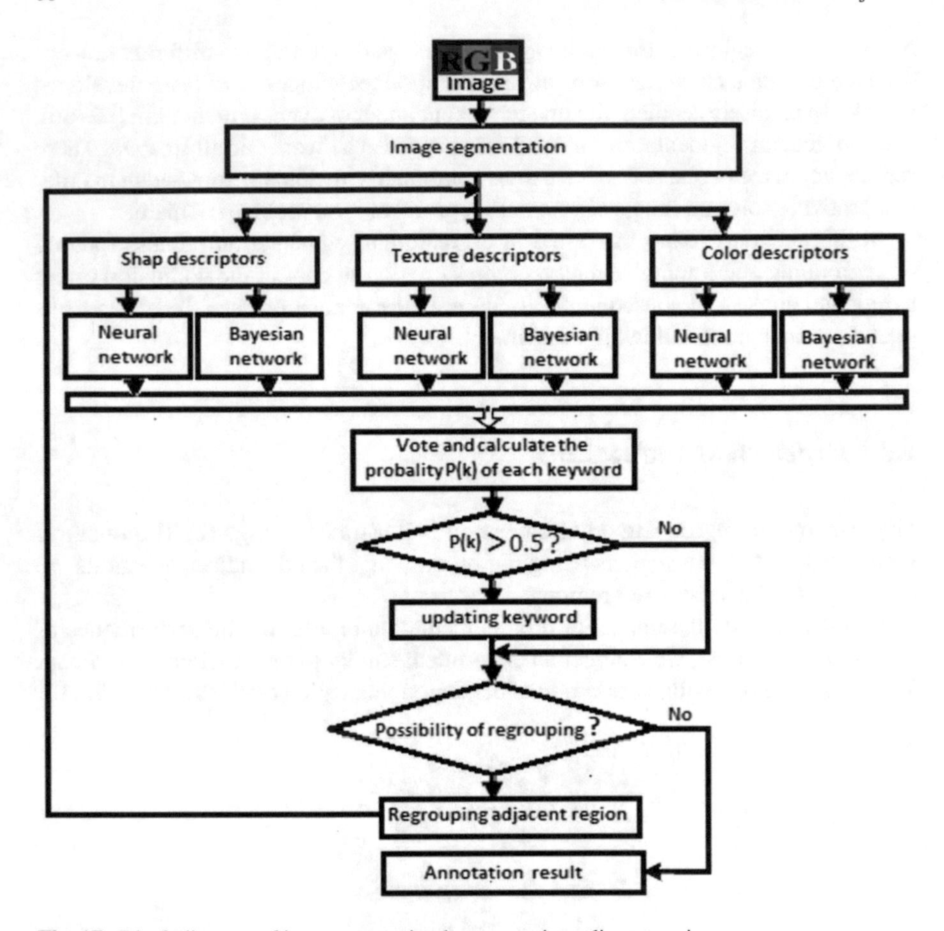

Fig. 17 Block diagram of image annotation by regrouping adjacent regions

Table 5 Average image annotation rate and error rate

Database	Global average Annotation rate	Error rate
ETH-80	97.50%	2.50%
COIL-100	92.50%	7.50%

The adjacent regions are regrouped iteratively and annotated by the appropriated keyword if the probability of a regrouped regions is higher than 0.5.

The results obtained in this experiment are presented in Table 5 and Fig. 18.

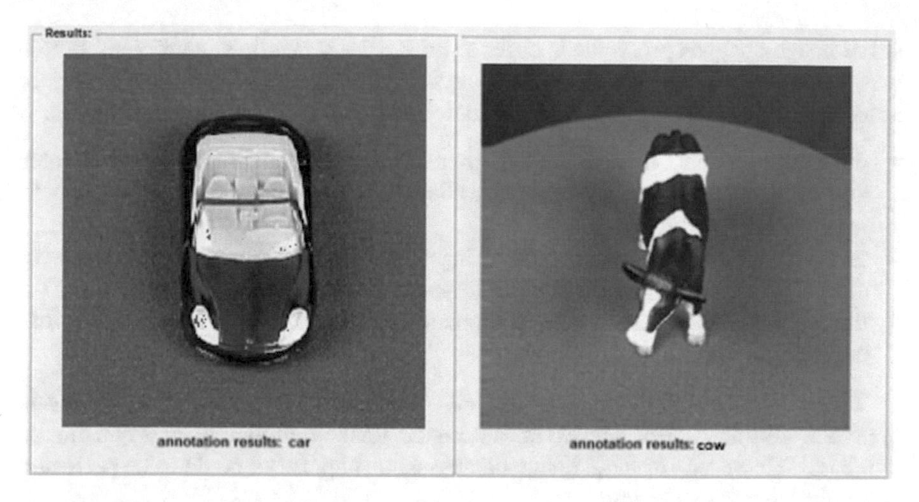

Fig. 18 Result annotation of Car and Cow obtained by regrouping adjacent regions

4.3 Analysis of Results

We can see from Table 5 that the annotation rates have considerably increased despite the increase in processing time. This result is obtained because the confusion problem between objects that have some similar segmented regions, is addressed as shown in Fig. 18 where Cow and Car are correctly annotated contrary to Fig. 15. So, it appears from this part of the study that the regrouping adjacent regions allow to get good compact regions that correspond well to the observed semantic contents in image.

4.4 Conclusion

In this part of the research study, we have examined the impact of regrouping segmented regions technique on image annotation accuracy. The results obtained in this study have shown that this technique helps to discriminate objects that could to be confused at the classification level. It is believed that in future research and exploration, there will be more image segmentation methods to be further developed and more widely used to gain benefit of semantic image analysis for automatic image annotation

5 General Conclusion

The ultimate objective of semantic image analysis is to extract high-level semantic information from images. To achieve this goal, great progress has been made to solve image analysis problems in order to bridge the semantic gap problem. In this research work, we have explored several aspects of image analysis process to perfect automatic image annotation. This exploration has revealed the following outputs:

- the image annotation accuracy is higher when the visual features texture, color and shape features are combined in parallel scheme instead of concatenating them into a single vector;
- an efficient classifier should combine a generative classifier with a discriminative classifier to gain the benefit of their complementarity;
- the regrouping adjacent regions of segmented image reduces the confusion errors at the classification level.

These outputs of the present study led to elaborate a semantic image analysis approach getting a clear success in automatic image annotation for small image databases. However, for large image databases, this approach could to be perfected by exploring and exploiting:

- other combination schemes that integrate the joint distribution of multiple low visual features;
- new hybrid classification schemes that seek to integrate the intra-class information from generative models and the complementary inter-class information from discriminative models;
- segmentation methods endowed with an ability to produce semantically significant compact regions;
- a disambiguation method based on a semantic similarity measure which integrates several sources of visual, contextual and spatial information.

References

1. Bouchakwa, M., Ayadi, Y., Amous, I.: Multi-level diversification approach of semantic-based image retrieval results. Progress Artif. Intell. **9**(1), 1–30 (2020)
2. Potapov, A., et al.: Semantic image retrieval by uniting deep neural networks and cognitive architectures. In: Proceedings of the 11th International Conference on Artificial General Intelligence (AGI), Prague, Czech Republic, pp. 196–206 (2018)
3. Pratap, R., et al.: A review study—in image retrieval bridging of semantic gap. J. Inf. Tech. Sci. **4**(3), 11–15 (2018)
4. Hirwane, R.: Semantic based image retrieval. Int. J. Adv. Res. Comput. Commun. Eng. **6**(4), 120–122 (2017)
5. Alzubi, A., Amira, A., Ramzan, N.: Semantic content-based image retrieval: a comprehensive study. J. Vis. Commun. Image Represent. **32**, 20–54 (2015)
6. Khodaskara, A., Ladhakeb, S.: Image analysis for intelligent image retrieval. In: International Conference on Intelligent Computing, Communication & Convergence (ICCC), Odisha, India, pp. 192–197 (2015)
7. Theodosiou, Z., Tsapatsoulis, N.: Image annotation: the effects of content, lexicon and annotation method. Int. J. Multimedia Inf. Retrieval **9**, 191–203 (2020)
8. Zhang, J.: Vision to keywords: automatic image annotation by filling the semantic gap. Thesis (2019)

9. Amiri, K., Farah, M.: Graph of concepts for semantic annotation of remotely sensed images based on direct neighbors in RAG. Can. J. Remote Sens. **44**(6), 551–574 (2018)
10. Bouzayani, A.: Automatic image annotation extension for search and classification. Thesis, Université de Lorraine (2018)
11. Nair, L.R., Subramaniam, K., PrasannaVenkatesan, G.K.D., Baskar, P.S., Jayasankar, T.: Essentiality for bridging the gap between low and semantic level features in image retrieval systems: an overview. J. Ambient Intell. Humanized Comput. (2020). https://doi.org/10.1007/s12652-020-02124-6
12. Kwasnicka, H., Jain, L.C.: Semantic gap in image and video analysis. In: Kwaśnicka, H., Jain, L. (eds.) Bridging the Semantic Gap in Image and Video Analysis, Chapter 1. Springer, Cham (2018)
13. Pardede, J., Sitohang, B.: Reduce semantic gap in content-based image retrieval. Adv. Sci. Lett. **23**(11), 10664–10671 (2017)
14. Anusree, B.S.: Reducing semantic gap in image retrieval by integrating high level query and low level facial features. Int. J. Sci. Res. (IJSR) **4**(3), 1415–1418 (2015)
15. Melani, S.G., Ruben, M.C., Jose, J.G.: Semantic and structural image segmentation for prosthetic vision. J. PLoS One **15**(1), (2020)
16. Li, W.: Semantic-aware image analysis, thesis, Heidelberg, German, December 2019
17. Bindhu, V.: Biomedical image analysis using semantic segmentation. J. Innov. Image Process. (JIIP) **1**(02), 91–101 (2019)
18. Oujaoura, M., Minaoui, B., Fakir, M.: A semantic hybrid approach based on grouping adjacent regions and a combination of multiple descriptors and classifiers for automatic image annotation. Int. J. Pattern Recogn. Image Anal. **26**(2), 316–335 (2016)
19. Ogiela, M.R., Hachaj, T.: Cognitive methods for semantic image analysis in medical imaging applications. In: Natural User Interfaces in Medical Image Analysis, Chapter 2 (2015)
20. Ajimi, A., Sree, K.: Efficient automatic image annotation using weighted feature fusion and its optimization using genetic algorithm. Commun. Appl. Electron. **1**(6), 15–19 (2015)
21. Ivan, D., Luciano, S.: Mixing low-level and semantic features for image interpretation. In: Proceedings of Computer Science. Computer Vision – Workshops, Zurich, Switzerland, vol. 8926, pp. 283–298 (2015)
22. Cong, J.: Automatic image annotation using feature selection based on improving quantum particle swarm optimization. Signal Process. J. **109**, 172–181 (2015)
23. Dongping, Z., Li, Y., Peng, H., Lu, Y.: Image annotation based on joint feature selection with sparsity. Inf. Tech. J. **13**, 102–109 (2014)
24. ping Tian, D.: A review on image feature extraction and representation. Tech. Multimedia Ubiquit. Eng. J. **8**(4), 385–395 (2013)
25. Fernando, B., Fromont, E., Muselet, D., Sebban, M.: Discriminative feature fusion for image classification. In: IEEE Conference Computer on Vision and Pattern Recognition (CVPR), pp. 3434–3441 (2012)
26. Zhang, R.: Combining visual features and contextual information for image retrieval and annotation. Theses and dissertations, Ryerson University, Toronto, Ontario, Canada (2011)
27. Zhang, R., Guan, L., Zhang, L., Xin-Jing, W.: Multi-feature pLSA for combining visual features in image annotation. In: Proceedings of the 19th ACM International Conference on Multimedia, Scottsdale, Arizona, USA, 1513–1516 (2011)
28. Jian, H., Zhang, B., Qi, N.-M., Yang, Y.: Evaluating feature combination in object classification. Adv. Vis. Comput. J. **6939**, 597–606 (2011)
29. Wang, M.T., Gong, S., Hua, X.: Combining global, regional and contextual features for automatic image annotation. Pattern Recogn. J. **42**(2), 259–266 (2009)
30. Minaoui, B., Oujouara, M., Fakir, M., Sajieddine, M.: Toward an effective combination of multiple visual features for semantic image annotation: TELKOMNIKA. Indonesian J. Electric. Eng. **15**(3), 401–408 (2015)
31. Haralick, R., Shanmugan, K., Dinstein, L.: Textural features for image classification. IEEE Trans. SMC, **3**(6), 610–621 (1973)
32. Chonga, C., Raveendranb, P., Mukundan, R.: Translation and scale invariants of Legendre moments. Pattern Recogn. J. **37**, 119–129 (2004)

33. ETH-80 database image. http://www.d2.mpi-inf.mpg.de/Datasets/ETH80
34. COIL-100 database image. http://www.cs.columbia.edu/CAVE/software/softlib/coil-100.php
35. Minaoui, B., Oujouara, M., Fakir, M.: Combining generative and discriminative classifiers for semantic automatic image annotation. Image Process. J. **8**(5), 225–244 (2014)
36. Shih, Y., Cheng, S.: Automatic seeded region growing for color image segmentation. Image Vis. Comput. J. **23**, 877–886 (2005)
37. Becker, A., Naim, P.: les réseaux bayésiens: modèles graphiques de connaissance. Eyrolles (1999)
38. Sabine, B.: Modèles graphiques probabilistes pour la reconnaissance de formes. Theses, Nancy 2 University (2009)
39. George, H., Langley, P.: Estimating continuous distributions in bayesian classifiers. In: The Eleventh Conference on Uncertainty in Artificial Intelligence (1995)
40. Leray, Ph.: Réseaux bayésiens: apprentissage et modélisation de systèmes complexes, These, Rouen University (2006)
41. Tom, M.: Generative and discriminative classifier: Naïve bayes and logistic regression. Machine learning. Draft (2010)
42. Li, Z., Shi, P., Liu, X., Shi, Z.: Automatic image annotation with continuous PLSA. In: Proceedings of the 35th IEEE International Conference on Acoustics, Speech and Signal Processing, p. 806–809 (2010)
43. Carneiro, G., Chan, A., Moreno, P., et al.: Supervised learning of semantic classes for image annotation and retrieval. IEEE Trans. Pattern Anal. Mach. Intell. **29**(3), 394–410 (2007)
44. Fan, J., Gao, Y., Luo, H.: Hierarchical classification for automatic image annotation. In: SIGIR Proceedings (2007)
45. Zhang, R., Zhang, Z., Li, M., et al.: A probabilistic semantic model for image annotation and multi-model image retrieval. In: Proceedings of the 10th IEEE International Conference on Computer Vision, pp. 846–851 (2005)
46. Lavrenko, V., Manmatha, R., Jeon, J.: A model for learning the semantics of, pictures. In: Proceedings of Advances in Neural Information Processing Systems, pp. 251–259 (2003)
47. Yakhnenko, O.: Learning from text and images: generative and discriminative models for partially labeled data. Thesis, Iowa State University Ames (2009)
48. Ng, A.Y., Jordan, M.: On discriminative vs. generative classifiers: a comparison of logistic regression and naïve Bayes. Neural Inf. Process. Syst. **14**, 841– 848 (2001)
49. Holub, L., Welling, M., Perona, P.: Hybrid generative-discriminative visual categorization. Inter. J. Comput. Vis. **77**(3), 239–258 (2008)
50. Ulusoy1, I., Bishop, M.: Comparison of generative and discriminative techniques for object detection and classification: toward category-level object recognition. In: Ponce, J., Hebert, M., Schmid, C., Zisserman, A. (eds.) Toward Category-Level Object Recognition. LNCS, vol. 4170, 173–195. Springer, Heidelberg (2006)
51. Lasserre, J., Bishop, C., Minka, P.: Principled hybrids of generative and discriminative models. In: Proceedings of the IEEE Computer Society Conf on Computer Vision and Pattern Recognition (CVPR), pp. 87–89 (2006)
52. Timothy, M., Shaogang, G., Xiang, T.: Finding rare classes: active learning with generative and discriminative models. IEEE Trans. Knowl. Data Eng. **25**(2), 374–386 (2013)
53. Cristani, A., Castellani, U., Murino, V.: A hybrid generative/discriminative classification framework based on free energy terms. In: ICCV (2009)
54. Anna, B., Andrew, Z., Xavier, M.: Scene classification using a hybrid generative/discriminative approach. IEEE Trans. Pattern Anal. Mach. Intell. **30**(4), 712–727 (2008)
55. Kelm, M., Pal, C., McCallum, A.: Combining generative and discriminative methods for pixel classification with multi-conditional learning. In: ICPR, pp. 828–832 (2006)
56. Bouchard, G., Triggs, B.: The trade-off between generative and discriminative classifiers. In: Proceedings of Computational Statistics Symposium. PhysicaVerlag, Springer, Heidelberg (2004)
57. Li, Z., Tang, Z., Zhao, W., Li, Z.: Combining generative/discriminative learning for automatic image annotation and retrieval. Int. J. Intell. Sci. 55–62 (2012)

58. Yang, S.H., Bian, J., Zha, H.: Hybrid generative/discriminative learning for automatic image annotation. In: Proceedings of the Uncertainly Artificial Intelligence (UAI), pp. 683–690 (2010)
59. Cao, Y., Liu, X., Bing, J., Song, L.: Using neural network to combine measures of word semantic similarity for image annotation. In: IEEE International Conference on Information and Automation (ICIA), pp. 833–837 (2011)
60. Simard, P., Steinkraus, D., Platt, J.: Best practices for convolutional neural networks applied to visual document analysis. In: ICDAR, pp. 958–962 (2003)
61. Lepage, R., Solaiman, B.: Les réseaux de neurones artificiels et leurs applications en imagerie et en vision par ordinateur, Ecole de technologie supérieure (2003)
62. Ivanovici, M., Coliban, R.-M., Hatfaludi, C., Nicolae, I.E.: Color image complexity versus over-segmentation: a preliminary study on the correlation between complexity measures and number of segments. J. Imaging **6**(16), 1–15 (2020)
63. Yao, Y., Wang, S.: Evaluating the effects of image texture analysis on plastic greenhouse segments via recognition of the OSI-USI-ETA-CEI pattern. J. Remote Sens. **11**(231), 36–57 (2019)
64. Aloun, M.S., Hitam, M.S., Wan Yussof, W.N., Abdul Hamid, A.A.K., Bachok, Z.: Modified JSEG algorithm for reducing over-segmentation problems in underwater coral reef images. Int. J. Electric. Comput. Eng. (IJECE) **9**(6), 5244–5252 (2019)
65. Sigut, J., Fumero, F., Nuñez, O.: Over- and under-segmentation evaluation based on the segmentation covering measure. In: 23rd International Conference in Central Europe on Computer Graphics, Visualization and Computer Vision, pp. 83–89 (2015)
66. Chen, B., Qiu, F., Wu, B., Du, H.: Image segmentation based on constrained spectral variance difference and edge penalty. J. Remote Sens. **7**(5), 5980–6004 (2015)

New Method for Data Replication and Reconstruction in Distributed Databases

Noreddine Gherabi

Abstract Data replication between servers is a critical and important step to share, or can be saved, the data between multiple remote sites in a distributed database. Data replication is used to replicate data to remote sites, improve system performance data access and reduce the connection time to the database.

This paper presents a new algorithm for dynamic data replication in distributed databases. The algorithm is adaptive in the sense that changes the schema objects then replicates these objects in the central scheme.

In this technique, using the strengths of communication between mobile agents, this communication can facilitate extraction, reconstruction and replication of schemes from remote sites to the central site. We show that the algorithm can be combined with the mechanisms of the distributed database management systems, because the performance of the algorithm is analyzed theoretically and experimentally.

Keywords Web semantic · Web services · Distributed database

1 Introduction

Most companies require decentralization of data, it is because companies today are more dispersed geographically and the database is distributed and appearing to applications as a single database. Unlike centralized databases, replication can provide data availability through remote site transactions and restoration can be the data in case of failures.

Data replication is an important area of research, either in theory or practice. In practical terms, data replication plays a very important role in several contexts: data backup, data distribution, increased scalability, parallel processing, etc.

The database replication is necessary in the case where there is a failure in the system; in this case a secondary copy of the database is required to retain

N. Gherabi (✉)
Sultan Moulay Slimane University, ENSA Khouribga, Lasti Laboratory, Khouribga, Morocco

N. Gherabi and J. Kacprzyk (eds.), *Intelligent Systems in Big Data, Semantic Web and Machine Learning*, Advances in Intelligent Systems and Computing 1344,
https://doi.org/10.1007/978-3-030-72588-4_5

data. A reconstitution service is needed to maintain data consistency across diverse environments.

Replication solutions remain a challenge for researchers. In recent years, database replication has experienced strong revolution because of the importance and the role it plays in large databases.

Replication is the process of creating and maintaining duplicate objects databases in a distributed database system [1, 2].

The replication role is improve the performance and increase the application availability by providing the options for accessing data in an easy and efficient way.

Database replication algorithms can provide an effective mechanism for replicating data and manage backups that can be used in case of problems.

This paper presents a survey replication technique in several sites using replication features based agents. In addition, they are traditional replication techniques that provide limited solutions and are not satisfactory.

2 Related Work

The data replication approach was used extensively in many areas, such as the Internet and distributed systems [3, 4]. Whatever the replication method used should meet some criteria [5–7], such as, data replication time, what data to be replicated, and what are the locations storage of these replicas.

However, there are not a lot of prototypes available in the literature for the replication of data. For example, PostgresR [8] is a replication solution based on a modified version of PostgreSQL. It can replicate large objects and uses a store and forward an asynchronous way.

Several methods have been developed in databases and object-relational databases, namely migration techniques and conversion [9, 10] optimization [11, 12] or may be replication. For example, in our work [13] we propose a new technique to convert a relational database to an OWL description, we have developed a prototype to create ontology from a relational database. This prototype extracts the schema of the database then transforms it into a canonical data model, after the system generates the structure of OWL; finally the data of RDB is stored in an OWL document.

The remainder of this paper is organized as follows:

First, I present an overview of the replication methods. Second, I describe in detail our proposal; this section explains the proposed method for extracting the schema and data of database and describes how to replicate it into global cheme. Finally, our conclusion and future work is given.

3 Replication Methods

Replication is a technique of creating and maintaining multiple replicated database objects in a distributed database system. The replication technique improves performance and make the applications more available by providing the alternative data access options.

With replication users access a local database in an easy and fast way instead of accessing a remote server, it reduces network traffic and provide location transparency. One important thing in replication is that the application can continue to function even parts of the distributed database are down. The database replication operation will be necessary in the case of a system failure, which can easily recover a secondary copy in case if a copy of the primary database is failed.

In the literature, there are several Open Source replication tool that helps the implementation of replication activities of a database. Postgres-R is set to a PostgreSQL database update which is a database replication technique very effective for data clusters while maintaining consistency [14, 15]. The Postgres-R tool is used to duplicate Big objects and uses a storage technique based asynchronous replication engine. In the case of a normal system, Postgres-R can help add or remove nodes. The bad nodes are automatically detected and removed without affecting the operations database system.

The basic architecture of Postgres-R is detailed in Fig. 1. the process and supported interfaces of PostgreSQL are available in Postgres-R.

Replication steps are based on several servers each one running an instance of Postgres-R [16]. When a client wants to access a database, it sends a request to the postmaster (server process). The postmaster starts a background process (backend) for each client, after a direct communication is established between the client and the backend. When a remote transaction is started, it is given to one of remote back end.

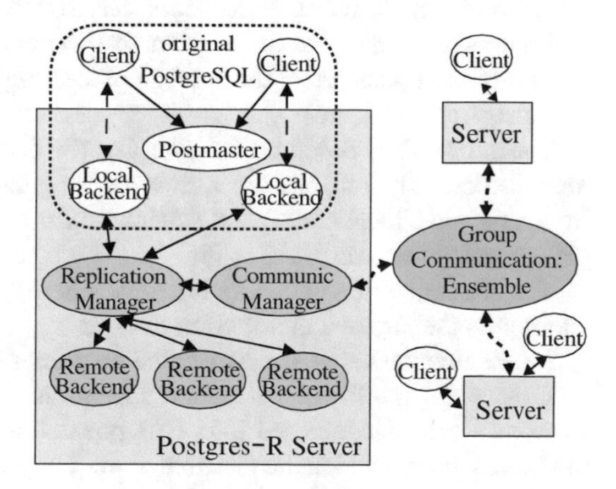

Fig. 1 Architecture of Postgres-R

Another replication tool called Slony-I [17]. Slony-I is a master–slave system that can replicate large databases using a limited number of slave systems. If the number of servers exceeds the specified threshold, the cost of communication can be increased.

The goal of Slony-I is to maintain data centers and backup sites where the sites are available and secure.

4 Our Approach

In this section, we present our proposed method for constrction and replication schemes of multiple sites using JADE technology.

In the first time, the system begins to create agents to execute the necessary tasks for constrcuction a global schema from multiple sites. the system creates an agent called interface agent that begins to analyze the databases found in sites and then creates an allocation scheme from local schemes existing in the remote databases that same agent analysis the fragmentation scheme then extract the global scheme.

In the second, the global scheme is transferred to the central site for replication by an other agent, called supervisor agent.

4.1 Extraction and Reconstruction of Global Schema

4.1.1 Construction of Global Scheme

The definition of the distribution pattern is the most delicate of the design phase of a distributed database part, because there is no quick way to find the optimal solution. The administrator must make decisions based on technical and organizational criteria with the aim to minimize the number of transfers between sites, transfer time, the volume of data transferred, the average processing time of applications, the number of copies fragments, etc. …

The approach is based on the fact that the distribution is already done, but we must successfully integrate the various existing databases into one global database. In other words, local conceptual schemas exist and must be passed to unify into a global conceptual scheme (See Fig. 2).

The allocation of a data base intervenes in the three levels of the architecture in addition to the physical distribution of data:

Externally: the views are distributed on users sites.

Conceptual level: the conceptual schema is associated, through the allocation structure (itself decomposed into a fragmentation pattern and allocation scheme), the local schemes which are distributed in multiple sites (physical sites).

Internally: the global internal schema has no real existence but is related to the local internal schemes located at different sites.

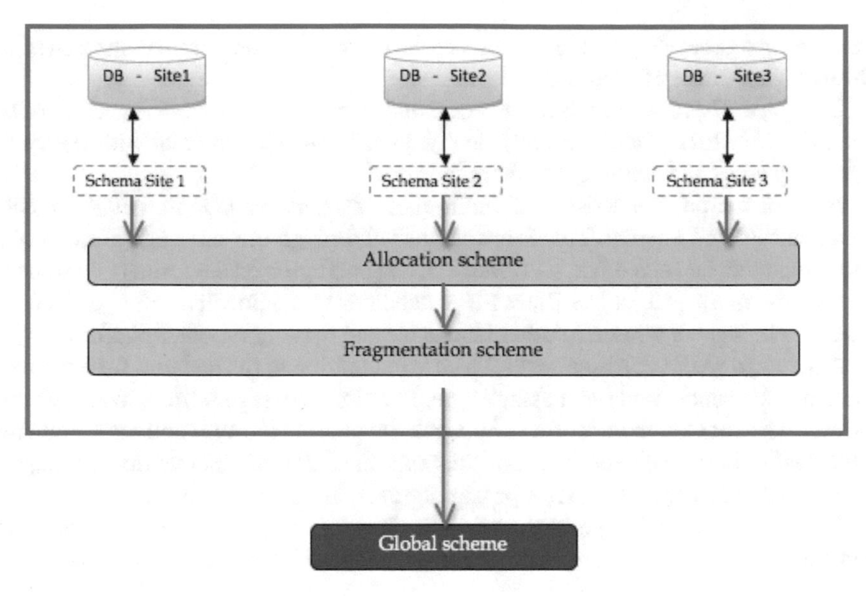

Fig. 2 Construction of global scheme

4.1.2 Algorithm for Extraction and Construction

In this section, I present the algorithm for extracting schemes for multiple sites. This algorithm is developed as follows:

Input: The RDB Sites
Output: The global scheme

Step 1: at each site, create an agent for extracting the local scheme.
Step 2: start an interface agent to construct an allocation scheme
Step 3: the interface agent create a fragmentation scheme
Step 4: the Supervisor Agent receives the global scheme
Step 5: the global scheme is replicated into central site
End

4.1.3 Multi-agent in JADE

A basic thing of multi-agent systems is that agents communicate and interact with easy and efficient way. For more JADE agents arriving to collaborate, they must exchange messages. Each JADE agent has a sort of mailbox that contains messages from other agents. These mailboxes are in a list that contains the messages in chronological order of their arrival.

Using JADE [18–20] in Java to create and communicate mobile agents. JADE complies with FIPA standards and, ideally, Jade agents may interact with agents in other languages and running on other platforms.

A fundamental charactistic of multi-agent systems is that individual agents communicate and interact. This is accomplished through the exchange of messages and, to understand each other, it is crucial that agents agree on the format and semantics of these messages. Jade follows FIPA standards so that ideally Jade agents could interact with agents written in other languages and running on other platforms.

A message JADE contains several pieces in addition to the content, for example: receivers, the sender and the message type. In JADE, messages strictly comply with the ACL (Agent Communication Language) that provides several options for coding the actual content. Specifically, Jade supports the FIPA SL (Semantic Language), actions and predicates. There is a possibility to serialize Java objects.

In this work, the message content is in the form of serialized objects, because the content of the message can be retrieved from objects of an object-oriented database.

4.2 Our Replication Strategy

4.2.1 Process for Replication

The proposed process is based on communication between agents as each agent executes a well-defined task, and sending the result to another agent to complete the process.

For a good replication schema and data, several techniques of distributed databases can be used to copy the schemes and data from a remote database namely materialized views, synonyms, stored procedures, or may be snapshots.

The replication mechanism starts with the creation of an agent with analysis of the global scheme, called the agent supervisor. Supervisor reads the global scheme produced by the interface agent and translates this scheme into serialized objects, these objects will be replicated to a central database.

The replication mechanism is detailed in Fig. 3.

4.2.2 Algorithm for Replication

This section presents our replication strategy. This strategy aims to ensure strong consistency between copies of a replicated object. Informally this amounts to ensure that the status of each copy is identical.

Replication being the result of the modification of the table in the central site, it is possible that the central site and the remote site are in the same state, in this case must be taken in replication.

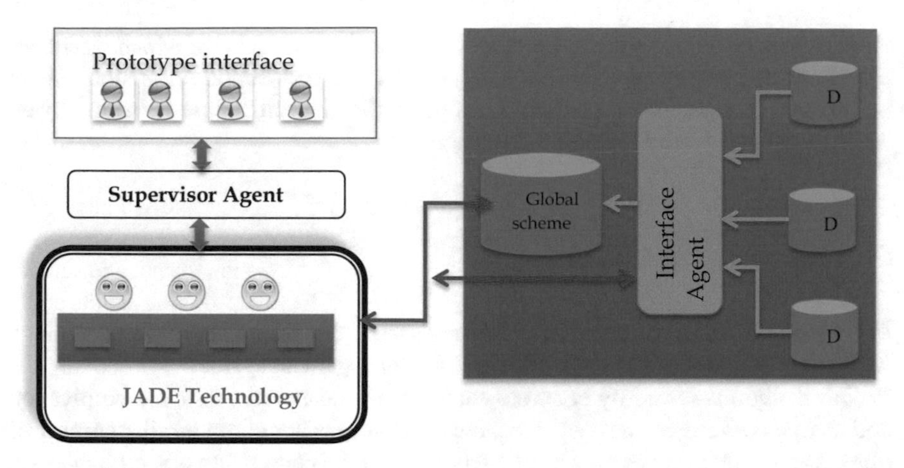

Fig. 3 Architecture of replication

If the objects exist in the central site, the supervisor agent appends the data of the object copied in the existing object data, else the supervisor agent creates the object in the global scheme and copies the data in this object.

The copy of the data is performed using the data transfer commands from local site to another.

The concept of datalink of a database is used to communicate the remote databases.

The algorithm to replication schemas and data is as follows:

```
CS = Central Site
For i=1 to n ( n sites)

Si← Scheme of Site number i
For j=1 to k ( k objects of site i)

      Oj← Object of Si
      COPY Oj FROM Si TO CS

If (Oj exists in a central site)

CREATE  O
 USING Select * from Oj@Si UION Select * from Oj@ CS

Else
 CREATE  O
       USING Select * from Oj@ Si
End j
End i
```

In this algorithm, the database is identified by datalink service which facilitates the communicate between remote sites. During the connection, a session should be initiated by a login and password. So the replication procedure is treated by an administrator with administrative previleges.

5 Test and Evaluation

For a good validation of our work, I have created a simulator using Java technology with remote databases in the Oracle database management system.

Our method is tested by concrete examples of databases. After the verification and analysis of remote sites schemes, the agent supervisor computes the number of objects and compares the objects of each site with the central site objects.

Figure 4 shows an overview of results generated by the system. The system displays the list of exported objects and also displays the behavior of agents during replication.

In this method, we use several mobile agents, each one execute a specific task. Figure 5 shows the JADE remote agent management containing agents used in the replication process.

For better appreciation of this approach, we compute the execution time in the processus of replication remotes sites and we compare this time with others methods without agents.

We compute the execution time of tasks effected by mobile agents in the replication process from multiple sites. The execution times are specified in the table in Fig. 6.

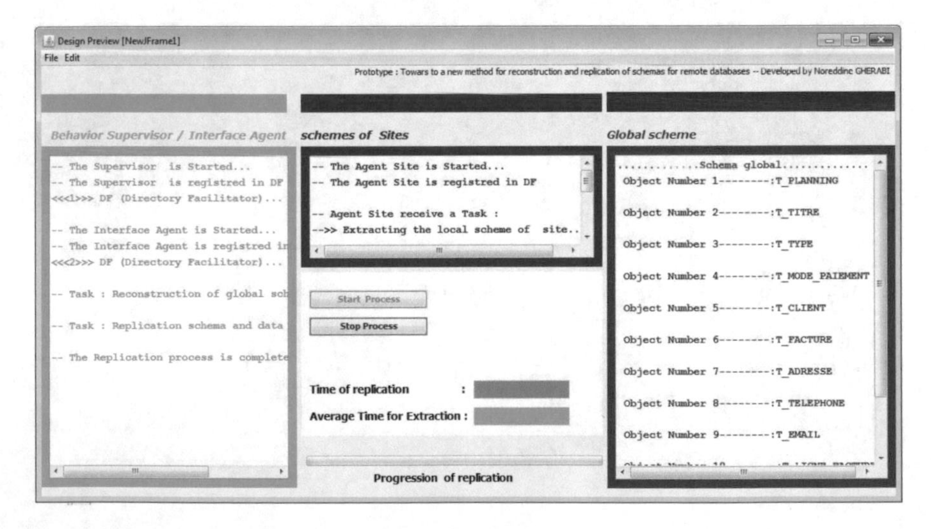

Fig. 4 The result of replication schema and data

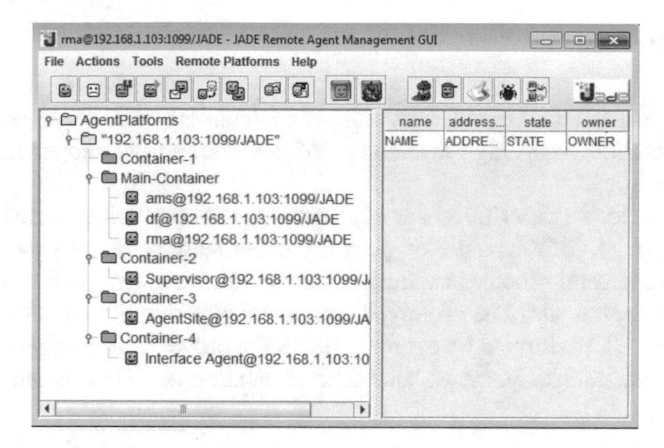

Fig. 5 JADE remote agent management

Number of sites		2	4	6
Average Time (Agent Sites)	Ms	**59**	**145**	**256**
Execution time (Interface Agent)		**125**	**261**	**342**
Replication time		**141**	**295**	**453**

Fig. 6 Execution time using the multi-agent

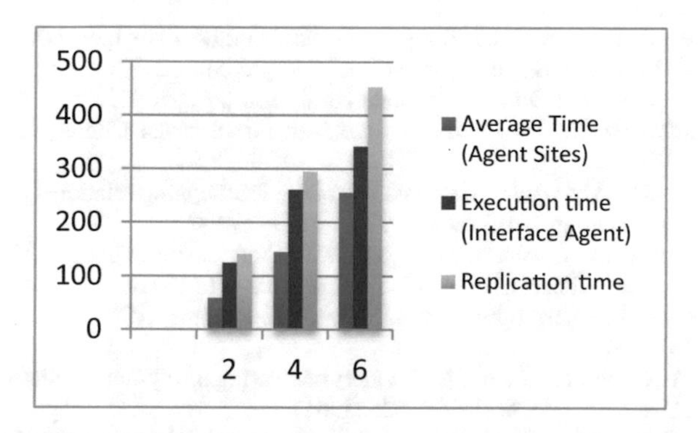

Fig. 7 Replication time using multi-agents

The system maintains better performance although with the increasing number of remote sites, we note in Fig. 7 that there is a small difference in terms of execution between the different cases.

6 Conclusion

The database replication is the process of maintaining a copy of one or more databases, aims to keep high availability of data and ensure proper maintenance in case of failures.

In this article, we describe an automated and optimal technique based on mobile agent technology, for this, a prototype was created helping us to extract the global schema from several schemes existing in remote sites and also replicate the global schema in a central site. The prototype compute replication time and the execution time of each task performed by agents. The results prove that our approach is practical and solves the problems of automatic replication of schemes and data in the distributed database.

References

1. Distributed database in P2P system using agent approach. Int. J. Comput. **4**(1) (2010)
2. Gherabi, N., Bahaj, M.: New technique for duplication the remote databases based on multi-agents. J. Theor. Appl. Inf. Technol. **73**, 465–472 (2015)
3. Wolfson, O., Jajodia, S., Huang, Y.: An adaptive data replication algorithm. ACM Trans. Database Syst **22**(4), 255–314 (1997)
4. Rabinovich, M., Rabinovich, I., Rajaraman, R.: Dynamic replication on the Internet, Technical report, HA6177000–980305-01-TM, AT&T Labs, March 1998
5. Lei, M., Vrbsky, S.V., Hong, X.: An on-line replication strategy to increase availability in data grids. Futur. Gener. Comput. Syst. **24**(2), 85–98 (2008)
6. Ranganathan, K., Foster, I.: Identifying dynamic replication strategies for a high performance data grid. In: Proceedings of the Second International Workshop on Grid Computing, Denver, CO, pp. 75–86, November 2001
7. Tang, M., Lee, B.-S., Yeo, C.-K., Tang, X.: Dynamic replication algorithms for the multi-tier data grid. Futur. Gener. Comput. Syst. **21**, 775–790 (2005)
8. Kemme, B., Alonso, G.: Don't be lazy, be consistent: postgres-R, a new way to implement database replication. In: Proceedings of the 26th International Conference on Very Large Databases (2000)
9. Gherabi, N., Bahaj, M.: Discovering new technique for mapping relational database based on semantic web technology. IEEE (2012). 978-1-4673-1949-2
10. Gherabi, N., Bahaj, M.: Outline matching of the 2D shapes using extracting XML data. Lecture Notes in Computer Science, LNCS, vol. 7340, pp. 502–512 (2012)
11. Jarke, M., Koch, J.: Query optimization in database systems. ACM Comput. Surv. (CSUR) **16**(2), 111–152 (1984)
12. Bernstein, P.A., Goodman, N., Wong, E.: Query processing in a system for distributed databases. ACM Trans. Database Syst. **6**(4), 602–625 (1981)
13. Gherabi, N., Bahaj, M., Addakiri, K.: Mapping relational database into OWL structure with data semantic preservation. Int. J. Comput. Sci. Inf. Secur. (IJCSIS) **10**, 42–47 (2012)
14. Wu, S., Kemme, B.: Postgres-R (SI): combining replica control with concurrency control based on snapshot isolation. In: 21st International Conference on Data Engineering (ICDE 2005) (2005)
15. Kemme, B., Alonso, G.: Database replication: a tale of research across communities. Proc. VLDB Endow. **3**(1), 5–12 (2010)

16. Kemme, B., Alonso, G.: Don't be lazy, be consistent: postres-R, a new way to implement database replication. In: Proceedings of the 26th International Conference on Very Large Databases (VLDB) (2000)
17. Weick, J.: Slony-I, a replication system for PostgreSQL (2016)
18. Bellifemine, F., Poggi, A., Rimassa, G.: JADE: a FIPA compliant agent framework. In: Practical Applications of Intelligent Agents and Multi-Agents (1999)
19. Bellifemine, F., Caire, G., Greenwood, D.: Developing Multi-Agent Systems with JADE. Wiley, New York (2007)
20. Ahuja, P., Sharma, V.: A JADE implemented mobile agent based host platform security. Comput. Eng. Intell. Syst. **3**(7), 8–20 (2012)

A Distributed Intrusion Detection Approach Based on Machine Leaning Techniques for a Cloud Security

Azidine Guezzaz, Ahmed Asimi, Younes Asimi, Mourade Azrour, and Said Benkirane

Abstract Nowadays, the rapid evolution of computer network and large amount of data which are generated by systems require an accurate monitoring and high security of data and resources. Thereby, the security task of personal information is highly requested. The cloud computing technology allows users and organizations several capabilities to store and process data in a privately owned cloud or on a third party server to make data accessing mechanisms more relevant. Then, it is useful to monitor and protect data housed on the cloud by detecting and stopping intrusive activities and attacks. For this reason, many security tools are available to monitor transactions in cloud and have been taken to prevent networks from intrusions. It is also very useful to choose a cloud provider that considers the security and protection like an essential goal. The intrusion detection and prevention is a recent technology used to secure data in cloud computing environments. In this paper, our main is to propose and design a new distributed intrusion detection system as a service to improve a cloud security. We describe all solutions that we suggest to validate various parts of proposed system for detecting intrusions and defending against attacks in cloud.

Keywords Security · Monitoring · Cloud computing · Intrusion detection · Machine learning

---

A. Guezzaz (✉) · S. Benkirane
Technology High School, Laboratoire MIMSC, Cadi Ayyad University, Marrakech, Morocco

A. Asimi
Faculty of Sciences, Ibn Zohr University, B.P 8106, Dakhla, Agadir, Morocco

Y. Asimi
High School of Technology Guelmim, Ibn Zohr University, Agadir, Morocco

M. Azrour
Faculty of Science and Technology, IDMS Team Errachidia, Moulay Ismail University, Errachidia, Morocco

N. Gherabi and J. Kacprzyk (eds.), *Intelligent Systems in Big Data, Semantic Web and Machine Learning*, Advances in Intelligent Systems and Computing 1344,
https://doi.org/10.1007/978-3-030-72588-4_6

1 Introduction

The cloud computing is a novel architecture of computing in which virtualized resources are provided as a service over the Internet. It is a development of distributed computing aiming to store and manage resources. It allows possibility to access data of persons and organizations from anywhere at any time [9, 11].The automatic intrusion detection is one of security measures which monitor network traffic to detect vulnerabilities, make a relevant decision and alert the existence of intrusions. It is a crucial tool that used to secure data in cloud [13–15]. The malicious persons develop attacks to infect systems and applications. Thereby, the research in intrusion detection is oriented towards on automatic response to intrusions. In second section, we present a background of intrusion detection and prevention and we cite different methods, techniques and types of some intrusion detection systems. For the third part, we discuss the cloud computing technology, its security challenges and certain attacks that try to infect the cloud resources. A new proposed approach of distributed intrusion detection system is described in fourth section. We sit up various solutions that aim to validate our contribution. This document is achieved with a conclusion and future work.

2 Intrusion Detection and Prevention

The intrusion detection and prevention is a novel security technology which used to detect vulnerabilities against applications or systems [1–4]. To anticipate intrusions, a number of solutions are available but any one is complete and satisfactory. Intrusion prevention adds the ability to block attacks in addition to just detecting them. There are two main intrusion detection techniques [1, 5, 6]:

- Anomaly detection based on behavior of users or applications.
- Misuse detection based on signatures of known attacks.

The goal of hybrid detection is to detect both known and unknown attacks [1, 2]. It is interesting to use a combination of these techniques to improve the performances of intrusion detection systems. The very used intrusion system tools are [5, 6]:

- Network intrusion detection systems have a role to control traffic within network.
- Host intrusion detection systems are implemented in level of host. They monitor the concerned host in the networks.
- Hybrid intrusion detection systems combine advantages and functions of network and host intrusion detection systems.
- Application intrusion detection systems receive data on application, generated file log by management software of database, web server or firewall…
- Wi-Fi intrusion detection systems are similar than Network intrusion detection systems. They analyze specifically Wi-Fi traffic.

The intrusion detection and prevention systems are the new security tools classified in various types according to used detection methods, monitoring level, use frequency or reaction state of detection (active and passive) [1, 3, 5]. The positive false are generated when a detection system identifies a normal activity like intrusion, then negative false corresponds to attacks that not detected. So, any alert is generated.

The intrusion prevention system is a new advanced technology applied by certain new intrusion detection system. It contains firewall filtering functions and intrusion detection analysis method. The intrusion prevention system stops an unwanted traffic and blocks attacks rapidly.

3 Cloud Computing and Its Security Challenges

This section presents a state of art for cloud computing models and its security challenges.

3.1 Cloud Computing Models

Following the advancement of information technologies, the cloud computing is a development delivery of computing services over the networks. It removes the needs to be in the same location that stores data. The cloud models allow access to data and resources from anywhere that a network connection is available. The cloud services allow to use software and hardware that include virtualization resources, bandwidth and on demand services. There are different types of models cloud that depend on needs. In general, the cloud model is composed of four deployment models and three service models while providing scalability, mobility and flexibility. For deployment models [10, 11]:

- Public cloud is accessed by anyone with an internet connection and access to the cloud space. It may be owned and managed by a business or organization.
- Private cloud is established for a specific group or organization and limits access to just that group.
- Community cloud is shared among two or more organizations that have similar cloud requirements.
- Hybrid Cloud is a combination of at least two clouds; it is a mixture of public, private, or community.

There are three services models of cloud that can differ in the amount of control over information [9, 11]:

- SaaS: the provider gives access to both resources and applications.
- PaaS: the provider gives access to the components that they require to develop and operate applications.

- IaaS: outsources the storage and resources, such as hardware and software that they need.

The cloud computing provides software and stores services on remote servers. It enables convenient on demand network access to configurable resources (networks, servers, storage, applications, and services) with a low cost and a minimal management [9–12]. Due to the distributed and open nature of cloud, resources, applications, and data are vulnerable to intruders. Thereby, the security provided by organizations is not adequate; this represents a big obstacle for users to adapt into the cloud computing systems. Hence, more security objectives, such as availability, confidentiality, data integrity, control, and audit should be taken into account to secure transaction in cloud [12, 14–16].

3.2 Security Challenges

In fact, it is confirmed that security and privacy tasks represent a crucial challenge in the cloud [15]. Thus, the cloud uses the recent technologies of intrusion detection and prevention to protect each host against intrusive events. As every cloud service model is different from others services models similarly, intrusion detection and prevention techniques used for each cloud services models also differs [3, 11, 15]. Therefore, to provide safe services is interesting. However, these services delivery models are vulnerable to amount of attacks, exploiting the existing vulnerabilities. Therefore, the services become unavailable due to various attacks:

- Port scanning is exploited to search for open ports on a cloud. It can attack available services over the scanning of ports.
- Deny of Services and Distributed Deny of Services attempt to render resources unavailable to users. The Distributed Deny of Services attack is a specialized form of the Deny of Service attack that is launched with the help of a huge number of distributed machines across the internet.
- BotNet attacks are the most well-known attacks on distributed network environment. They consist of a number of bots which generate large volume of spam or launch Distributed Deny of Services attacks from compromised hosts.
- Several attacks cloud services depend on internet protocols such as man in the middle attack, IP spoofing, RIP attacks, and flooding.

The security of the cloud is becoming increasingly important due to the trend of data storage. The service delivery differs according to the service models, so the detection techniques used to control these models are also different. Intrusions detection technology can be used in various layers of cloud computing. There are some major types of intrusion detection and prevention systems which are used in cloud [12–14]:

- Network Intrusion Detection and Prevention Systems observe, examine and analyze the traffic to and from all the devices on the network.

- Host Intrusion Detection and Prevention Systems analyze the machines which contain the important information.
- Hypervisor based intrusion detection system implements the abstract layer between the underlying host and the guest operating system.

The studies which we carried on [1, 17, 18] leads to highlight certain limits and constraints of existing intrusion detection and prevention systems. The majority of whom suffer a wide of positive false and no detection of some negative false, it influences negatively on operational functioning of these tools. Certain packets are not treated and some attacks are not detected during a monitoring of high traffic. Another important aspect is how many data that analyst can analyze effectively and efficiently, the quantity of data that will be developed rapidly. It is impossible to have a perfect intrusion detection and prevention systems because of decision taken by these systems due to number of positive false that are produced when the system detects by error on anomaly and number of negative false when not wanted activity is not detected.

There is no standard detection system able to overcome all vulnerabilities of actual systems. The down researches in intrusion detection try to minimize in maximum those vulnerabilities by proposing optimal architectures integrating a sophistical techniques and algorithms.

4 Our Contribution

This section will cite a new contribution for intrusion detection to improve a cloud environment security. We describe in detail the proposed solutions to validate our proposed approach.

4.1 *The Distributed Approach of Intrusion Detection System for a Cloud*

To protect efficiently a cloud environment, a simple intrusion blocking of not authorized data activities is not satisfactory. The collected activities have to be preprocessed, analyzed and inspected. The detection systems are used to alert a detected intrusion and inform security administrator to take adequate decision. To design an efficient intrusion detection and prevention system cloud, we require reliable techniques. The proposed distributed system in [1] consists of various parts and databases to manage gathered data (Fig. 1).

For this, we implement a specific sniffer called PcapSockS to collect events from cloud taking into account some sophistical functions [7, 8] that intercepts packets. The preprocessing is carried out to transform data evens into a new representation before being presented to a neural network inputs. A classifier based on multilayer

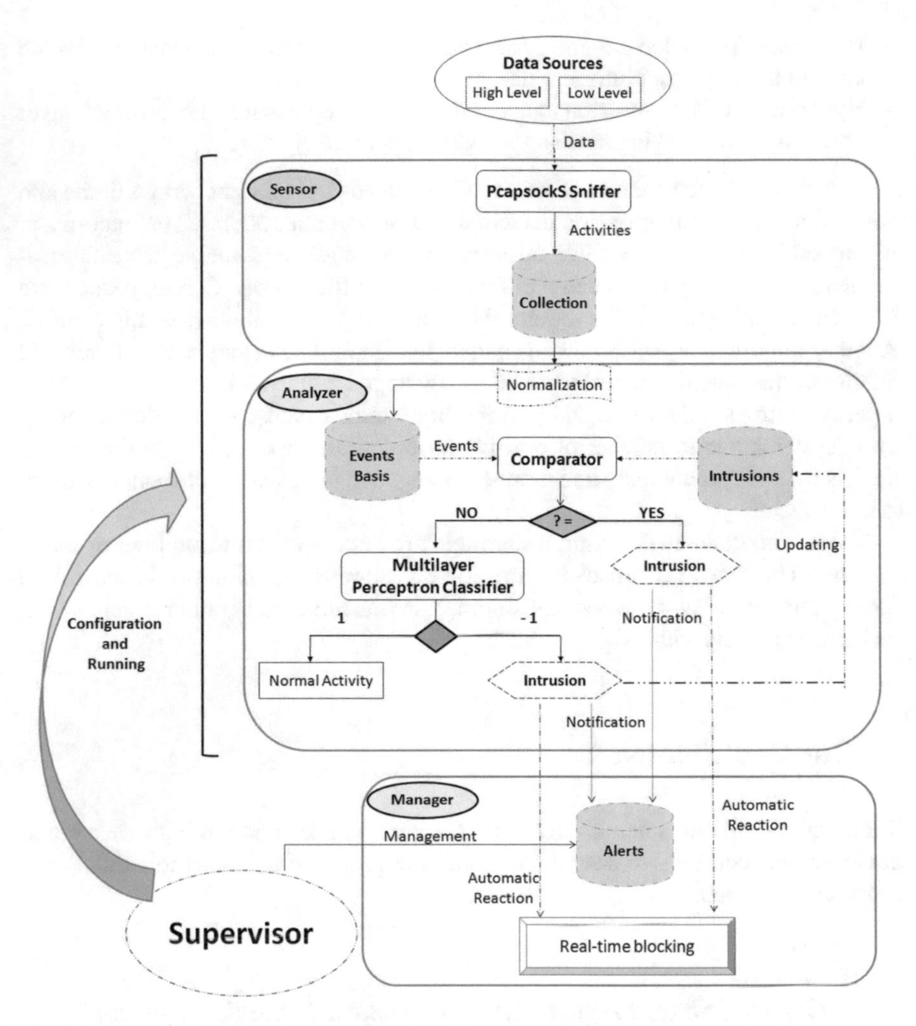

Fig. 1 Proposed IDPS based on PcapSockS Sniffer and multi-layer perceptron classifier

perceptron algorithm to classify events and finally take a convenient decision. It aims to improve the accuracy of detection. We propose a rigorous algorithm for training and recognition. Thus, we use the multilayer perceptron. The proposed machine learning algorithms used for training phase and recognition phase to make an accurate classifier [1, 17, 18]:

- Training phase algorithm

Initialize weights $W^{(0)} = \left(w_{i,0}\right)_{i=1..n}$ such as $w_{i,0} \leq 10^{-3}$ for $**i^{**} = 1..n$ and $w_{0,i} = 1$.

For i from 1 to n do

1 Present the inputs $X_i = \left(x_{i,j}\right)_{j=1...m}$.

2 Calculate $W_i^{(op)}$ *and* ε_i:

$$\varepsilon_i = \min_{a_i}(1 - y(a_i))$$

$$\begin{cases} a_i = \sum_{j=1}^{m} w_{i,j} x_{i,j} + w_{0,i}; \\ y(a_i) = f(a_i); \\ w_{0,i} = w_{0,i} + [1 - y(a_i)]; \\ for\ j\ from\ 1 \text{ to } m\ do \\ w_{i,j} = w_{i,j-1} + [1 - y(a_i)]x_{i,j}; \\ endfor \end{cases}$$

3 End For

- Recognition phase algorithm

1 New input $X = \left(x^{(j)}\right)_{j=1..m}$, final output d, activation state a, calculated result $y(a)$.

2 Computing of output

$$a = \sum_{j=1}^{m} w_j^{(\max)} x^{(j)} + w_0^{(\max)};\ y(a) = f(a);$$

3 Classification of activities

$$\begin{cases} if\ (1 - y(a) \leq \varepsilon)\ then \\ \quad d = 1 // Normal\ activity. \\ Else \\ \quad d = -1 // Intrusion. \\ End\ if \end{cases}$$

The Manager is the person who administers various parts of the proposed system. He is responsible for lunching management and monitoring several steeps of implementation.

4.2 Implementation of Proposed Approach and Description of Solutions

The role of our proposed distributed intrusion detection and prevention system in cloud is interesting because it provides an additional layer for protection. It allows watching what happen in a cloud environment. The implementation of our intrusion detection approach cloud is shown in the figure below (Fig. 2):

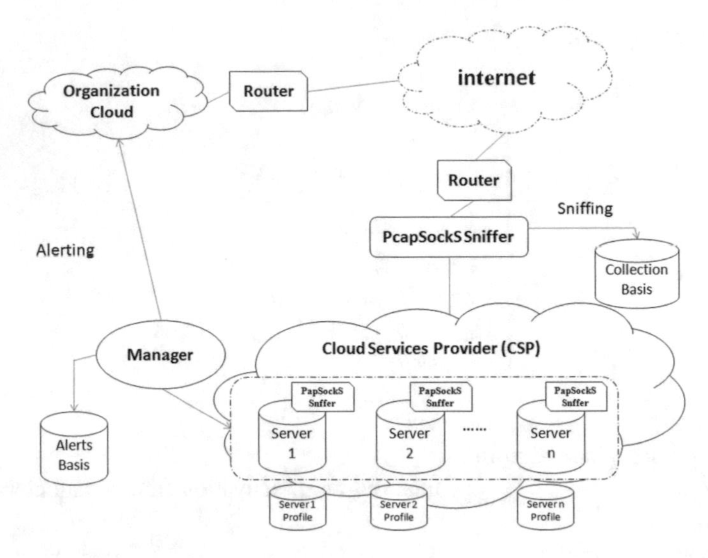

Fig. 2 Implementation of our proposed IDS cloud

The proposed intrusion detection and prevention system is installed at Cloud Solution Provider level. The system components are distributed according to the distributed aspect of a cloud. The PcapSockS Sniffer is installed just after the router that connects Cloud Solution Provider with the internet; it is also implemented in every monitored server. The intrusion detection and prevention system manager is an independent Cloud Solution Provider component, able to administer other intrusion detection and prevention system components; it is responsible to run sniffing, training to build a specific profile for each server, managing databases…

The first detection is based on signature; the collected packet at network level is normalized and is compared by the content of signature basis. If an intrusion is detected, it is recorded. The manager alerts user cloud and the provider about the existence of a threat. This is the role of Network Intrusion Detection and Prevention cloud (Fig. 3).

The second detection is behavioral and implemented inside the Cloud Solution Provider. Each PcapSockS Sniffer is placed at the server to be monitored. Each server is represented by a profile that characterizes its reference state using parameters such as CPU consumption time, memory size, peak hours, etc. The profile is built during the training phase. The intercepted activity is compared with this profile. If there is a deviation, the intrusion is detected. The proposed approach is one of the practical approaches to defend various vulnerabilities (Fig. 4).

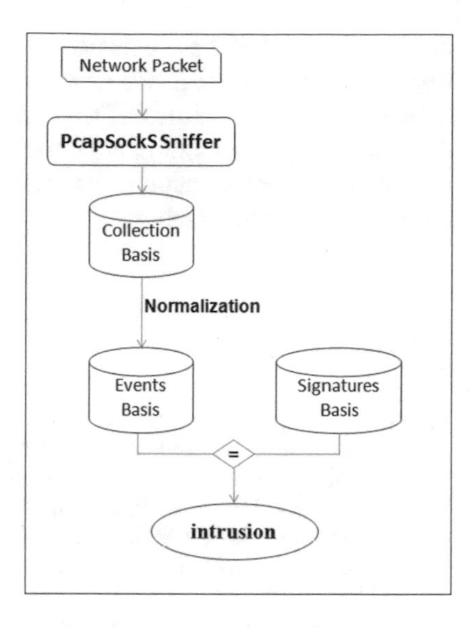

Fig. 3 Network intrusion detection and prevention system for cloud environment based on misuse detection

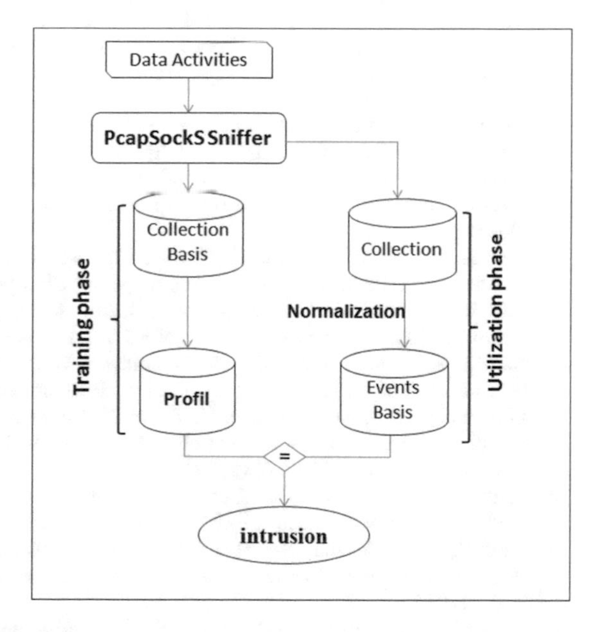

Fig. 4 Host intrusion detection and prevention system for cloud environment based on anomaly detection

5 Conclusion and Future Work

Security and privacy issues present a strong barrier for users to adapt into cloud. It is important to ensure the security and privacy of data outsourced to the cloud as well as the security of cloud services and cloud infrastructures. This paper proposed an optimal approach and presents the alternative methods to incorporate intrusion detection techniques in cloud. We have explored the locations in cloud service delivery models where these Intrusion Detection and Prevention can be deployed for efficient detection and prevention. In the future work, we will develop an in depth, advanced protection system to monitor the activities of different layers of cloud to ensure integrity and confidentiality.

Reference

1. Guezzaz, A., Asimi, A., Batou, Z., Asimi, Y., Sadqi, Y.: A global intrusion detection system using PcapSockS Sniffer and multilayer perceptron classifier. Int. J. Netw. Secur. (IJNS) **21**(3), 438–450 (2019). https://doi.org/10.6633/IJNS.201905_21(3).10
2. Farhaoui, Y., Asimi, A.: Performance assessment of tools of the intrusion detection and prevention systems. IJCSIS **10**, 7 (2012)
3. Gupta, A., Kumar, M., Rangra, A., Tiwari, V.K., Saxena, P.: Network intrusion detection types and analysis of their tools. Department of Computer Science and Information Technology, Jaypee University of Information Technology, India (2012)
4. Day, D.J., Burns, B.: A performance analysis of snort and suricata network ıntrusion detection and prevention engines. In: The Fifth International Conference on Digital Society (2011)
5. China, R., Avadhani, P.: A comparison of two intrusion detection systems. IJCST **4**(1), 316–319 (2013). ISSN 0976-8491
6. Santos, B., Chandra, T., Ratnakar, M., Baba, S., Sudhakar, N.: Intrusion detection system-types and prevention. Int. J. Comput. Sci. Inf. Technol. **4**(1), 77–82 (2013)
7. Guezzaz, A., Asimi, A., Sadqi, Y., Asimi, Y., Tbatou, Z.: A new hybrid network sniffer model based on Pcap language and sockets (Pcapsocks). Int. J. Adv. Comput. Sci. Appl. **7**(2), 207–214 (2016)
8. Guezzaz, A., Asimi, A., Asimi, Y.: A hybrid NIPS based on PcapSockS Sniffer and neural MLP. In: Advances in Intelligent Systems and Computing. Springer (2017). ISBN 978-3-319-64718-0
9. Mell, P., Grance, T.: The NIST definition of cloud computing. Recommendations of the National Institute of Standards and Technology. NIST Special Publication 800-145
10. Kumar, S., Goudar, R.H.: Cloud computing – research ıssues, challenges, architecture, platforms and applications: a survey. Int. J. Future Comput. Commun. **1**(4), 356 (2012)
11. El Balmany, C., Asimi, A., Tbatou, Z., Asimi, Y., Guezzaz, A.: Openstack: launch a secure user virtual machine image into a trust public cloud IaaS environment. In: 4th World Conference on Complex Systems (WCCS) (2019)
12. Zouhair, C., et al.: Review of intrusion detection systems in cloud computing. LIMSAD Labs Faculty of Sciences, Hassan II University of Casablanca, Casablanca, Morocco (2019)
13. Iqbal, S., et al.: On cloud security attacks: a taxonomy and ıntrusion detection and prevention as a service. J. Netw. Comput. Appl. S1084-8045(16)30177-1. https://doi.org/10.1016/j.jnca.2016.08.016
14. Shelke, P., et al.: Intrusion detection system for cloud computing. Int. J. Sci. Technol. Res. **1**(4), 67–71 (2012). ISSN 2277-8616
15. Parekh, D.H., Sridaran, R.: An analysis of security challenges in cloud computing. Int. J. Adv. Comput. Sci. Appl. (IJACSA) **4**(1), 38–46 (2013)
16. Achbarou, O., et al.: Cloud security: a multi agent approach based ıntrusion detection system. Indian J. Sci. Technol. **10**(18) (2017). https://doi.org/10.17485/ijst/2017/v10i18/109044
17. Guezzaz, A., Asimi, A., Asimi, Y., Tbatou, Z., Sadqi, Y.: A lightweight neural classifier for ıntrusion detection. Gen. Lett. Math. **2**(2), 57–66 (2017)
18. Guezzaz, A., Asimi, A., Azrour, M., Batou, Z., Asimi, Y.: A multilayer perceptron classifier for monitoring network traffic. In: Big Data and Networks Technologies, January 2020

Cb2Onto: OWL Ontology Learning Approach from Couchbase

Sajida Mhammedi, Hakim El Massari, and Noreddine Gherabi

Abstract Big Data are collections of large data sets of both unstructured and structured data characterized by volume, variety and velocity. These characteristics accentuate the heterogeneity and complexity of data, which exceeds the capacity of traditional systems to cope with it. Due to the aforementioned characteristics, it is crucial to have a unified conceptual view of the data, as well as, an efficient representation of knowledge for big data management. Since the ontology has the maturity we need to understand and provide the meaning to process big data. However, Construction of ontology by hand is an incredibly challenging and error-prone process. Learning Ontology from existing resource gives a reasonable alternative. Therefore, this paper proposes an approach to learn OWL ontology from data in Couchbase database by the application of six mapping rules, we use Ontop reasoner to evaluate the consistency of extracted ontology.

Keywords Ontology learning · Transformation rules · OWL · NoSQL database · Couchbase

1 Introduction

With the rapid advancement of technology, the amount of available data is increasing continuously, which has defined big data as a huge volume of both unstructured and structured data set. This dramatic explosion of data from different sources exceeds the use of conventional software and database techniques. Moreover, the ability to share and reuse knowledge remains difficult issues in big data applications especially when data is unstructured. To overcome this challenge, semantic integration in big data seems to be the right solution. In order to better manage big data, it is crucial for representing these data as knowledge [1].

Ontologies are at the heart of the semantic Web technology, they define a formal description of knowledge as a set of concepts within a domain and the relationships between them that make global interoperability possible. The primary aim of

S. Mhammedi · H. El Massari (✉) · N. Gherabi
Sultan Moulay Slimane University, ENSA Khouribga, Lasti Laboratory, Khouribga, Morocco

N. Gherabi and J. Kacprzyk (eds.), *Intelligent Systems in Big Data, Semantic Web and Machine Learning*, Advances in Intelligent Systems and Computing 1344,
https://doi.org/10.1007/978-3-030-72588-4_7

ontologies is to collect knowledge about a particular domain and to provide common vocabulary as well as an accepted conceptual representation for reuse and share.

The process of manual ontology extraction is vulnerable to errors and very time consuming especially in big data since data are voluminous, of a large variety, and of a wide velocity. Hence the need for algorithms capable of producing automated semantic models of data is becoming paramount. At this level, ontology learning field has emerged to propose a solution to this issue [12, 13]. It focusses on automatically deriving ontology from current data [2]. Nowadays, the latest attempts are geared towards the automated extraction of ontologies from NOSQL databases[1] due to the proliferation of these databases in storage.

Big data applications increasingly use NoSQL databases as they provide a flexible schema design, support easy replication, have a simple API and support large volumes of data. Four types of NoSQL data models are available, namely key-value model, document-oriented model, graph model and columnar data model. We choose to deal with couchbase which is a document-oriented database.

This paper proposes an automatic ontology learning approach, which requires OWL ontology based on data in Couchbase[2] database. The resulting ontology is to be used as a conceptual view over the data stored in the Couchbase database.

The reminder of this paper is organized as follows: We start in Sect. 2 by giving related work drawing from prior researches in the associated areas. Section 3, describes about NoSQL databases besides an overview of Couchbase. Our approach to construct an ontology for Big Data integration is stated in Sect. 4. In Sect. 5, we perform experiments and discuss evaluation results. Finally, Sect. 6 presents future work and draws the conclusions.

2 Related Work

We focus on ontology-based NOSQL databases approaches since these databases are in the basis of storage of Big Data and concentrate on recent solutions addressing ontology-learning and Big Data. In this section we briefly review the most popular ontology learning approaches taking as input NOSQL databases.

In [3], authors proposes a tool (M2Ont) for ontology learning from MongoDB to address the stored data semantically. They presented transformation rules for all possible cases in the MongoDB database, in order to build an OWL ontology. The final ontology extracted is used as a conceptual view over the data stored in the MongoDB database.

In [4], authors develop a data integration framework, in order to get schema represented as an OWL ontology from the NOSQL sources correspond to databases, namely MongoDB and Cassandra. The first step is to use an inductive approach to generate a local schema for each integrated source. The second step allows to define

[1] https://hostingdata.co.uk/nosql-database/.

[2] https://www.couchbase.com/.

a global ontology based on the local ontologies created for each data source. To build the global ontology, authors enhance the two ontologies to be aligned by introducing subsumption relationships, to detect both simple and complex correspondences.

With the aim of representing unstructured data Jabbari et al. proposes ontology engineering method for MongoDB [5]. They used formal concept analysis (FCA) as the main component of ontology development. After the generation of the formal context from MongoDB, they apply FCA to obtain a concept lattice from that formal context. Finally, the proposed approach transforms the concept lattice obtained into ontology according to its mapping rules.

To solve the issue of semantic heterogeneity and structural heterogeneity in the big data integration process, Li et al. [6] presented ontology-based approach for constructing big data semantic models. This study analyses big data model and then develops ontology-based big data model in accordance with the Key/Value framework by applying their proposed approach. to implement and verify their approach they used HBase.

3 NoSQL Database

NoSQL databases "not only SQL" are non-tabular, and store data differently than relational tables. It come in a variety of types based on their data model. The main types are wide-column, key-value, document, and graph [7]. They provide flexible scale and schemas easily with high user loads and large amounts of data.

- Wide-column stores: Data are stored in records with an ability to hold very large numbers of dynamic columns. Cassandra and HBase are examples of wide-column stores.
- key-value model: a simpler type of database where each item contains keys and values. The query speed is higher than relational database, support mass storage and high concurrency, etc. Redis and DynamoDB are two of the most popular key-value databases.
- Document databases: store data in documents similar to JSON objects or XML format, that support semi-structured and hierarchical formats. Each document contains pairs of fields and values (Examples: MongoDB, Couchbase).
- Graph databases: They are schema-less databases. Store data in nodes and edges. Neo4j, InfoGrid, Infinite Graph and JanusGraph are popular Graph databases.

We choose to deal with document databases, they contain documents, which are records that describe the data in the document, as well as the actual data. Documents can contain nested structures. This allows developers to express one-to-many as well as one-to-one relationships without requiring a reference or junction table. Thus, a complex document can be extracted or saved without using joins. We are particularly interested to Couchbase as a document NOSQL database.

3.1 Overview of Couchbase

Couchbase Server is a scale-out NoSQL database with an architecture designed to simplify building modern applications with a flexible data model, powerful SQL-based query language, and a secure core database platform that provides high availability, scalability, and performance. Surely providing these features in production is not a trivial thing, but Couchbase achieves this in a simple and easy manner [8].

Couchbase stores data as individual documents, comprised of a key and value, where the value is a JSON-formatted. Couchbase introduces multiple ways and methods to access data. These methods include basic key-value operations, SQL querying, full-text searching, real-time analytics, server-side eventing, and mobile application synchronization.

With the research focus on performance, the authors in [9] compare the durability and performance trade-offs of several state-of-the-art NoSQL systems. Their results firstly show that Couchbase well performed in-memory, and that MongoDB and Cassandra lagged behind in bulk loading capabilities. Regarding durability. In their second paper [10], the authors study failover characteristics. Their results allow for many conclusions, but the entire tend to indicate that Couchbase give strong availability guarantees.

Data model of Couchbase is organized as follow: Cluster, bucket, document. Where cluster consists of one or more nodes running Couchbase Server, bucket hold JSON document, and document structure consists of its inner arrangement of attribute-value pairs [11].

4 The Proposed Ontology Learning Approach

The proposed approach consists of six mapping rules to learn ontology from big data in Couchbase. Our approach uses OWL as the ontology representation language since it is the standard recommended by the World Wide Web Consortium (W3C)[3]. OWL provide additional vocabulary along with formal semantics and facilitates greater machine interoperability of web content. The mapping rules are as follows:

Rule 1. Learning classes.

All buckets in Couchbase are transformed into owl:Class.

Algorithm 1 allow to learn ontology classes.

[3]https://www.w3.org/TR/owl-features/.

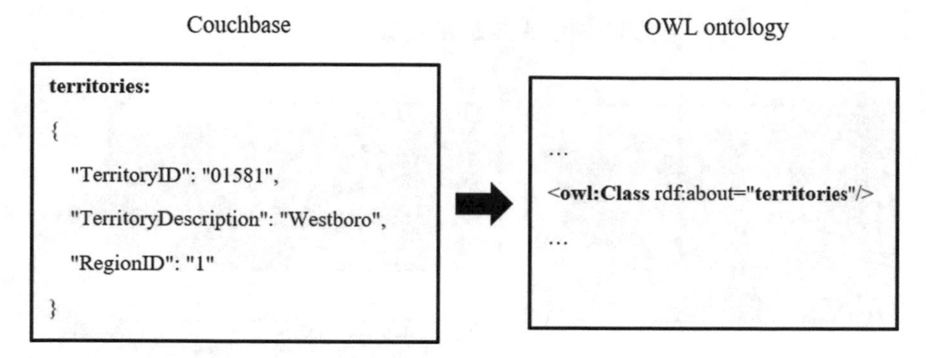

Fig. 1 A Couchbase bucket mapped to an OWL class

Algorithm 1: Learning classes.

1 *Browse the whole cluster*
2 **foreach** *bucket Bi* **do**
3   Extract the bucket name
4   Associate to *Bi* an OWL class *CLi*
5   Add *CLi* to the ontology
6 *Save ontology*

Figure 1 provides a bucket in Couchbase (left) and its corresponding OWL class (right).

Rule 2. Learning objectProperties from embedded document.

The value of field may have different data types, collection types (map, set, list) and basic data types (string, int, etc.).

An *objectProperty* relationship is extracted either from *embedded docments* (rule 2-1) or from *referenced with Referenced_ID* (rule 2-2) Fig. 2.

- Rule 2-1: Let's consider a document docB embedded into a document docA. The relation of that embedment is extracted into owl:objectProperty in which the *"domain"* is the class corresponding to the embedding document (docA) and the *"range"* is the class corresponding to the embedded document (docB). The name of the objectProperty relationship is the concatenation of the word "has" with the name of the docB.

Algorithm 2 enable to learn *objectProperties* from embedded documents.

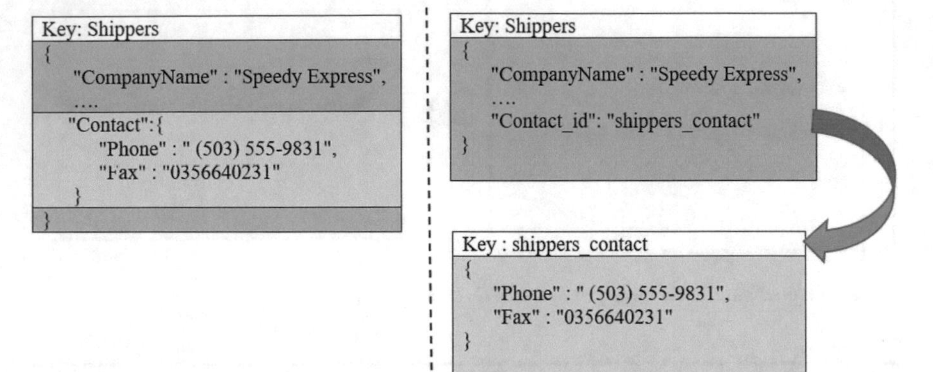

Fig. 2 Embedded versus referenced documents

Algorithm 2: Learning objectProperties from embedded document

---

1 *Browse the whole cluster*
2 **foreach** *bucket Bi* **do**
3   **foreach** *document Dij* **do**
4   **foreach** *field F* **do**
5   Extract its type (basic, embedded document, Referenced_ID, List)
6   **if** *the type of the field is embedded document* **then**
7   DocA = *Dij*
8   DocB = the referenced document
9   ObjectProperty name = "has" + name of DocB
10   Domain = the class corresponding to current document (DocA)
11   Range = the class extracted from the nameDocument_ID field (DocB)
12   Save ObjectProperty
13 *Save ontology*

According to rule 2-1 an example of a learned objectProperty from embedded document is given in Fig. 3.

- Rule 2-2: To reference another document the item key is using. Referencing allows Couchbase to cache, store, and retrieve the documents independently. For example, let's consider a DocA containing a *reference_ID* to a DocB by the use of the *Referenced_ID* construct. *Referenced_ID* are references from one document to another utilizing the value of nameDocument_ID field, *Referenced_ID* allow documents located in multiple collections to be more easily linked with documents from a single collection. The nameDocument_ID field retains the name of the collection where the referenced document resides.
  The *Referenced_ID* reference is mapped to an *objectProperty* for which the *"domain"* is the class corresponding to the referencing document (docA) and the *"range"* is the referenced class (corresponding to docB). The name of the

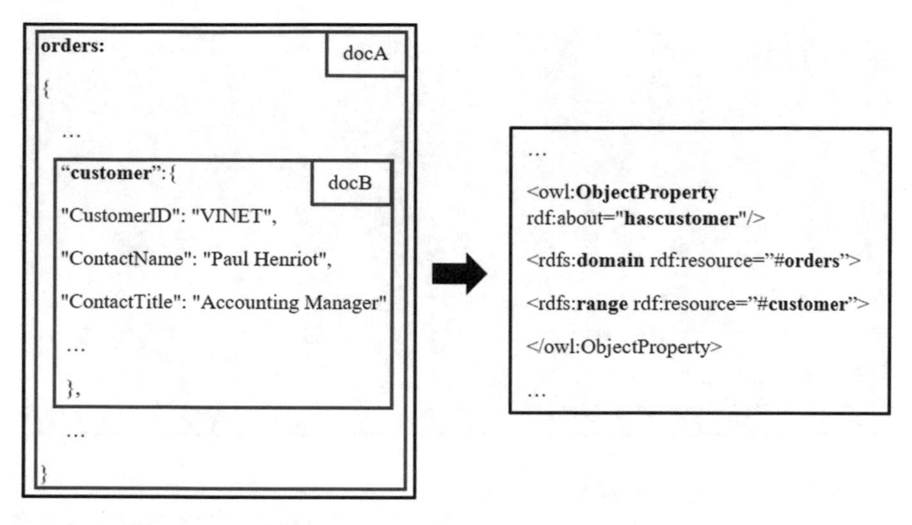

Fig. 3 Embedded document mapped to an objectProperty

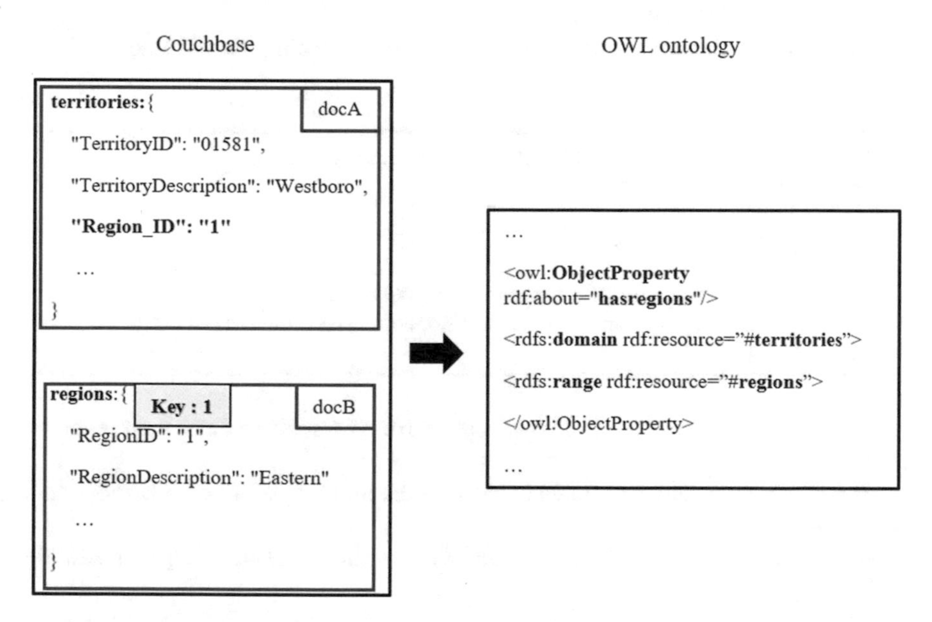

Fig. 4 Referenced document mapped to an objectProperty

objectProperty relationship is the concatenation of the word "has" with the name of the document B.

Algorithm 3 explains how to learn objectProperties from *Referenced_ID* references. As example Fig. 4 show how to learn objectProperty from references with *Refrenced_ID*.

Algorithm 3: Learning objectProperties from references with Referenced_ID.
1 *Browse the whole cluster*
2 **foreach** *bucket Bi* **do**
3   **foreach** *document Di j* **do**
4   **foreach** *field F* **do**
5   Extract its type (basic, embedded document, *Referenced_ID*, List)
6   **if** *the type of the field is Refereced_ID* **then**
7   DocA = *Dij*
8   DocB = the referenced document
9   ObjectProperty name = "has" + name of docB
10   Domain = the class corresponding to current document (DocA)
11   Range = the class extracted from the nameDocument_ID field (DocB)
12   Save ObjectProperty
13 *Save ontology*

Rule 3. Learning dataTypeProperties.

Each *basic field* in a document *Dij* is mapped into owl:dataTypeProperty which "domain" is the class corresponding to the collection containing this document and "range" is limited to that basic data type.

Algorithm 4 demonstrates *dataTypeProperties* learning from basic fields in Couchbase database.

Algorithm 4: Learning dataTypeProperties
1 *Browse the whole cluster*
2 **foreach** *bucket Bi* **do**
3   **foreach** *document Dij* **do**
4   Extract its fields
5   Basic fields are transformed into DatatypeProperties
6   Add DatatypeProperties to the corresponding class into the ontology
7 *Save ontology*

An example of a learned dataTypeProperty from Couchbase basic field is shown in Fig. 5.

Rule 4. All data values of fields in each document is transformed into individual in the ontology.

Algorithm 5 allows to learn individuals. Figure 6 illustrates an example of learning individuals.

Algorithm 5: Defining individual from field's values
1 *Browse the whole cluster*
2 **foreach** *bucket Bi* **do**
3   **foreach** *document Dij* **do**
5   Create an individual I with the values of its fields
6   Associate the individual I to the ontology class corresponding to the document *Dij*
7   Save individual
13 *Save ontology*

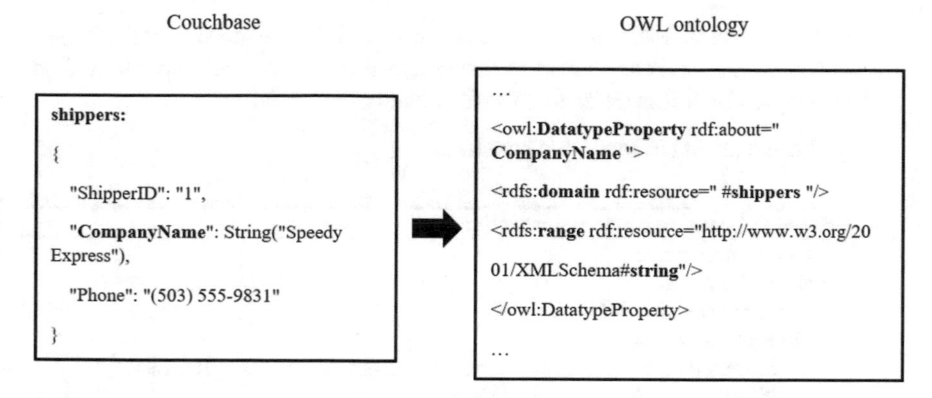

Fig. 5 Basic field mapped to a dataTypeProperty

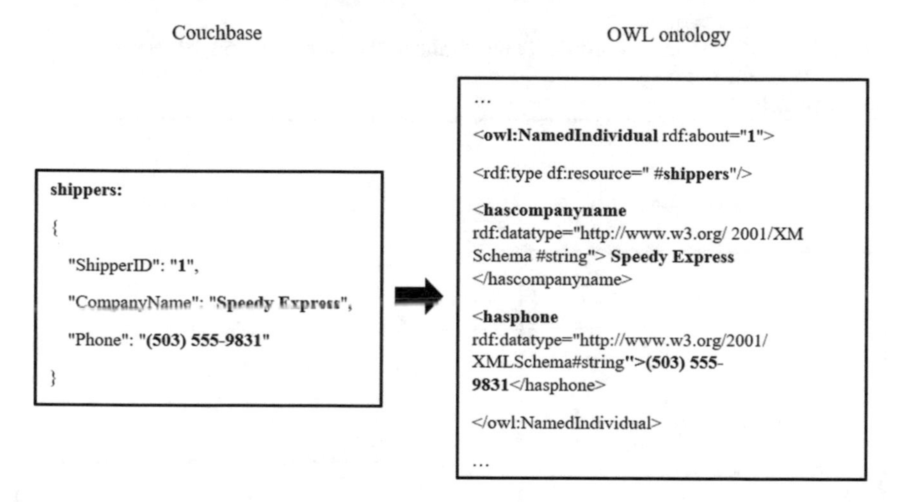

Fig. 6 Mapping values to individuals

Rule 5. Learning property restrictions.

In the OWL language, restrictions or constraints are divided into cardinality constraints and value constraints. Cardinality constraints are *owl:maxCardinality*, *owl:minCardinality* and *owl:cardinality*. For the value constraints, we cope with the value constraint *owl:allValuesFrom*.

The maxCardinality restriction describes that the concerned property must have at most N semantically distinct values, where N is the value of the cardinality constraint. In the same way, the minCardinality restriction defines that the concerned property must have at least N semantically distinct values.

- *Rule 5-1.* The *maxCardinality* constraint is concluded from the size of the biggest list in a List field. It is applied to the property that match the corresponding field. As example Fig. 7 illustrates the maxCardinality constraint.

  Algorithm 6 allows to learn this constraint.

Algorithm 6: Learning property restrictions

1 *Browse the whole cluster*
2 **foreach** *bucket Bi* **do**
3 **foreach** *document Dij* **do**
4 **foreach** *field F* **do**
5 Extract its type (basic, embedded document, Referenced_ID, List)
6 **if** *the type of the field is List* **then**
7 Extract the size of the list
8 The size of the biggest list for the concerned field is transformed into max cardinality
9 Add the max cardinality constraint to the corresponding property into the ontology
10 *Save ontology*

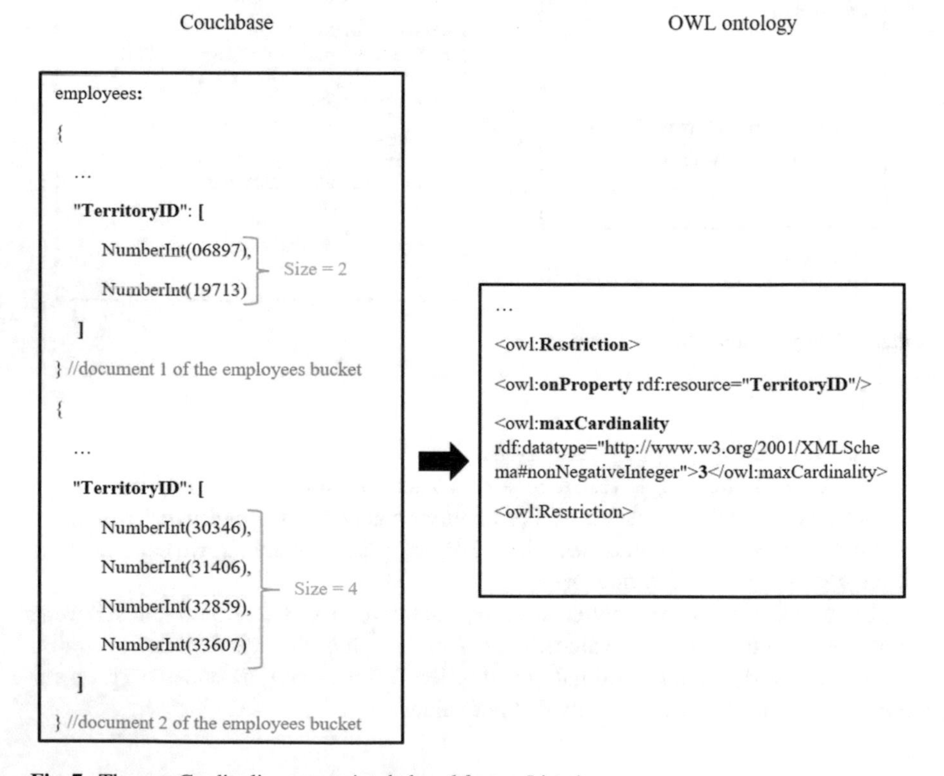

Fig. 7 The maxCardinality constraint deduced from a List size

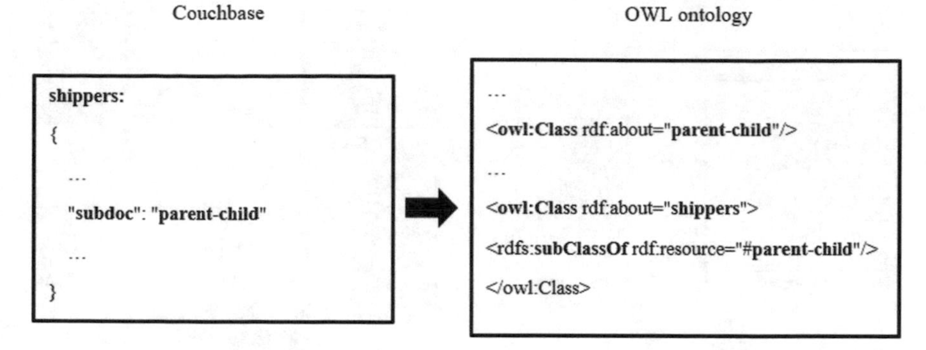

Fig. 8 A parent field mapped to a class hierarchy

- *Rule 5-2.* The *minCardinality* constraint is extracted from the size of the smallest list in a List field. It is applied to the property that match the corresponding field.

Rule 6. Learning classes hierarchies.

A *Superclass/subclass* is a relationship (parent/child), extracted from the field *"subdoc"* in every document *Dij* belonging to a bucket *Bi*. Therefore, an OWL class C is associated with the document *Dij* and *Bi* becomes a *"subClassOf"* the class C.

Algorithm 7: Learning classes hierarchies.

1 *Browse the whole cluster*
2 **foreach** *bucket Bi* **do**
3 Extract the name of the current bucket
4 **foreach** document *Dij* **do**
5 Extract the value of the field "subdoc"
6 Associate to it an OWL class C
7 Save the class corresponding to the document *Dij* as a "subClassOf" the class C in the ontology
8 *Save ontology*

As example Fig. 8 provide the learned classes hierarchy from Couchbase.

5 Experiment and Evaluation

Our proposed approach focuses on learning ontology from Couchbase database using mapping rules. This work is done by Java programming language and OWL API 5.1.14 version.[4] Couchbase-java-client:3.0.4 is used to access Couchbase. For the graphical representation, we used "WebVOWL[5]" a web application for the interactive

[4]https://owlcs.github.io/owlapi/.

[5]https://vowl.visualdataweb.org/webvowl.html.

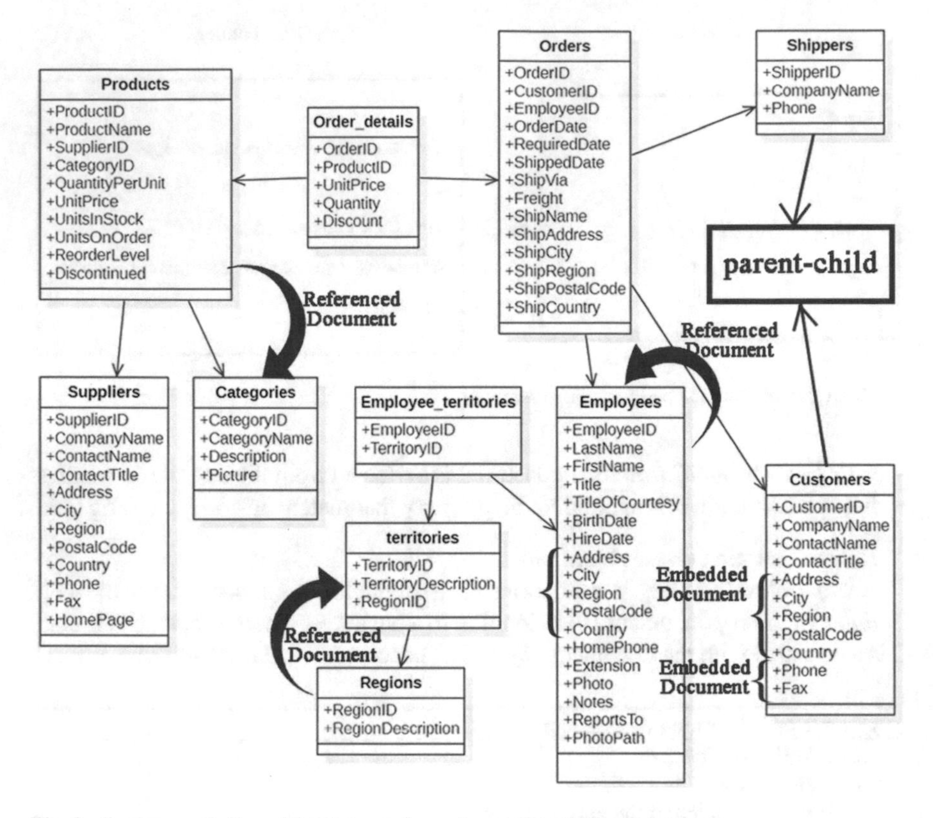

Fig. 9 Summary of all modifications performed over "NorthWind" database

visualization of ontologies. We first converted the output OWL ontologie file to VOWL-JSON format using owl2vowl 0.3.7 version, then we uploaded the ontology file to webvowl 1.1.7 version.

To realize the proposed approach, "NorthWind" database is imported to cluster named "northwind" in Couchbase. The "Northwind" database is available in the CSV format. We developed a java method to dump "NorthWind[6]" data from CSV to Couchbase. We were obliged to make changes over the database design with the aim of covering all Couchbase constructs. It consists simply to transformed relational tables to buckets and ignores joins and integrity constraints which have to be transformed into references and embedded documents. Moreover, to represent the «parent/child» relationship, we created a bucket "parent-child" and added the field "parent" in the buckets "customers" and "shippers". These modifications performed over the relational "Northwind" are shown in Fig. 9.

The execution of our Java Program (Cb2Onto: Couchbase to ontology) over the "northwind" database gives a consistent Owl ontology as output (.owl). Figure 10 provides the resulting OWL file. Figure 11 shows Converted file (.json) generated

[6]https://github.com/tmcnab/northwind-mongo.

```
<?xml version="1.0"?>
<rdf:RDF xmlns="http://www.semanticweb.org/hakim/ontologies/2020/7/Northwind#"
     xml:base="http://www.semanticweb.org/hakim/ontologies/2020/7/Northwind"
     xmlns:owl="http://www.w3.org/2002/07/owl#"
     xmlns:rdf="http://www.w3.org/1999/02/22-rdf-syntax-ns#"
     xmlns:xml="http://www.w3.org/XML/1998/namespace"
     xmlns:xsd="http://www.w3.org/2001/XMLSchema#"
     xmlns:rdfs="http://www.w3.org/2000/01/rdf-schema#">
    <owl:Ontology rdf:about="http://www.semanticweb.org/hakim/ontologies/2020/7/Northwind"/>



    <!--
    ///////////////////////////////////////////////////////////////////////////////////////
    //
    // Object Properties
    //
    ///////////////////////////////////////////////////////////////////////////////////////
     -->




    <!-- http://www.semanticweb.org/hakim/ontologies/2020/7/Northwind#hascategories -->

    <owl:ObjectProperty rdf:about="http://www.semanticweb.org/hakim/ontologies/2020/7/Northwind#hascategories">
        <rdfs:domain rdf:resource="http://www.semanticweb.org/hakim/ontologies/2020/7/Northwind#products"/>
        <rdfs:range rdf:resource="http://www.semanticweb.org/hakim/ontologies/2020/7/Northwind#categories"/>
    </owl:ObjectProperty>



    <!-- http://www.semanticweb.org/hakim/ontologies/2020/7/Northwind#hascustomers -->

    <owl:ObjectProperty rdf:about="http://www.semanticweb.org/hakim/ontologies/2020/7/Northwind#hascustomers">
        <rdfs:domain rdf:resource="http://www.semanticweb.org/hakim/ontologies/2020/7/Northwind#orders"/>
        <rdfs:range rdf:resource="http://www.semanticweb.org/hakim/ontologies/2020/7/Northwind#customers"/>
    </owl:ObjectProperty>



    <!-- http://www.semanticweb.org/hakim/ontologies/2020/7/Northwind#hasemployee_territories -->

    <owl:ObjectProperty rdf:about="http://www.semanticweb.org/hakim/ontologies/2020/7/Northwind#hasemployee_territories">
```

Fig. 10 Visualization of resulting OWL file with the Cb2Onto

```
{
  "_comment" : "Created with OWL2VOWL (version 0.3.7), http://vowl.visualdataweb.org",
  "header" : {
    "baseIris" : [ "http://www.semanticweb.org/hakim/ontologies/2020/7/Northwind", "http://www.w3.org/2002/
    "prefixList" : {
      "owl" : "http://www.w3.org/2002/07/owl#",
      "rdf" : "http://www.w3.org/1999/02/22-rdf-syntax-ns#",
      "xsd" : "http://www.w3.org/2001/XMLSchema#",
      "" : "http://www.semanticweb.org/hakim/ontologies/2020/7/Northwind#",
      "xml" : "http://www.w3.org/XML/1998/namespace",
      "rdfs" : "http://www.w3.org/2000/01/rdf-schema#"
    },
    "iri" : "http://www.semanticweb.org/hakim/ontologies/2020/7/Northwind"
  },
  "namespace" : [ ],
  "class" : [ {
    "id" : "3",
    "type" : "rdfs:Datatype"
  }, {
    "id" : "7",
    "type" : "owl:Class"
  }, {
    "id" : "9",
    "type" : "owl:Class"
  }, {
    "id" : "16",
    "type" : "owl:Class"
  }, {
    "id" : "17",
    "type" : "owl:Class"
  }, {
    "id" : "28",
    "type" : "rdfs:Datatype"
  }, {
    "id" : "29",
    "type" : "rdfs:Datatype"
  }, {
    "id" : "19",
    "type" : "rdfs:Datatype"
  }, {
    "id" : "37",
    "type" : "rdfs:Datatype"
  }, {
```

Fig. 11 Visualization of converted OWL file to JSON

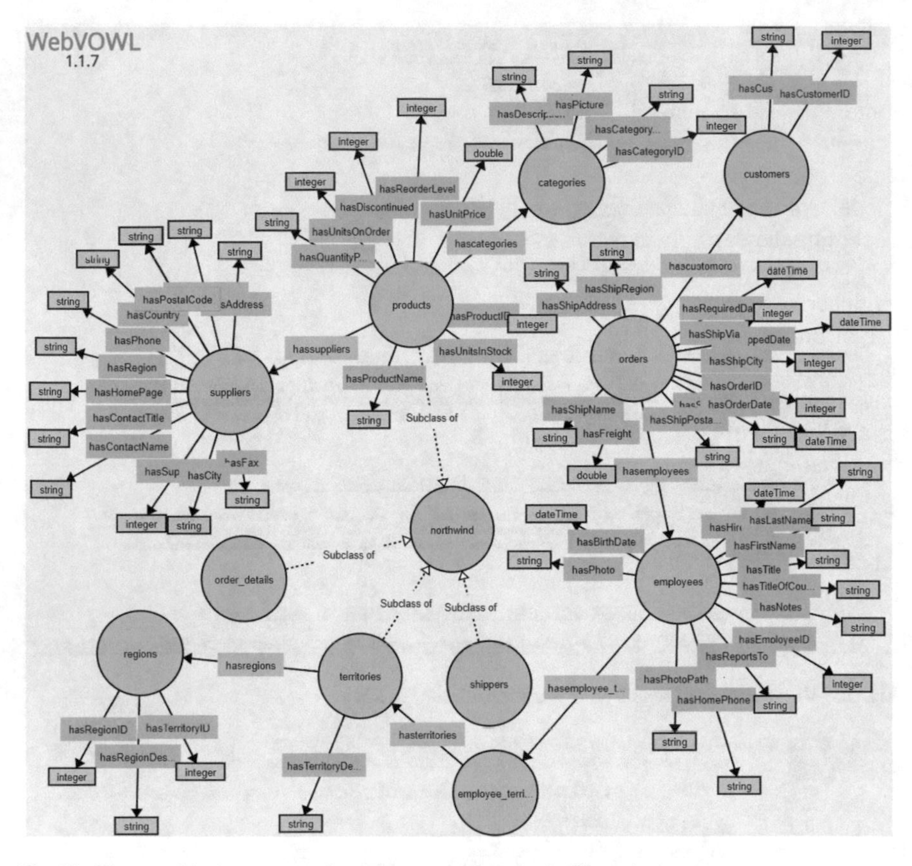

Fig. 12 The graphical representation of extracted "northwind" ontology

from OWL file and Fig. 12 illustrates the graphical representation of the "northwind" ontology.

Blue circles represent all classes and subclasses. Yellow rectangles present types of data value. DatatypeProperties are presented by the solid arrows that link the class and data types, marked by green rectangles. ObjectProperties are described by the solid arrows that link between two classes, marked by blue rectangles.

We used Ontop[7] reasoner to evaluate the consistency of extracted ontology. To realize that, protégé 5.5 version with Ontop plugin (it.unibz.inf.ontop.protege-3.0.1.jar) is used.

[7]https://github.com/ontop/ontop/releases/tag/ontop-3.0.1.

6 Conclusion and Future Work

Ontologies play an important role to represent knowledge extracted from data. Further, it helps to fetch out the hidden semantics which are not directly available from the data sources. The basic issue is to automatically build the ontology model from Big Data sources and to provide a shared model. Research on ontology is becoming increasingly extensively used in the computer science. Ontology learning is a good alternative to minimize mass of handwork encountered by the ontology engineer. In this paper we address the semantic aspect of the stored data through the use of ontology learning from Couchbase database. The aim of our approach is to propose an automatic construction of OWL ontology based on data stored in Couchbase database. It can be seen that our approach can sufficiently extract knowledge as a hierarchy of concepts from data in Couchbase. Consequently, these data can be managed in a simple manner for later processes. Furthermore, the time-consuming problem of manual ontology building from Couchbase can be solved by our approach.

As future works, we attend to develop a system that allows for querying Couchbase database through the conceptual representation of the domain of interest, provided in terms of ontology learning by using Ontop.

we envisage also to propose an approach to update the resulting ontology taking into account the updates of the Couchbase database.

References

1. Al-Aswadi, F.N., Chan, H.Y., Gan, K.H.: Automatic ontology construction from text: a review from shallow to deep learning trend. Artif. Intell. Rev. **53**, 3901–3928 (2020). https://doi.org/10.1007/s10462-019-09782-9
2. Abbes, H., Gargouri, F.: Big data integration: a MongoDB database and modular ontologies based approach. Procedia Comput. Sci. **96**, 446–455 (2016). https://doi.org/10.1016/j.procs.2016.08.099
3. Abbes, H., Gargouri, F.: M2Onto: an approach and a tool to learn OWL ontolo-gy from MongoDB database. In: Madureira, A.M., Abraham, A., Gamboa, D., Novais, P. (eds.) Intelligent Systems Design and Applications, pp. 612–621. Springer, Cham (2017)
4. Curé, O., Lamolle, M., Duc, C.L.: Ontology based data integration over document and column family oriented NOSQL. arXiv:1307.2603 [cs] (2013)
5. Jabbari, S., Stoffel, K.: Ontology extraction from MongoDB using formal concept analysis. In: 2017 2nd International Conference on Knowledge Engineering and Applications (ICKEA), pp. 178–182. IEEE, London (2017)
6. Kang, L., Yi, L., Dong, L.: Research on construction methods of big data semantic model, p. 6 (2014)
7. Bhogal, J., Choksi, I.: Handling big data using NoSQL. In: 2015 IEEE 29th International Conference on Advanced Information Networking and Applications Workshops, pp. 393–398. IEEE, Gwangiu (2015)
8. Ostrovsky, D., Haji, M., Rodenski, Y.: Getting started with Couchbase server. In: Pro Couchbase Server, pp. 3-18. Apress, Berkeley (2015). https://doi.org/10.1007/978-1-4842-1185-4_1
9. Nelubin, D., Engber, B.: Ultra-high performance NoSQL benchmarking: analyzing durability and performance tradeoffs, p. 43 (2013)

10. Nelubin, D., Engber, B.: NoSQL failover characteristics: aerospike, cassandra, couchbase, MongoDB, p. 19 (2013)
11. Couchbase Server: CouchDocs. https://docs.couchbase.com/server/current/introduction/intro.html
12. Daoui, A., Gherabi, N., Marzouk, A.: A new approach for measuring semantic similarity of ontology concepts using dynamic programming. J. Theor. Appl. Inf. Technol. **95**(17), 4132–4139 (2017)
13. Daoui, A., Gherabi, N., Marzouk, A.: An enhanced method to compute the similarity between concepts of the ontology. In: Noreddine, G., Kacprzyk, J. (eds.) International Conference on Information Technology and Communication Systems. Advances in Intelligent Systems and Computing, vol. 640, pp. 95–107. Springer (2018)

Data Profiling over Big Data Area

A Survey of Big Data Profiling: State-of-the-Art, Use Cases and Challenges

Bahaa Eddine Elbaghazaoui, Mohamed Amnai, and Abdellatif Semmouri

Abstract Before consuming datasets for any application, we need to understand the dataset at hand and its metadata. Discovering metadata process known as data profiling. Data profiling focus on examining the data sets and collecting metadata such as statistics or informative summaries about that data. In this chapter, we will discuss the importance of data profiling and shed light on the area of data profiling in big data. In addition, we will detail data profiling use cases and reviewing the state-of-the-art data profiling systems and techniques. Finally, we conclude with directions and challenges for future research in the area of data profiling.

Keywords Data profiling · Data quality rules · Big data profiling · Data lake · Linked data

1 Introduction

The world of the data scale is growing over time, many data sources produce data stream of records as never-end. The data sets must be structured, have actual content and with good quality. Researchers and data scientists will be in high demand to use their skills to manage and understand datasets.

Data is an important asset to a company or an organization, improving quality of data can be added value. Data profiling is one of the techniques that could be used to ensure data quality. In order to analyze efficiently a given data set. It contains a range of methods that could uncover and extract metadata such as data types and value patterns, uniqueness of columns, completeness, primary and foreign keys, dependencies and association rules.

B. E. Elbaghazaoui (✉) · M. Amnai
Laboratory of Computer Sciences, Faculty of Sciences Kenitra, Ibn Tofail University, Kenitra, Morocco
e-mail: bahaaeddine.elbaghazaoui@uit.ac.ma

M. Amnai
e-mail: mohamed.amnai@uit.ac.ma

A. Semmouri
Faculty of Sciences and Techniques, Lab. TIAD, Sultan Moulay Slimane University, Beni Mellal, Morocco
e-mail: abd_semmouri@yahoo.fr

N. Gherabi and J. Kacprzyk (eds.), *Intelligent Systems in Big Data, Semantic Web and Machine Learning*, Advances in Intelligent Systems and Computing 1344,
https://doi.org/10.1007/978-3-030-72588-4_8

The aim of data Profiling measure data consistency and accuracy, detect data duplications in order to obtain the correct value of data that could be used in decision-making purposes.

Existing work on this axis become difficult to applied in large datasets due to their different nature. To overcome this gap, we explain in this chapter the different tasks of data profiling, how could we use data profiling in relational and non-relational data, what are our propositions to use data profiling in big data and we conclude with an outlook on the future challenges of data profiling.

2 Profiling Metadata

Most of the features and services become or have the vision to becoming digital. Applications store transaction information requests during time. This data helps us make business decisions in most aspects such as activity market data, customer information, billing details, sales records, accounting information, production details, collection details, personnel records, salary records, and so on.

Data profiling is a process used to analyze content, quality and structure of data sets. Jack Olson [1] explains so clearly that data profiling is a necessary design precursor for any type of system to use this data.

Data profiling involves [19]:

- Collecting descriptive statistics, data types, keywords, descriptions or categories.
- Performing data quality assessment.
- Discovering metadata and assessing its accuracy.
- Identifying distributions, key candidates, foreign-key candidates, functional dependencies, embedded value dependencies, and performing inter-table analysis.

2.1 Data Profiling Types

The data profiling process comprises structure discovery, data discovery and relationship discovery.

2.1.1 Structure Discovery

The data inconsistency makes it possible to have problems at the level of the structure. Among these problems are also due to old data sources which are still in use or which have been migrated to a new application.

The discovery of the structure consists in examining complete columns or tables of data, and determining if the data in these columns or tables is adaptable to the expectations for this data. There are three main structure discovery techniques as shown in the figure:

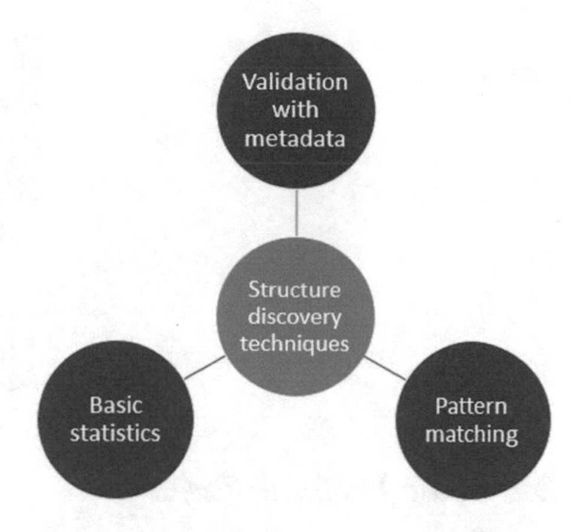

Fig. 1 Main structure discovery techniques

In the first place, metadata validation analyzes the data and indicates that the length of the field is appropriate, there are fields with missing values and determine if the data collected is in accordance with the original plan or if there are discrepancies. Then, pattern matching also determines whether the data values in a field are consistent for the data schema and whether the information is in the expected format. Finally, basic statistics present statistical information such as min, max, mean, median, mode, and standard deviation, to highlight outliers from normal values.

2.1.2 Content Discovery

Data discovery or content discovery allows you to examine the data elements individually. However, in this step, the following data discovery techniques that used are (Fig. 1):

- uncover non-standard data;
- find data elements that dont make sense using outlier detection and frequency counts.

The data of an organization is distributed and comes from different sources. In fact, Standardization uncovers inconsistencies in the data and then provides a solution for the inconsistency. And the frequency count examines the link between the values across the occurrences of data and thus detects the outliers.

2.1.3 Relationship Discovery

The last step of the data profiling process is discovering relationships. Indeed, it provides information about the relationships between data records. To more clarify, this method determines key relationships using metadata, provide primary and foreign keys, and identifies pending records that do not adhere to the relationship.

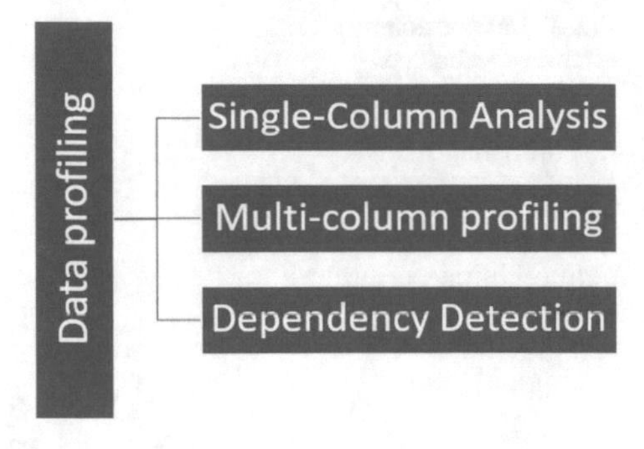

Fig. 2 Classification of data profiling tasks

2.2 *Data Profiling Tasks*

By definition, data profiling includes large and diverse set of tasks. They can be classified according to the type of data they process. Ziawasch Abedjan et al. [2], illustrates that data profiling tasks are composed to single column, multi-columns and dependencies profiling.

2.2.1 Single-Column Analysis

The basic step of data profiling is the columns analysis individually in a given table. Indeed, the metadata generated included various counts, such as number of values, number of unique values and number of non-null values. In addition, discover the minimum and maximum values and data type. Thus, Data profiling give ability to create value distributions histograms and identify typical patterns in data values in the form of regular expressions [13], and also predict semantic data type or domain of a given column (Fig. 2).

2.2.2 Multi-column Profiling

The second class of profiling tasks is multicolumn profiling, which covers several columns simultaneously. The several columns profiling is generalized the tasks on unique column and identify the dependencies between the values of the columns. This task consists to identify correlations between values through frequent patterns or association rules. In addition, multicolumn profiling and through clustering approaches give the possibility to discover coherent subsets of data records and outliers. In addition, multicolumn profiling generates summaries and sketches of large data sets.

2.2.3 Dependency Detection

Dependencies are metadata that describe the relationship between columns. A lot of research has been invested in examining a large set of columns. The big problem that remains with these researches is the automatization of detecting dependencies in a given data sets. The aim of Dependency discovery or dependency detection is to define Unique column combinations (UCCs) [22], Functional dependencies (FDs) [21] or Inclusion dependencies (INDs) [20].

- Unique column combinations are sets of columns that fulfill the uniqueness constraint in a relational database table [5].
- Functional Dependency states that values in one set of columns (the dependent column) functionally determine the value of another column (the determinant column).
- Inclusion dependencies reports the extent to which the values from one set of columns also appear in the other set of columns.

3 Profiling Linked Data

Data must be analyzed and understood at least at the basic level. In order to benefit from the massive amount of data, W3C talked about the Semantic Web project based on the concept of linked data in 2006. Linked data refers to publishing data on the web. In fact, this data can be linked to URIs from other data sources, using standards such as RDF without being available under an open license.

In order to profile, analyze and understand a huge linked data such as Linked Open Data or LOD [23], traditional data profiling methods do not suffice. In this section, we explore an overview of profiling approaches to understand the semantics of linked datasets.

3.1 Large Scale Linked Data

Over the past years, large number of data sources has been published as a part of the web of data. For example, LOD starting with 12 datasets in 2007, currently contains 1,255 datasets with 16,174 links. Furthermore, the range of domains and topics covered by these datasets has also increased. In general, most algorithms exist for data profiling cannot deal with large datasets. As well as, running data profiling tasks can take a long time for large linked datasets.

Simple requirements are littered with performance challenges. A basic profiling approach is to summarize each columns values and across pairs of columns for drill-down. For large linked datasets, profiling application deal with query optimization and answering, data cleansing, schema induction and also data mining.

3.2 *Semantic Profiling*

The Semantic Web is to automate the machine understand and process of content, metadata and the web documents to make their meanings explicit [17]. The principal goal in semantic web is giving content the ability to understand and present itself in the most useful forms matched to a customers need.

Semantic arts explain semantic profiling as a technique using ontologies and semantic-based tools in order to understand data and manipulated in an existing system. Semantic profiling describes the semantic meaning of data in a machine-readable format and observe how data sets are tagged and profiled [18]. Moreover, it enable the creation of relations with linking data sets.

4 Profiling Data Lake

The concept of Data Lake was created by James Dixon [4], he defined data lake as a method of storing data within a system or repository in its natural format. Similarly, that facilitates the collocation of data in various schemata and structural forms, usually object blobs or files. In fact, data lake is a centralized storage location that contains big data from a large number of sources in a raw and granular format.

Accordingly, to reduce initial integration costs for data lakes, current systems work to provide a more flexible way of data integration and analysis. However, a deep semantic integration is done only after the initial loading of the data into the data lake. Thereby, data can be restructured, transformed and aggregated as required.

4.1 *Metadata Extraction in Data Lakes*

Metadata is frequently pointed out as the key to describe and navigate through the massive content of a data lake, but details about the functional requirements and design of a metadata management component in data lakes are missing. Therefore, Data profiling solutions are responsible for structuring and managing data lake in order to unlock their full potential and offer valuable knowledge.

4.2 *Profiling Non-relational Data*

Generally, the most approaches published in data profiling focused to relational and structured data. However, big data need powerful algorithms and processes to analyze also non-relational and unstructured data. For this reason, profiling non-relational data is becoming a critical issue.

The rapid growth of data has put an emphasis on semi-structured data and unstructured data.

4.2.1 Semi-structured Data

Semi-structured data does not reside in a relational database, but that have some standards and organizational properties that make it easier to analyze. Semi-structured data such as XML and RDF does not constrain by a fixed schema, but many applications such as web services use XML and RDF to their transactions.

Profiling semi-structured data comprises a well-established set of basic operations. Advanced techniques include detecting seasonality, frequent patterns and identify which data are similar, which can be done through clustering or nearest-neighbor search.

4.2.2 Unstructured Data

Unstructured data does not have a pre-defined data model or schema, it can be a headache for many data scientists and IT personnel [24]. Current profiling approaches focus to extract essential metadata from unstructured data such as texts, social media, emails, call transcriptions and etc. Profiling unstructured data must processes all forms of files and document types, creating a searchable index and extracting key terms.

5 Data Profiling with Machine Learning

A good data profiling system can process a huge source of data and can uncover all sort of issues. Profiling is enabled by progress in powerful material resources, the huge availability of data, the faster internet and especially smart algorithms to process, understand and employ the data. These algorithms stem from the field of artificial intelligence, including machine learning and data mining.

In order to understand machine learning approaches to profiling, we need to distinguish data and behavior. In fact, in the context of user profiling, we talk more specifically about user data and user behavior. User data are any set of information pieces that can be attributed to the same entity, eg account details in social media. User behavior is defined similarly, but now the pieces of information are actions, eg click behavior on a web page. In 2010, De Vries calls these pieces informational shibboleths [3]. Generally, user profiling evolved with the data mining and machine learning approach. Its root can be found in the knowledge data discovery model and many of the steps in KDD (Knowledge discovery in databases) model resemble to the steps involved in the user profiling process.

Through data profiling and machine learning, you determine whether the dataset is complete and accurate enough to solve a practical problem. In addition, the automated way version of data profiling need using machine learning. Automated procedures for profiling include analyzing behavioral patterns, clustering data and analytics reporting in real time.

6 Use Cases

Data profiling can be implemented where data quality is important, it can use in various use cases. In fact, applying the data profiling process to data warehousing, business intelligence or linked data can help to identify potential issues and corrections that need to process. Additionally, data profiling is crucial in data conversion or data migration, it can help to identify data quality issues that may get lost in translation or adaptations.

In this section, we will provide an overview of popular use cases for data profiling and the exact moment when we need to apply it.

6.1 Data Exploration

Before any formal analysis of the data, the analyst must know the cases number in the data set, the variables found, the number of missing observations and the general assumptions that the data suggests. To answer all these questions, a first data set exploration allowing to analysts to familiarize with the data they will handle. In fact, Database exploration helps to identify important database properties, whether it is data of interest or data quality problems [6].

6.2 Data Cleaning

The general process of extracting metadata begins with analysis, followed by cleansing. Data cleaning is the process of detecting and correcting corrupt or inaccurate records from a data sets and identify incomplete and then modifying the dirty or coarse data. To ensure data quality, we need to repeat some of the profiling and cleansing processes to be sure all errant data is caught and fixed.

6.3 *Data Integration*

Data profiling used to help understand and prepare data for integration. In general, the present invention provides for systems and methods that aid in efficiently and effectively gathering and storing information about data being profiled for potential integration opportunities and for actually integrating the data [7]. For example, two data sources may comprise similar data.

6.4 *Query Optimization*

Exact data profiling results are directly applicable to query optimization. Effectively, data profiling keeps the query text along with its results to let queries optimized. Profiling is an important part of performance and optimization queries to keep the database running smoothly (Query profiling). Data profiling should include executing the query to collect runtime measurements such as the amount of memory consumed, typical duration, locks required while the query executes, number of file handles used, etc.

6.5 *Reverse Engineering*

The activity which consists in studying an object to determine its internal functioning or the manufacturing method is reverse engineering. Data profiling helps to determine more information for effective questioning and interrogation. For example, profiling the victim could provide fruitful about the way a victim was done and extracting the leaves traces of the attacker [25].

7 Data Profiling Tools

Data profiling tools into big data sources help to create a deep insight into the data quality. Number of data-profiling tools exist in the web, with varying degrees of effectiveness, efficacy and powerful. One of them will satisfy certainly our requirements for database archiving projects.

So far, most of the profiling tools are part of a larger software suite and often embedded in data quality or data integration systems. In this section, we will focus specially on data profiling tools. We first give an overview tools created as part of a research project. Next, we give a brief overview of the vast set of business profiling tools.

7.1 Prototype Search

Achieve data quality starting generally with data profiling. Every research axis need data that is accurate, is reliable, is validated before using it as business intelligence or analytics. Many tools have emerged that support data scientists and IT professionals in this task.

Metanome is an open framework and an extensible profiling platform that incorporates many state-of-the-art profiling algorithms and the datasets as external resources. Moreover, researchers and IT professionals can add their own algorithms. Current version of Metanome project support many algorithms such as DUCC [8], SPIDER [9], BINDER [10], etc.

ProLOD++ is a profiling framework for exploring specially LOD and RDF (Resource Description Framework) datasets [11]. Open RDF datasets contain a lack of meta-information necessary to understand.

Data profiling useful in many fields to analyze data and upgrade their quality value. Many research projects are existing for profiling these fields. For example, Bellman [12] project focus on data quality browser. Potters Wheel [13] for detecting data types and syntactic patterns. Data Auditor [14] and RuleMiner [15] are two research prototype systems that perform rule discovery to some degree. In addition, MADLib toolkit [16] for scalable in-database analytics, and moreover.

7.2 Commercial Tools

Generally, commercial data quality or data cleansing tools often support a set of profiling tasks. In business intelligence or analytics, most ETL (Extract Transform Load) tools have some profiling capabilities. Many commercial tools such as IBM, SAP, Attacama, or Informatica focus to solving some profiling tasks.

8 Data Profiling Challenges

Data profiling is often difficult to apply due to the sheer big volume of data. In fact, A bad profiling system might have years to return acceptable results. Big Data is reported to raise many challenges when it comes to data profiling. Such as, profiling could be useless in assessing usefulness of data when not understand. In addition, profiling tools and techniques challenge big data in terms of computational complexity and memory requirements. Also, we need to re-think how data profiling needs to be carried out in the context of Big Data and how data profiling system can support different data source types.

Big data profiling topics could classify as shown in the following chart maps (Fig. 3).

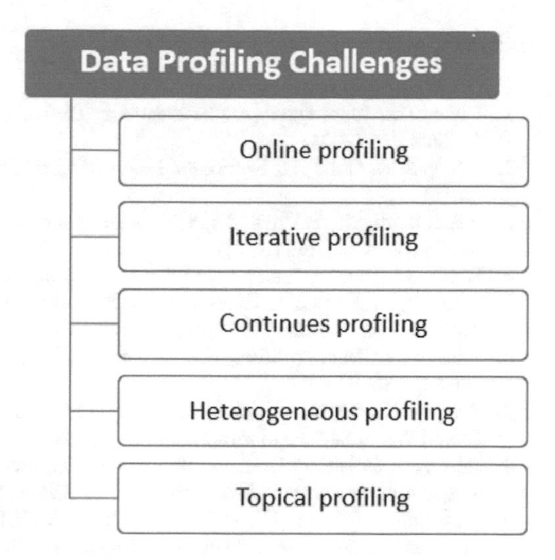

Fig. 3 Big research challenges for Data profiling

Data source grow every instance, Big Data potential focus to analyze data and display results in real time. For this reason, online profiling is triggered. It is necessary to incrementally update the outputs without processing the complete dataset again. In addition, the Big Data velocity characteristic is growing in real time, data profiling should apply upon changing data with periodic timespans and also performe upon data while it is being created or updated.

Profiling for big data would be to determine the common properties or level of heterogeneity of different datasets. In fact, heterogeneous profiling focus mainly to find inconsistent formatting between data, and structural heterogeneity, which is about unmatched schemata. Finally, most of social media datasets are unknown. Data profiling should recognizing the domain of this unknown data and research upon how topical profiling could be implemented efficiently.

9 Conclusion and Future Work

Data profiling has been of major importance since the growing importance of data. In this chapter, we provided an overview into the state-of-the-art in data profiling, we define the set of activities and processes to determine metadata from different database types, we detail data profiling use cases and we describe most know research prototype and commercial tools of data profiling.

Profiling will continue to be an important data management problem in research and practice. To facilitate interpretation of metadata, further work needs to focus on profiling many types of data, supporting new data management architectures, and interpreting and visualizing data profiling results in real time.

References

1. Olsen, J.E.: Data Quality: The Accuracy Dimension. Morgan Kaufmann Publishers. ISBN 1558608915 (2003)
2. Abedjan, Z., Golab, L., Naumann, F.: Profiling relational data: a survey. VLDB J. **24**, 557–581 (2015)
3. Hildebrandt, M., de Vries, K.: Privacy, Due Process and the Computational Turn, 43 (58 / 271). Routledge, New York (2013)
4. Dixon, J.: Pentaho, Hadoop, and Data Lakes. James Dixon's Blog (2010)
5. Abedjan, Z., Naumann, F.: Advancing the Discovery of Unique Column Combinations. Universittsverlag Potsdam (2011). ISBN 978-3-86956-148-6
6. Johnson, T.: Data Profiling, Encyclopedia of Database Systems, pp. 604–608. Springer, Heidelberg (2009)
7. Suereth, R., Ennis, W., Clavens, G.: Systems and methods of profiling data for integration, United Parcel Service of America Inc., US7912867B2, US12/036,611 (2008)
8. Heise, A., Quiané-Ruiz, J.A., Abedjan, Z., Jentzsch, A., Naumann, F.: Scalable discovery of unique column combinations. Proc. VLDB Endow. **7**(4) (2013)
9. Bauckmann, J., Leser, U., Naumann, F., Tietz, V.: Efficiently detecting inclusion dependencies. In: International Conference on Data Engineering (ICDE 2007), Istanbul, Turkey (poster paper, to appear)
10. Papenbrock, Thorsten., Kruse, Sebastian., Quian-Ruiz, Jorge-Arnulfo, Naumann, Felix: Divide and conquer-based inclusion dependency discovery. Proc. VLDB Endow. **8**(7), 774–785 (2015)
11. Abedjan, Z., Grütze, T., Jentzsch, A., Naumann, F.: Profiling and mining RDF data with ProLOD++. In: Proceedings of the International Conference on Data Engineering (ICDE) (2014)
12. Dasu, T., Johnson, T., Muthukrishnan, S., Shkapenyuk, V.: Mining database structure; or, how to build a data quality browser. In: Proceedings of the International Conference on Management of Data (SIGMOD), pp. 240–251 (2002)
13. Raman, V., Hellerstein, J.M.: Potters wheel: an interactive data cleaning system. In: Proceedings of the International Conference on Very Large Databases (VLDB), Rome, Italy, pp. 381–390 (2001)
14. Golab, L., Karloff, H., Korn, F., Srivastava, D.: Data auditor: exploring data quality and semantics using pattern tableaux. Proc. VLDB Endow. **3**(12), 16410–1644 (2010)
15. Chu, X., Ilyas, I., Papotti, P., Ye, Y.: RuleMiner: data quality rules discovery. In: Proceedings of the International Conference on Data Engineering (ICDE), pp. 1222–1225 (2014)
16. Hellerstein, J.M., Christopher, R., Schoppmann, F., Wang, D.Z., Fratkin, E., Gorajek, A., Ng, K.S., Welton, C., Feng, X., Li, K., Kumar, A.: The MADlib analytics library or MAD skills, the SQL. Proc. VLDB Endow. **5**(12), 1700–1711 (2012)
17. Mohamed, F.S., Bellahsene, B.E.Z., Todorov, K.: Towards semantic dataset profiling. In: (2014)
18. Shoaib, M., Basharat, A.: Ontology based knowledge representation and semantic profiling in personalized semantic social networking framework. In: 2010 3rd International Conference on Computer Science and Information Technology. IEEE (2010)
19. Gangadharan, S.P.: Digital inclusion and data profiling. First Monday **17**(5) (2012). https://doi.org/10.5210/fm.v17i5.3821
20. Bauckmann, J., Leser, U., Naumann, F., Tietz, V.: Efficiently detecting inclusion dependencies. In: Proceedings of the International Conference on Data Engineering (ICDE), Istanbul, Turkey, pp. 1448–1450 (2007)
21. Papenbrock, T., Ehrlich, J., Marten, J., Neubert, T., Rudolph, J.-P., Schönberg, M., Zwiener, J., Naumann, F.: Functional dependency discovery: an experimental evaluation of seven algorithms. Proc. VLDB Endow. **8**(10), 1082–1093 (2015)
22. Heise, A., Quian-Ruiz, J.A., Abedjan, Z., Jentzsch, A., Naumann, F.: Scalable discovery of unique column combinations. Proc. VLDB Endow. **7**, 301–312 (2013)
23. Bohm, C., Naumann, F., Abedjan, Z., Grutze, D.F.T., Hefenbrock, D., Pohl, M., Sonnabend, D.: Profiling linked open data with ProLOD. In: IEEE 26th International Conference on Data Engineering Workshops (ICDEW) (2010)

24. Buneman, P., Davidson, S., Fernandez, M., Suciu, D.: Adding structure to unstructured data. In: International Conference on Database Theory ICDT 1997: Database Theory ICDT 1997, pp. 336-350 (2005)
25. Bruinsma, G., Weisburd, D. (eds.) Encyclopedia of Criminology and Criminal Justice. Springer, New York (2014)

Generalization of the Fuzzy Conformable Differentiability with Application to Fuzzy Fractional Differential Equations

Atimad Harir, Said Melliani, and L. Saadia Chadli

Abstract In the present paper, we study generalized conformable differentiability concepts for fuzzy valued functions. Existence of the solutions of fuzzy fractional differential equations involving generalized conformable differentiability is studied. Also, some concrete applications to ordinary fuzzy fractional differential equations with fuzzy initial values.

Keywords Fuzzy fractional differential equation · Conformable fractional derivative · Fuzzy number · Fuzzy conformable fractional derivative

1 Introduction

Roshdi [13] discussed the properties of conformable fractional derivative, This definition seems to be a natural extension of the usual derivative, and it satisfies some properties. Definition coincides with the known fractional derivatives on polynomials (up to a constant multiple). In [8] discussed the properties of conformable fractional derivative fuzzy valued mappings by means of the concept of H-conformable differentiability due to Puri and Ralescu [15]. Seikkala [17] defined the fuzzy derivative which is a general-wxization of the Hukuhara derivative in [15] and the fuzzy integral which is the same as that proposed by Dubois and Prade [5, 6].

In this paper, by the use of the concept of H-conformable differentiability, we study the problem of fuzzy fractional differential equations for the fuzzy valued mappings of a real variable whose values are normal, convex, uppersemicontinuous, and compactly supported fuzzy sets R. Several important results are obtained by applying the embedding theorem in [12] which is a generalization of the classical Radstrom embedding results [16]. As preliminaries we recall some basic results on fuzzy number and differentiability properties for the fuzzy set-valued mappings in [11]. We list several definitions and theorems for conformable fractional derivative.

A. Harir (✉) · S. Melliani · L. S. Chadli
Laboratory of Applied Mathematics and Scientific Computing, Sultan Moulay Slimane University, Beni-Mellal, Morocco
e-mail: s.melliani@usms.ma

N. Gherabi and J. Kacprzyk (eds.), *Intelligent Systems in Big Data, Semantic Web and Machine Learning*, Advances in Intelligent Systems and Computing 1344,
https://doi.org/10.1007/978-3-030-72588-4_9

We show the solution to the conformable Cauchy problem of the fuzzy fractional differential equation, and furthermore, in Sect. 4, we prove the existence theorem for a solution to the fuzzy fractional differential equation.

2 Preliminnaries

Let us denote by $R_{\mathcal{F}}$ the class of fuzzy subsets of the real axis $y : R \to [0, 1]$ satisfying the following properties:

(i) y is normal, i.e. $\exists\, x_0 \in R$ with $y(x_0) = 1$,
(ii) y is convex fuzzy set i.e.

$$y(\lambda x + (1-\lambda)t) \geq \min\{y(x), y(t)\}, \ \forall x, t \in R, \ 0 < \lambda \leq 1$$

(iii) y is upper semicontinuous on R
(iv) $[y]^0 = cl\{x \in R \mid y(x) > 0\}$ is compact.

Then $R_{\mathcal{F}}$ is called the space of fuzzy numbers see [7]. Obviously, $R \subset R_{\mathcal{F}}$. If y is a fuzzy set, we define $[y]^\alpha = \{x \in R \mid y(x) \geq \alpha\}$ the α-level (cut) sets of y, with $0 < \alpha \leq 1$. Also if $y \in R_{\mathcal{F}}$ then α-cut of u denoted by $[y]^\alpha = \left[y_1^\alpha, y_2^\alpha\right]$

Definition 1 [7, 14]**.** We represent an arbitrary fuzzy number by an ordered pair of functions $[y]^\alpha = \left[y_1^\alpha, y_2^\alpha\right], \alpha \in [0, 1]$, which satisfy the following requirements :

1. y_1^α is an increasing function over [0,1],
2. y_2^α is a decreasing function on [0,1]
3. y_1^α and y_2^α are bounded left continuous on $(0, 1]$, and right continuous at $\alpha = 0$,
4. $y_1^\alpha \leq y_2^\alpha$, for $0 \leq \alpha \leq 1$

Lemma 1. *see [10] Let $y, z : R_{\mathcal{F}} \to [0, 1]$ be the fuzzy sets. Then $y = z$ if and only if $[y]^\alpha = [z]^\alpha$ for all $\alpha \in [0, 1]$.*

For $y, z \in R_{\mathcal{F}}$ and $\lambda \in R$ the sum $y + z$ and the product λy are defined by

$$[y+z]^\alpha = \left[y_1^\alpha + z_1^\alpha, y_2^\alpha + z_2^\alpha\right]$$

$$[\lambda y]^\alpha = \lambda[y]^\alpha = \begin{cases} \left[\lambda y_1^\alpha, \lambda y_2^\alpha\right] & if\, \lambda \geq 0 \\ \left[\lambda y_2^\alpha, \lambda y_1^\alpha\right] & if\, \lambda < 0 \end{cases}$$

$\forall \alpha \in [0, 1]$. Additionally if we denote $\hat{0} = \chi_{\{0\}}$ then $\widehat{0} \in R_{\mathcal{F}}$ is neutral element with respert to $+$.
Let $d : R_{\mathcal{F}} \times R_{\mathcal{F}} \to R_+ \cup \{0\}$ by the following equation:

$$d(y, z) = \sup_{\alpha \in [0,1]} d_H\left([y]^\alpha, [z]^\alpha\right), \quad for\ all\ y, z \in R_{\mathcal{F}}$$

where d_H is the Hausdorff metric defined as.

$$d_H\left([y]^\alpha, [z]^\alpha\right) = \max\left\{\left|y_1^\alpha - z_1^\alpha\right|, \left|y_2^\alpha - z_2^\alpha\right|\right\}$$

The following properties are well-known see [18].

$$\begin{aligned} d(y+w, z+w) &= d(y,z) \text{ and } d(y,z) = d(z,y),\ y, z, w \in \mathrm{R}_{\mathcal{F}} \\ d(ky, kz) &= |k| d(y,z),\ k \in \mathrm{R} \text{ and } y, z \in \mathrm{R}_{\mathcal{F}} \\ d(y+z, w+e) &\leq d(y,w) + d(z,e),\ \ y, z, w, e \in \mathrm{R}_{\mathcal{F}} \end{aligned} \tag{1}$$

Definition 2. Let $u, v \in \mathrm{R}_{\mathcal{F}}$. If there exists $w \in \mathrm{R}_{\mathcal{F}}$ such as $u = v + w$ then w is called the H-difference of u, v and it is denoted $u \ominus v$

Theorem 1 *[1].*

(i) Let we denote

$$\overline{0} = \begin{cases} 1 & t = 0, \\ 0 & t \neq 0. \end{cases}$$

then $\overline{0} \in \mathrm{R}_{\mathcal{F}}$ be a neutral element with respect to $+$ i.e. $y + \overline{0} = \overline{0} + y \quad y \in \mathrm{R}_{\mathcal{F}}$

(ii) With respect to $\overline{0}$, none of $y \in \mathrm{R}_{\mathcal{F}} \backslash \mathrm{R}$, has opposite in $\mathrm{R}_{\mathcal{F}}$.

(iii) For any $a, b \in \mathrm{R}$ with $a, b \geq 0$ or $a, b \leq 0$ and any $y \in \mathrm{R}_{\mathcal{F}}$, we have $(a + b) \cdot y = a \cdot y + b \cdot y$, for general $a, b \in \mathrm{R}$ the alpove property does not hold.

(iv) For any $\lambda \in \mathrm{R}$ and any $y, z \in \mathrm{R}_{\mathcal{F}}$, we have $\lambda.(y + z) = \lambda \cdot y + \lambda \cdot z$.

(v) For any $\lambda, \nu \in \mathrm{R}$ and any $y \in \mathrm{R}_{\mathcal{F}}$, we have $\lambda.(\nu.y) = (\lambda.\nu).y$

Let $\mathcal{C}[0, 1] = \{F : [0, 1] \to \mathrm{R}$, F bounded on $[0, 1]$, left continuous for any $t \in (0, 1]$, right continuous on 0 and F has right limit for any $t \in (0, 1]\}$. Endowed with the norm $\|F\| = \sup\{|F(t)|, t \in [0, 1]\}$, $\mathcal{C}[0, 1]$ is a banach space.

It is known the following result which embeds $\mathrm{R}_{\mathcal{F}}$ into $\mathcal{C}[0, 1] \times \mathcal{C}[0, 1]$ isometrically and isomorphically.

Theorem 2. *If we define $j : \mathrm{R}_{\mathcal{F}} \to \mathcal{C}[0, 1] \times \mathcal{C}[0, 1]$ by $j(y) = (y_1, y_2)$, where $y_1, y_2 : [0, 1] \to \mathrm{R}$, y_1^α, y_2^α, then $j(\mathrm{R}_F)$ is a closed convex cone with vertex 0 in $\mathcal{C}[0, 1] \times \mathcal{C}[0, 1]$ (here $\mathcal{C}[0, 1] \times \mathcal{C}[0, 1]$ is a banach space with the norm $\|(f, g)\| = \max\{\|F\|, \|G\|\}$) and j satisfies:*

(i) $j(s.y + t.z) = sj(y) + tj(z), \forall y, z \in \mathrm{R}_{\mathcal{F}}, s, t \geq 0$;

(ii) $d(y, z) = \|j(y) + j(z)\|$.

3 Fuzzy Conformable Fractional Calculus

Definition 3. Let $\gamma \in (0, 1]$, a function $F : [0, a] \to R_{\mathcal{F}}$ is called H-differentiable on $t_0 \in [a, b]$ if for $\varepsilon > 0$ sufficiently small there exist the H-differences $F\left(t_0 + \varepsilon t_0^{1-\gamma}\right) \ominus F(t_0)$, $F(t_0) \ominus F\left(t_0 - \varepsilon t_0^{1-\gamma}\right)$ and an element $T_\gamma(F)(t_0) \in R_{\mathcal{F}}$ such that

$$\lim_{\varepsilon \to 0^+} d\left(\frac{F\left(t_0 + \varepsilon t_0^{1-\gamma}\right) \ominus F({}_0t)}{\varepsilon}, T_\gamma(F)(t_0)\right) =$$

$$\lim_{\varepsilon \to 0^+} d\left(\frac{F(t_0) \ominus F\left(t_0 - \varepsilon t_0^{1-\gamma}\right)}{\varepsilon}, T_\gamma(F)(t_0)\right) = 0 \tag{2}$$

Definition 4. A fuzzy function $F : [0, a] \to R_{\mathcal{F}}$ with $a > 0$. γ^{th} order "fuzzy conformable fractional derivative" of F is defined by

$$T_\gamma(F)(t) = \lim_{\varepsilon \to 0^+} \frac{F\left(t + \varepsilon t^{1-\gamma}\right) \ominus F(t)}{\varepsilon}$$

for all $t > 0$, $\gamma \in (0, 1)$.
If F is γ-differentiable in some $[0, a]$, and $\lim_{t \to 0^+} T_\gamma(F)(t)$ exists, then

$$T_\gamma(F)(0) = \lim_{t \to 0^+} T_\gamma(F)(t)$$

and the limits (in the metric d)

Definition 5. Let $F : (0, a) \to R_{\mathcal{F}}$ with $a > 0$ and $\gamma \in (0, 1]$. We say that F is strongly generalized γ-differentiable at t_0, if there exists an element $T_\gamma(F)(t_0) \in R_{\mathcal{F}}$ such that either:

(*i*) For all $\varepsilon > 0$ sufficiently close to 0, the $F\left(t_0 + \varepsilon t_0^{1-\gamma}\right) \ominus F(t_0)$, $F(t_0) \ominus F\left(t_0 - \varepsilon t_0^{1-\gamma}\right)$ exist and

$$T_\gamma(F)(t_0) = \lim_{\varepsilon \to 0^+} \frac{F\left(t_0 + \varepsilon t_0^{1-\gamma}\right) \ominus F(t_0)}{\varepsilon} = \lim_{\varepsilon \to 0^+} \frac{F(t_0) \ominus F\left(t_0 - \varepsilon t_0^{1-\gamma}\right)}{\varepsilon}$$

for all $t_0 > 0$, $\gamma \in (0, 1)$.

(*ii*) For all $\varepsilon > 0$ sufficiently close to 0, the $F(t_0) \ominus F\left(t_0 + \varepsilon t_0^{1-\gamma}\right)$, $F\left(t_0 - \varepsilon t_0^{1-\gamma}\right) \ominus F(t_0)$ exist and

$$T_\gamma(F)(t_0) = \lim_{\varepsilon \to 0^+} \frac{F(t_0) \ominus F\left(t_0 + \varepsilon t_0^{1-\gamma}\right)}{-\varepsilon} = \lim_{\varepsilon \to 0^+} \frac{F\left(t_0 - \varepsilon t_0^{1-\gamma}\right) \ominus F(t_0)}{-\varepsilon}$$

for all $t_0 > 0, \gamma \in (0, 1)$.

(iii) For all $\varepsilon > 0$ sufficiently close to 0, the $F\left(t_0 + \varepsilon t_0^{1-\gamma}\right) \ominus F(t_0), F\left(t_0 - \varepsilon t_0^{1-\gamma}\right) \ominus F(t_0)$ exist and

$$T_\gamma(F)(t_0) = \lim_{\varepsilon \to 0^+} \frac{F\left(t_0 + \varepsilon t_0^{1-\gamma}\right) \ominus F(t_0)}{\varepsilon} = \lim_{\varepsilon \to 0^+} \frac{F\left(t_0 - \varepsilon t_0^{1-\gamma}\right) \ominus F(t_0)}{-\varepsilon}$$

for all $t_0 > 0, \gamma \in (0, 1)$.

(iv) For all $\varepsilon > 0$ sufficiently close to 0, the $F(t_0) \ominus F\left(t_0 + \varepsilon t_0^{1-\gamma}\right), F(t_0) \ominus F\left(t_0 - \varepsilon t_0^{1-\gamma}\right)$ exist and

$$T_\gamma(F)(t_0) = \lim_{\varepsilon \to 0^+} \frac{F(t_0) \ominus F\left(t_0 + \varepsilon t_0^{1-\gamma}\right)}{-\varepsilon} = \lim_{\varepsilon \to 0^+} \frac{F(t_0) \ominus F\left(t_0 - \varepsilon t_0^{1-\gamma}\right)}{\varepsilon}$$

for all $t_0 > 0, \gamma \in (0, 1)$, where ε and $-\varepsilon$ at the denominators mean $\frac{1}{\varepsilon}$ and $-\frac{1}{\varepsilon}$, respectively. The limit is taken in the metric space $(R_{\mathcal{F}}, d)$. Moreover, let F be differentiable at each point $t \in (a, b)$, then F is said to be γ-differentiable on (a, b)

Theorem 3 [8]. *Let $\gamma \in (0, 1]$*

(i) *If F is (1)-differentiable and F is $\gamma_{(1)}$-differentiable then*

$$T_{\gamma_{(1)}} F(t) = t^{1-\gamma} D_1^1 F(t)$$

(ii) *If F is (2)-differentiable and F is $\gamma_{(2)}$-differentiable then*

$$T_{\gamma_{(2)}} F(t) = t^{1-\gamma} D_2^1 F(t)$$

Note that the definition of (n)-differentiable or $\left(D_n^1\right)$ for $n \in 1, 2$ see [2–4].

Theorem 4. *Let $\gamma \in (0, 1]$ and $F : (0, a) \to R_{\mathcal{F}}$ be such that $[F(t)]^\alpha = \left[f_1^\alpha, f_2^\alpha\right]$. Suppose that the functions f_1^α and f_2^α are real-valued functions, differentiable w.r.t. t uniformly w.r.t. $\alpha \in [0, 1]$. Then the function $F(t)$ is generalized conformable differentiable at a fixed $t \in (a, b)$ if and only if one of the following two cases holds:*

1. $\left(f_1^\alpha\right)^{(\gamma)}(t)$ is increasing, $\left(f_2^\alpha\right)^{(\gamma)}(t)$ is decreasing as functions of α and

$$\left(f_1^1\right)^{(\gamma)}(t) \le \left(f_2^1\right)^{(\gamma)}(t)$$

or

2. $\left(f_1^\alpha\right)^{(\gamma)}(t)$ is decreasing, $\left(f_2^\alpha\right)^{(\gamma)}(t)$ is increasing as functions of α and

$$\left(f_2^1\right)^{(\gamma)}(t) \le \left(f_1^1\right)^{(\gamma)}(t)$$

Also, $\forall\alpha \in [0, 1]$ *we have*

$$\left[F^{(\gamma)}(t)\right]^{\alpha} = \left[\min\left\{\left(f_1^{\alpha}\right)^{(\gamma)}(t), \left(f_2^{\alpha}\right)^{(\gamma)}(t)\right\}, \max\left\{\left(f_1^{\alpha}\right)^{(\gamma)}(t), \left(f_2^{\alpha}\right)^{(\gamma)}(t)\right\}\right] \tag{3}$$

Proof. Let $\gamma \in (0, 1]$ and F be H-differentiable and assume that f_1^{α} and f_2^{α} are differentiable. So we have

$$\left[F^{(\gamma)}(t)\right]^{\alpha} = \left[\min\left\{\left(f_1^{\alpha}\right)^{(\gamma)}(t), \left(f_2^{\alpha}\right)^{(\gamma)}(t)\right\}, \max\left\{\left(f_1^{\alpha}\right)^{(\gamma)}(t), \left(f_2^{\alpha}\right)^{(\gamma)}(t)\right\}\right]$$

Now suppose that for fixed $t \in (a, b)$, the differences $\left(f_2^{\alpha}\right)^{(\gamma)}(t) - \left(f_1^{\alpha}\right)^{(\gamma)}(t)$ change sign at a fixed $\beta \in (0, 1)$. Then $\left[F^{(\gamma)}(t)\right]^{\beta}$ is a singleton and, for all α such that $\beta \leq \alpha \leq 1$, also $\left[F^{(\gamma)}(t)\right]^{\alpha}$ is a singleton because $\left[F^{(\gamma)}(t)\right]^{\alpha} \subseteq \left[F^{(\gamma)}(t)\right]^{\beta}$; it follows that, for the same values of α, $\left(f_2^{\alpha}\right)^{(\gamma)}(t) - \left(f_1^{\alpha}\right)^{(\gamma)}(t) = 0$, which is a contradiction with the fact that $\left(f_2^{\alpha}\right)^{(\gamma)}(t) - \left(f_1^{\alpha}\right)^{(\gamma)}(t)$ changes sign. We then conclude that $\left(f_2^{\alpha}\right)^{(\gamma)}(t) - \left(f_1^{\alpha}\right)^{(\gamma)}(t)$ cannot change sign with respect to $\alpha \in [0, 1]$.

To prove our conclusion, we distinguish three cases according to the sign of $\left(f_2^{\alpha}\right)^{(\gamma)}(t) - \left(f_1^{\alpha}\right)^{(\gamma)}(t)$.

- If $\left(f_1^{1}\right)^{(\gamma)}(t) < \left(f_2^{1}\right)^{(\gamma)}(t)$, then $\left(f_2^{\alpha}\right)^{(\gamma)}(t) - \left(f_1^{\alpha}\right)^{(\gamma)}(t) \geq 0$ for every $\gamma \in (0, 1]$ and $\alpha \in [0, 1]$ and

$$\left[F^{(\gamma)}(t)\right]^{\alpha} = \left[\left(f_1^{\alpha}\right)^{(\gamma)}(t), \left(f_2^{\alpha}\right)^{(\gamma)}(t)\right].$$

  Since F is generalized conformable differentiable, the intervals $\left[\left(f_1^{\alpha}\right)^{(\gamma)}(t), \left(f_2^{\alpha}\right)^{(\gamma)}(t)\right]$ should from a fuzzy number , i.e., for any $\alpha > \beta$, $\left[\left(f_1^{\alpha}\right)^{(\gamma)}(t), \left(f_2^{\alpha}\right)^{(\gamma)}(t)\right] \subseteq \left[\left(f_1^{\beta}\right)^{(\gamma)}(t), \left(f_2^{\beta}\right)^{(\gamma)}(t)\right]$ which shows that $\left(f_1^{\alpha}\right)^{(\gamma)}(t)$ is increasing and $\left(f_2^{\alpha}\right)^{(\gamma)}(t)$ is decreasing as a function of $\alpha \in [0, 1]$.
- If $\left(f_1^{1}\right)^{(\gamma)}(t) > \left(f_2^{1}\right)^{(\gamma)}(t)$, then $\left(f_2^{\alpha}\right)^{(\gamma)}(t) - \left(f_1^{\alpha}\right)^{(\gamma)}(t) \leq 0$ for every $\gamma \in (0, 1]$ and $\alpha \in [0, 1]$ and

$$\left[F^{(\gamma)}(t)\right]^{\alpha} = \left[\left(f_2^{\alpha}\right)^{(\gamma)}(t), \left(f_1^{\alpha}\right)^{(\gamma)}(t)\right].$$

  So, for any $\alpha > \beta$, $\left[\left(f_2^{\alpha}\right)^{(\gamma)}(t), \left(f_1^{\alpha}\right)^{(\gamma)}(t)\right] \subseteq \left[\left(f_2^{\beta}\right)^{(\gamma)}(t), \left(f_1^{\beta}\right)^{(\gamma)}(t)\right]$ which shows that $\left(f_1^{\alpha}\right)^{(\gamma)}(t)$ is decreasing and $\left(f_2^{\alpha}\right)^{(\gamma)}(t)$ is increasing as a function of $\alpha \in [0, 1]$
- If $F^{(\gamma)}(t) \in \mathrm{R}$ is a crisp number, we have $\left(f_1^{1}\right)^{(\gamma)}(t) = \left(f_2^{1}\right)^{(\gamma)}(t)$, the conclusion is obvious; if this not case, then we may have either $\left(f_1^{0}\right)^{(\gamma)}(t) < \left(f_2^{0}\right)^{(\gamma)}(t)$ or

$\left(f_1^0\right)^{(\gamma)}(t) > \left(f_2^0\right)^{(\gamma)}(t)$, taking

$$\beta = \inf\left\{\alpha \mid \left(f_1^\alpha\right)^{(\gamma)}(t) = \left(f_2^\alpha\right)^{(\gamma)}(t)\right\},$$

we have corresponding that $\left(f_1^\alpha\right)^{(\gamma)}(t) \geq \left(f_2^\alpha\right)^{(\gamma)}(t)$ or $\left(f_1^\alpha\right)^{(\gamma)}(t) \leq \left(f_2^\alpha\right)^{(\gamma)}(t)$ for all $\gamma \in (0, 1]$ and $\alpha \in (0, 1)$ because the differences cannot change sign w.t.r α. We conclude that $\left(f_1^\alpha\right)^{(\gamma)}(t)$ and $\left(f_2^\alpha\right)^{(\gamma)}(t)$ are monotonic w.t.r α. Reciprocally, by Theorem 2 we have

$$j(F(t)) = (f_1(t), f_2(t)) = \left\{\left(f_1^\alpha(t), f_2^\alpha(t)\right) \mid \alpha \in [0, 1]\right\},$$

is an isometric embedding. Assuming that, for all α, the two functions $f_1^\alpha(t)$ and $f_2^\alpha(t)$ are differentiable with respect to t, the limits

$$\left(f_1^\alpha\right)^{(\gamma)}(t) = \lim_{\varepsilon\to 0^+} \frac{f_1^\alpha\left(t + \varepsilon t^{1-\gamma}\right) - f_1^\alpha(t)}{\varepsilon}$$
$$\left(f_2^\alpha\right)^{(\gamma)}(t) = \lim_{\varepsilon\to 0^+} \frac{f_2^\alpha\left(t + \varepsilon t^{1-\gamma}\right) - f_2^\alpha(t)}{\varepsilon}$$

let $h = \varepsilon t^{1-q}$ and using Theorem so (3), we have

$$\left(f_1^\alpha\right)^{(\gamma)}(t) = t^{1-\gamma} \lim_{h\to 0^+} \frac{f_1^\alpha(t+h) - f_1^\alpha(t)}{h}$$
$$\left(f_2^\alpha\right)^{(\gamma)}(t) = t^{1-\gamma} \lim_{h\to 0^+} \frac{f_2^\alpha(t+h) - f_2^\alpha(t)}{h}$$

exist uniformly for all $\alpha \in [0, 1]$. Taking a sequence $h_n \to 0$, we will have

$$\left(f_1^\alpha\right)^{(\gamma)}(t) = t^{1-\gamma} \lim_{n\to\infty} \frac{f_1^\alpha(t+h_n) - f_1^\alpha(t)}{h_n}$$
$$\left(f_2^\alpha\right)^{(\gamma)}(t) = t^{1-\gamma} \lim_{n\to\infty} \frac{f_2^\alpha(t+h_n) - f_2^\alpha(t)}{h_n}$$

i.e., $\left(f_1^\alpha\right)^{(\gamma)}(t), \left(f_2^\alpha\right)^{(\gamma)}(t)$ are uniform limits of sequences of left continuous functions at $\alpha \in (0, 1]$, so they are themselves left continuous for $\alpha \in (0, 1]$. Similarly the right continuity at 0 can be deduced. Assuming that, for a fixed $t \in (a, b)$, the function $\left(f_1^\alpha\right)^{(\gamma)}(t)$ is increasing and the function $\left(f_2^\alpha\right)^{(\gamma)}(t)$ is decreasing as functions of α, and that $\left(f_1^1\right)^{(\gamma)}(t) \leq \left(f_2^1\right)^{(\gamma)}(t)$, then also $\left(f_1^\alpha\right)^{(\gamma)}(t) \leq \left(f_2^\alpha\right)^{(\gamma)}(t)$ for all $\alpha \in [0, 1]$ and it is easy to see that the pair of functions $\left(f_1^\alpha\right)^{(\gamma)}(t), \left(f_2^\alpha\right)^{(\gamma)}(t)$ fulfill the conditions in Definition 1 and the intervals $\left[\left(f_1^\alpha\right)^{(\gamma)}(t), \left(f_2^\alpha\right)^{(\gamma)}(t)\right]$, $\alpha \in [0, 1]$ determine a fuzzy number.

Now we observe that the following limit uniformly exists

$$\lim_{\varepsilon\to 0+}\frac{\left[F\left(t+\varepsilon t^{1-\gamma}\right)\ominus F(t)\right]^{\alpha}}{\varepsilon}=\lim_{\varepsilon\to 0+}\left[\frac{f_1^{\alpha}\left(t+\varepsilon t^{1-\gamma}\right)-f_1^{\alpha}(t)}{\varepsilon},\frac{f_2^{\alpha}\left(t+\varepsilon t^{1-\gamma}\right)-f_2^{\alpha}(t)}{\varepsilon}\right]$$
$$=\left[\left(f_1^{\alpha}\right)^{(\gamma)}(t),\left(f_2^{\alpha}\right)^{(\gamma)}(t)\right],\alpha\in[0,1]$$

and it is a fuzzy number, then F is generalized conformable differentiable. If $\left(f_1^{\alpha}\right)^{(\gamma)}(t)$ is decreasing and the function $\left(f_2^{\alpha}\right)^{(\gamma)}(t)$ is increasing as functions of α and $\left(f_2^{1}\right)^{(\gamma)}(t)\leq\left(f_1^{1}\right)^{(\gamma)}(t)$, then also $\left(f_2^{\alpha}\right)^{(\gamma)}(t)\leq\left(f_1^{\alpha}\right)^{(\gamma)}(t)$, $\alpha\in[0,1]$, for all
$\gamma\in(0,1]$ and by Definition 1, the intervals $\left[\left(f_2^{\alpha}\right)^{(\gamma)}(t)\leq\left(f_1^{\alpha}\right)^{(\gamma)}(t)\right],\alpha\in[0,1]$ determine a fuzzy number. Observing that the following limit exists uniformly

$$\lim_{\varepsilon\to 0+}\frac{\left[F\left(t+\varepsilon t^{1-\gamma}\right)\ominus F(t)\right]^{\alpha}}{\varepsilon}=\lim_{\varepsilon\to 0+}\left[\frac{f_2^{\alpha}\left(t+\varepsilon t^{1-\gamma}\right)-f_2^{\alpha}(t)}{\varepsilon},\frac{f_1^{\alpha}\left(t+\varepsilon t^{1-\gamma}\right)-f_1^{\alpha}(t)}{\varepsilon}\right]$$
$$=\left[\left(f_2^{\alpha}\right)^{(\gamma)}(t),\left(f_1^{\alpha}\right)^{(\gamma)}(t)\right],\alpha\in[0,1]$$

and it is a fuzzy number, then F is generalized conformable differentiable.

Example 1. Consider $F:]0,1[\to \mathrm{R}_{\mathcal{F}}$ defined y the α-cuts, and $\gamma\in(0,1]$ (it is 0 symmetric)

1. If

$$[F(t)]^{\alpha}=\left[\frac{-1}{\left(1+\frac{t\gamma}{\gamma}\right)(1+\alpha)},\frac{1}{\left(1+\frac{t}{\gamma}\right)(1+\alpha)}\right]\tag{4}$$

i.e. $f_1^{\alpha}=\frac{-1}{\left(1+\frac{t\gamma}{\gamma}\right)(1+\alpha)}$ and $f_2^{\alpha}=\frac{1}{\left(1+\frac{t\gamma}{\gamma}\right)(1+\alpha)}$ the level sets.

$$\left(f_1^{\alpha}\right)^{(\gamma)}=\frac{1}{\left(1+\frac{t\gamma}{\gamma}\right)^2(1+\alpha)}\quad \textit{and}\quad \left(f_2^{\alpha}\right)^{(\gamma)}=\frac{-1}{\left(1+\frac{t\gamma}{\gamma}\right)^2(1+\alpha)}$$

Then for all $\alpha\in[0,1]$, both $\left(f_1^{\alpha}\right)^{(\gamma)}$ and $\left(f_2^{\alpha}\right)^{(\gamma)}$ are γ-differentiable and satisfy conditions of Theorem 4 and $\left[F^{(\gamma)}(t)\right]^{\alpha}=\left[\left(f_2^{\alpha}\right)^{(\gamma)},\left(f_1^{\alpha}\right)^{(\gamma)}\right]$.

2.

$$[F(t)]^{\alpha}=\left[\frac{-1}{\left(1-\frac{t\gamma}{\gamma}\right)(1+\alpha)},\frac{1}{\left(1-\frac{t\gamma}{\gamma}\right)(1+\alpha)}\right]\tag{5}$$

i.e. $f_1^{\alpha}=\frac{-1}{\left(1-\frac{t\gamma}{\gamma}\right)(1+\alpha)}$ and $f_2^{\alpha}=\frac{1}{\left(1-\frac{t\gamma}{\gamma}\right)(1+\alpha)}$ the level sets.

$$\left(f_1^{\alpha}\right)^{(\gamma)} = \frac{-1}{\left(1-\frac{t\gamma}{\gamma}\right)^2(1+\alpha)} \quad \textit{and} \quad \left(f_2^{\alpha}\right)^{(\gamma)} = \frac{1}{\left(1-\frac{t\gamma}{\gamma}\right)^2(1+\alpha)}$$

Then for all $\alpha \in [0,1]$, both $\left(f_1^{\alpha}\right)^{(\gamma)}$ and $\left(f_2^{\alpha}\right)^{(\gamma)}$ are γ -differentiable and satisfy conditions of Theorem 4 and $\left[F^{(\gamma)}(t)\right]^{\alpha} = \left[\left(f_1^{\alpha}\right)^{(\gamma)}, \left(f_2^{\alpha}\right)^{(\gamma)}\right]$.

Theorem 5. *Let $\gamma \in (0,1]$ and $F : (0,a) \to \mathrm{R}_{\mathcal{F}}$ be strongly generalized conformable derivative on each point $t \in (a,b)$ in sense of Definition 5 (iii) or (iv). Then $F^{(\gamma)}(t) \in \mathrm{R}$ for all $t \in (a,b)$.*

Proof. Suppose that F is γ-differentiable on t as Definition (5) (iii). Then the H−differences $F\left(t+\varepsilon t^{1-\gamma}\right) \ominus F(t)$, $F\left(t-\varepsilon t^{1-\gamma}\right) \ominus F(t)$ exist for $\varepsilon > 0$ sufficiently small. Then we have

$$F\left(t+\varepsilon t^{1-\gamma}\right) = F(t) + w\left(t, \varepsilon t^{1-\gamma}\right) \tag{6}$$

and

$$F\left(t-\varepsilon t^{1-\gamma}\right) = F(t) + v\left(t, \varepsilon t^{1-\gamma}\right) \tag{7}$$

for $\varepsilon > 0$ sufficiently small. If we take in (7) $t = t + \varepsilon t^{1-\gamma}$ we obtain:

$$F(t) = F\left(t+\varepsilon t^{1-\gamma}\right) + v\left(t+\varepsilon t^{1-\gamma}, \varepsilon t^{1-\gamma}\right) \tag{8}$$

for $\varepsilon > 0$ sufficiently small. By adding (6) and (8) it follows that $w\left(t, \varepsilon t^{1-\gamma}\right) + v\left(t+\varepsilon t^{1-\gamma}, \varepsilon t^{1-\gamma}\right) = \bar{0}$.
By Theorem 1 we obtain $w\left(t, \varepsilon t^{1-\gamma}\right), v\left(t+\varepsilon t^{1-\gamma}, \varepsilon t^{1-\gamma}\right) \in \mathrm{R}$ for $\varepsilon > 0$ sufficiently small. Then it is easy to see that $F^{(\gamma)}(t) = \lim_{\varepsilon \to 0^+} \frac{w\left(t, \varepsilon t^{1-\gamma}\right)}{\epsilon} \in \mathrm{R}$ If F is γ-differentiable on x as in Definition 5 (iv), the reasonings are similar.

Remark 1. (a) Let $F : (0,a) \to E$, where $(E, \|.\|)$ is real normed space, if there exists a linear continuous mapping, $F^{(\gamma)}(t_0) : (0,a) \to E$ such that

$$\lim_{\varepsilon \to 0^+} \frac{\left\| F\left(t_0+\varepsilon t^{1-\gamma}\right) - F(t_0) - F^{(\gamma)}(t_0)(\varepsilon) \right\|}{|\varepsilon|} = 0$$

is called Fréchet conformable (left) fractional derivative of order γ on to.

(b) If F is strangly generalized conformable fractional derivative of γ on $t_0 \in (a,b)$ by Definition 5 (i), then by Lemme 3.2 in [18] it follows that $j \circ F : (a,b) \to C[0,1] \times C[0,1]$ is Fréchet conformable differentiable at t_0 and $(j \circ F)^{(\gamma)}(t_0) = j\left(F^{(\gamma)}(t_0)\right)$

(c) $\tilde{j} : \mathrm{R}_{\mathcal{F}} \to C[0,1] \times C[0,1]$, $\tilde{j}(y) = j(-1.y)$, $y \in \mathrm{R}_{\mathcal{F}}$ It is easy to prove the following properties : $\|\tilde{j}(y) - \tilde{j}(z)\| = d(y,z)$

$$\tilde{j}\left(\mathrm{R}_{\mathcal{F}}\right) = j\left(\mathrm{R}_{\mathcal{F}}\right) \textit{ and } \tilde{j}(s.y+t.z) = s.\tilde{j}(y) + t.\tilde{j}(z) \quad \forall y, z \in \mathrm{R}_{\mathcal{F}}, t, s \geq 0$$

Theorem 6. *Let $F:(a,b)\longrightarrow \mathrm{R}_{\mathcal{F}}$ and $t_0\in(a,b)$.*

(i) *If F is generalized γ-differentiable on t_0 according to Definition 5 (ii), then $j\circ F$ is Fréchet γ-differentiable on t_0 and*

$$(j\circ F)^{(\gamma)}(t_0)=-\tilde{j}\left(F^{(\gamma)}(t_0)\right)$$

(ii) *If F is generalized γ-differentiable on t_0 according to Definition 5 (iii), then $j\circ F$ is Fréchet γ-differentiable at right and at left on t_0 and*

$$(j\circ F)_r^{(\gamma)}(t_0)=j\left(F^{(\gamma)}(t_0)\right),\ (j\circ F)_l^{(\gamma)}(t_0)=-\tilde{j}\left(F^{(\gamma)}(t_0)\right)$$

(iii) *If F is generalized γ-differentiable on to according to Definition 5 (iv), then $j\circ F$ is Fréchet γ-differentiable at right and at left on t_0 and*

$$(j\circ F)_r^{(\gamma)}(t_0)=-\tilde{j}\left(F^{(\gamma)}(t_0)\right),\ (j\circ F)_l^{(\gamma)}(t_0)=j\left(F^{(\gamma)}(t_0)\right)$$

Proof. The proofs of (i), (ii) and (iii) are similar, to have a sample, here we will prove, for example, (ii).
Let $\varepsilon>0$ and $\varepsilon\longrightarrow 0$, we have

$$\left\|\frac{(j\circ F)\left(t_0+\varepsilon t_0^{1-\gamma}\right)-(j\circ F)(t_0)}{\varepsilon}-j\left(F^{(\gamma)}(t_0)\right)\right\|$$

$$\leq\left\|\frac{(j\circ F)\left(t_0+\varepsilon t_0^{1-\gamma}\right)-(j\circ F)(t_0)}{\varepsilon}-j\left(\frac{F\left(t_0+\varepsilon t_0^{1-\gamma}\right)\ominus F(t_0)}{\varepsilon}\right)\right\|$$
$$+\left\|j\left(\frac{F\left(t_0+\varepsilon t_0^{1-\gamma}\right)\ominus F(t_0)}{\varepsilon}\right)-j\left(F^{(\gamma)}(t_0)\right)\right\|.$$

By $F\left(t_0+\varepsilon t_0^{1-\gamma}\right)=F(t_0)+y$ we get $j\left(F\left(t_0+\varepsilon t_0^{1-\gamma}\right)\right)=j(F(t_0))+j(y)$, i.e. $j(y)=j\left(F\left(t_0+\varepsilon t_0^{1-\gamma}\right)\ominus F(t_0)\right)=j\left(F\left(t_0+\varepsilon t_0^{1-\gamma}\right)\right)-j(F(t_0))$, which immediately implies that

$$j\left(\frac{F\left(t_0+\varepsilon t_0^{1-\gamma}\right)\ominus F(t_0)}{\varepsilon}\right)=j\left(t_0^{\gamma-1}\frac{F(t_0+h)\ominus F(t_0)}{h}\right)$$
$$=t_0^{\gamma-1}\frac{(j\circ F)(t_0+h)-(j\circ F)(t_0)}{h}$$

$$= \frac{(j \circ F)\left(t_0 + \varepsilon t_0^{1-\gamma}\right) - (j \circ F)(t_0)}{\varepsilon}$$

where $h = \varepsilon t_0^{1-\gamma}$ then $\varepsilon = h t_0^{\gamma-1}$. Also, for all $\gamma \in (0, 1]$

$$\left\| j\left(\frac{F\left(t_0 + \varepsilon t_0^{1-\gamma}\right) \ominus F(t_0)}{\varepsilon}\right) - j\left(F^{(\gamma)}(t_0)\right)\right\|$$

$$= d\left(\frac{F\left(t_0 + \varepsilon t_0^{1-\gamma}\right) \ominus F(t_0)}{\varepsilon}, F^{(\gamma)}(t_0)\right) \longrightarrow_{\varepsilon \to 0} 0$$

So

$$\lim_{\varepsilon \to 0} \left\| \frac{(j \circ F)\left(t_0 + \varepsilon t_0^{1-\gamma}\right) - (j \circ F)(t_0) - j\left(F^{(\gamma)}(t_0)\right) \cdot \varepsilon}{\varepsilon} \right\|$$

Therefore, passing to $h \to 0$, it follows

$$= \lim_{h \to 0} \left\| t_0^{\gamma-1} \frac{(j \circ F)(t_0 + h) - (j \circ F)(t_0) - j\left(F'(t_0)\right) \cdot h}{h} \right\| = 0$$

i.e. $j \circ F$ is frechet γ-differentiable at right on t_0, and $(j \circ F)_r^{(\gamma)}(t_0) = j\left(F^{(\gamma)}(t_0)\right)$
On the other hand, for the same $\varepsilon \longrightarrow 0$, we have for all $\gamma \in (0, 1]$

$$\left\| \frac{(j \circ F)\left(t_0 - \varepsilon t_0^{1-\gamma}\right) - (j \circ F)(t_0)}{-\varepsilon} - \left(-\tilde{j}\left(F^{(\gamma)}(t_0)\right)\right) \right\|$$

$$= \left\| \frac{(j \circ F)\left(t_0 - \varepsilon t_0^{1-\gamma}\right) - (j \circ F)(t_0)}{-\varepsilon} + \left(\tilde{j}\left(F^{(\gamma)}(t_0)\right)\right) \right\|$$

$$\leq \left\| \frac{(j \circ F)\left(t_0 - \varepsilon t_0^{1-\gamma}\right) - (j \circ F)(t_0)}{-\varepsilon} + \tilde{j}\left[\frac{F\left(t_0 - \varepsilon t_0^{1-\gamma}\right) \ominus F(t_0)}{-\varepsilon}\right] \right\|$$

$$+ \left\| \tilde{j}\left[\frac{F\left(t_0 - \varepsilon t_0^{1-\gamma}\right) \ominus F(t_0)}{-\varepsilon}\right] - \left(\tilde{j}\left(F^{(\gamma)}(t_0)\right)\right) \right\|.$$

But

$$\left\| \tilde{j}\left[\frac{F\left(t_0-\varepsilon t_0^{1-\gamma}\right)-F\left(t_0\right)}{-\varepsilon}\right]\ominus\left(\tilde{j}\left(F^{(\gamma)}\left(t_0\right)\right)\right)\right\|$$

$$= d\left(\frac{F\left(t_0-\varepsilon t_0^{1-\gamma}\right)\ominus F\left(t_0\right)}{-\varepsilon}, F^{(\gamma)}\left(t_0\right) \longrightarrow_{\varepsilon\to 0} 0\right.$$

Also,

$$\tilde{j}\left[\frac{F\left(t_0-\varepsilon t_0^{1-\gamma}\right)\ominus F\left(t_0\right)}{-\varepsilon}\right]=j\left[\frac{F\left(t_0-\varepsilon t_0^{1-\gamma}\right)\ominus F\left(t_0\right)}{\varepsilon}\right]$$

$$=\frac{j\left(F\left(t_0-\varepsilon t_0^{1-\gamma}\right)\right)-j\left(F\left(t_0\right)\right)}{\varepsilon}$$

because by $F\left(t_0-\varepsilon t_0^{1-\gamma}\right)-F\left(t_0\right)=y$, we get $F\left(t_0-\varepsilon t_0^{1-\gamma}\right)=F\left(t_0\right)+y$, $j\left(F\left(t_0- \varepsilon t_0^{1-\gamma}\right)\right)-j\left(F\left(t_0\right)\right)=j(y)$, and $j(\lambda y)=\lambda j(y)$ for $\lambda\geq 0$. As a conclusion,

$$\left\|\frac{(j\circ F)\left(t_0-\varepsilon t_0^{1-\gamma}\right)-(j\circ F)\left(t_0\right)}{-\varepsilon}+\tilde{j}\left(\frac{F\left(t_0-\varepsilon t_0^{1-\gamma}\right)\ominus F\left(t_0\right)}{-\varepsilon}\right)\right\|=0$$

denoting $\eta=-\varepsilon$,

$$\lim_{\eta\to 0}\left\|\frac{(j\circ F)\left(t_0+\eta t_0^{1-\gamma}\right)-(j\circ F)\left(t_0\right)}{\eta}-\left(-\tilde{j}F^{(\gamma)}\left(t_0\right)\right)\right\|=0$$

i.e. $(j\circ F)_l^{(\gamma)}=-\tilde{j}\left(F^{(\gamma)}\left(t_0\right)\right)$.

Remark 2. It is easy to show that $j\left(F^{(\gamma)}\left(t_0\right)\right)=-\tilde{j}\left(F^{(\gamma)}\left(t_0\right)\right)$ if and only if $F^{(\gamma)}\left(t_0\right)\in \mathrm{R}$.

4 Fuzzy Fractional Conformable Differential Equations

A fuzzy fractional differential equation is an equation of the form

$$u^{(\gamma)}=F(t,u), u\left(t_0\right)=u_0\in \mathrm{R}_{\mathcal{F}} \tag{9}$$

where we search for solutions $u : \mathrm{R} \longrightarrow \mathrm{R}_{\mathcal{F}}$ which are strongly generalized differentiable on any interval (a, b) with at most a finite set of points in which u is $(iii) + (iv)$-γ-differentiable. The following Theorem transforms the fuzzy fractional differential equation into integral equations.

Theorem 7. *For $t_0 \in \mathrm{R}$, the fuzzy fractional differential equation (9) where $F : \mathrm{R} \times \mathrm{R}_{\mathcal{F}} \longrightarrow \mathrm{R}_{\mathcal{F}}$ is supposed to be continuous, is equivalent to one of the integral equations for all $\gamma \in (0, 1]$:*

$$u(t) = u_0 + \int_{t_0}^{t} x^{\gamma-1} F(x, u(x))dx, \ \forall t \in [t_0, t_1]$$

or

$$u_0 = u(t) + (-1) \int_{t_0}^{t} x^{\gamma-1} F(x, u(x))dx, \ \forall t \in [t_0, t_1]$$

on some interval $(t_0, t_1) \subset \mathrm{R}$, depending on the strongly differentiability considered, (i) or (ii), respectively. Here the equivalence between two equations means that any solution of an equation is a solution too for the other one.

Proof. Suppose that F is γ_1 -differentiable on (t_0, t_1). Then we have by Theorem 3.6 [18] and definition of conformable integral [9, 13]

$$\int_{t_0}^{t} x^{\gamma-1} F^{(\gamma)}(x, u(x))dx = F(t) - F(t_0) \tag{10}$$

By the Newton-Leibniz formula for functions with values in Banach space see Theorem 3.5 [18] we have

$$(j \circ F)(t) = (j \circ F)(t_0) + \int_{t_0}^{t} x^{\gamma-1} (j \circ F)^{(\gamma)}(x)dx$$

By Theorem 6 there exists $(j \circ F)^{(\gamma)}(x)$ and we get

$$(j \circ F)(t) = (j \circ F)(t_0) + \int_{t_0}^{t} x^{\gamma-1} \left(-\tilde{j}\left(\left(F^{(\gamma)}(x)\right)\right)\right) dx \,.$$

Since the embedding j commutes with the integral (see Theorem 3.4 [18]) we obtain

$$(j \circ F)(t) = (j \circ F)(t_0) - \tilde{j}\left(\int_{t_0}^{t} x^{\gamma-1} \left(F^{(\gamma)}(x)dx\right)\right.$$

Then it follows that

$$\tilde{j}\left(\int_{t_0}^{t} x^{\gamma-1} \left(F^{(\gamma)}(x)dx\right) + (j \circ F)(t) = (j \circ F)(t_0)\right. .$$

By the definition of $\tilde{j}$ we obtain

$$j\left(-1\int_{t_0}^{t} x^{\gamma-1}F^{(\gamma)}(x)dx\right)+j(F(t))=j\left(F\left(t_0\right)\right).$$

By the additivity of the embedding j we have

$$-1\int_{t_0}^{t} x^{\gamma-1}F^{(\gamma)}(x)dx=F(t)-F\left(t_0\right)$$

and finally

$$\int_{t_0}^{t} x^{\gamma-1}F^{(\gamma)}(x)dx=-\left(F(t)-F\left(t_0\right)\right) \tag{11}$$

By adding (10) and (11) the required result is obtained.

References

1. Anastassiou, G.A., Gal, S.G.: On a fuzzy trigonometric approximation theorem of Weierstrass-type. J. Fuzzy Math. **9**, 701–708 (2001)
2. Bede, B., Rudas, I.J., Bencsik, A.L.: First order linear fuzzy differential equations under generalized differentiability. Inf. Sci. **177**, 1648–1662 (2007)
3. Harir, A., Melliani, S., Chadli, L.S.: Hybrid fuzzy differential equations. AIMS Math. **5**(1), 273–285 (2020). https://doi.org/10.3934/math.2020018
4. Chadli, L.S., Harir, A., Melliani, S.: Fuzzy Euler differential equation. SOP Trans. Appl. Math. **2**(1) (2015). https://doi.org/10.15764/AM.2015.01001
5. Dubois, D., Prade, H.: Towards fuzzy differential calculus. Part 1. Integration of fuzzy mappings. Fuzzy Sets Syst. **8**, 1–17 (1982)
6. Dubois, D., Prade, H.: Towards fuzzy differential calculus. Part 2. Integration on fuzzy intervals. Fuzzy Sets Syst. **8**, 105–116 (1982)
7. Diamond, P., Kloeden, P.E.: Metric Spaces of Fuzzy Sets: Theory and Applications. World Scienific, Singapore (1994)
8. Harir, A., Melliani, S., Chadli, L.S.: Fuzzy generalized conformable fractional derivative. Adv. Fuzzy Syst. **2020**, 7 (2019). Article ID: 1954975
9. Harir, A., Melliani, S., Chadli, L.S.: Fuzzy fractional evolution equations and fuzzy solution operators. Adv. Fuzzy Syst. **2019**, 10 (2019). Article ID: 5734190
10. Goo, H.Y., Park, J.S.: On the continuity of the Zadeh extensions. J. Chungcheong Math. Soc. **20**(4), 525–533 (2007)
11. Kaleva, O.: Fuzzy differential equations. Fuzzy Sets Syst. **24**, 301–317 (1987)
12. Kaleva, O.: The Cauchy problem for fuzzy differential equations. Fuzzy Sets Syst. **35**, 389–396 (1990)
13. Khalil, R., Al Horani, M., Yousef, A., Sababheh, M.: A new definition of fractional derivative. J. Comput. Appl. Math. **264**, 65–70 (2014)
14. Ma, M., Friedman, M., Kandel, A.: A new fuzzy arithmetic. Fuzzy Sets Syst. **108**, 83–90 (1999)
15. Puri, M.L., Ralescu, D.A.: Differentials of fuzzy functions. J. Math. Anal. Appl. **91**, 552–558 (1983)

16. Radstrom, H.: An embedding theorem for spaces of convex set. Proc. Am. Math. Soc. **3**(1), 165–169 (1952)
17. Seikkala, S.: On the fuzzy initial value problem. Fuzzy Sets Syst. **24**, 319–330 (1987)
18. Wu, C., Gong, Z.: On Henstock integral of fuzzy-number-valued functions I. Fuzzy Sets Syst. **120**, 523–532 (2001)

Artificial Intelligence for Healthcare: Roles, Challenges, and Applications

Said El Kafhali and Mohamed Lazaar

Abstract With the use of an intelligent technology-based healthcare technique, there may be a real opportunity to improve medical care quality and effectiveness, thereby increasing patient wellness. Around the world, with rising healthcare costs and the onset of many illnesses, it has become necessary to focus on the people-centered environment, not just the hospital. The future of healthcare may change completely using artificial intelligence (AI) that change how we prevent, diagnose, and cure health conditions. However, the potential of AI is hard to ignore. It is a decision-making machine that can exponentially increase the efficiency of the healthcare organization. Recently, many published papers use the AI technology to monitor and controls the spread of COVID-19 (Coronavirus) pandemic. There are not only the right set of circumstances by using AI in healthcare but also many obstacles and barriers. Data integration is complex, trust issues, time, and energy limitations are some of the barriers to implementing AI in healthcare. Hence, this chapter provides a survey of AI-driven healthcare and identifies proposed models, which health staff is using to bring AI solutions for health applications. It identifies existing approaches to designing models for AI healthcare. The readers can benefit from the chapter by understanding the roles, challenges, applications, and future opportunities of AI for healthcare.

Keywords Artificial intelligence · Machine learning · Deep learning · Healthcare

1 Introduction

AI dates back to the 1950s as a university field and has been developing rapidly in recent years with recent advances and innovations in information storage and processing in which enabled intelligent systems using AI to revolutionize industries in different fields [1]. The goal of AI is to develop technology that permits machines

S. El Kafhali (✉)
Faculty of Sciences and Techniques, Computer, Networks, Mobility and Modeling Laboratory: IR2M, Hassan First University of Settat, 26000 Settat, Morocco
e-mail: said.elkafhali@uhp.ac.ma

M. Lazaar
ENSIAS, Mohammed V University in Rabat, Rabat, Morocco
e-mail: mohamed.lazaar@um5.ac.ma

N. Gherabi and J. Kacprzyk (eds.), *Intelligent Systems in Big Data, Semantic Web and Machine Learning*, Advances in Intelligent Systems and Computing 1344,
https://doi.org/10.1007/978-3-030-72588-4_10

and computers to operate intelligently. The key principle of AI is machine learning (ML), or the capability of a machine to improve upon its skills by permanently analyzing its interactions with the real world. AI evenly refers to situations in which computers can simulate human minds in learning and analysis and therefore can work on problem-solving [2]. According to the types of problems that we want to solve, elementary ML algorithms can be divided into two classes: supervised machine learning (SML) and unsupervised machine learning (USML) algorithms [3]. SML algorithms work by collecting a big number of training cases that contain inputs and the required output labels such as support vector machines, random forest, logistic regression, most artificial neural networks, decision tree methods, fuzzy mathematical theory, and so on. USML algorithms provide untagged data as the algorithms try to identify the optimal parameters in the models to make sense by extracting functionality for the learning case such as the K-means clustering, k nearest neighbors, self-organization mapping model, principal component analysis, hierarchical cluster analysis, and so forth.

Another approach used in AI is deep learning (DL), which is based on the construction of artificial neural networks [4]. These networks, made up of thousands, even millions of neurons, are inspired by the human brain. DL is often applied to much larger amounts of data than ML and learns from this mass of data and in certain cases obtains much better results than ML. It is particularly effective for working with voice data. These voice data must be interpreted and translated into text before a result can be found. The reinforcement learning (RL) model is an AI in which machines identify the actions producing the highest probability of the result. It can be formed by a set of trial and error sequences of events, exposing the model to expert a combination of these strategies. This happens in a Markov decision process, consisting of an ensemble of states, an ensemble of actions, the probability that some action in some state will advance to a new state, and the reward that results from the new state. Using the RL model, the machine establishes a policy that determines the choice or action with the highest likelihood of the desired outcome, assessing the total rewards attributable to the multiple actions performed over time and the relative importance present and future rewards. However, AI habitually refers to a machine that learns from raw data with some degree of autonomy, as occurs with ML, DL, and RL as shown in Fig. 1. With ML, DL, and RL, AI makes it possible to solve problems that were thought to be reserved for human intelligence, such as interpreting natural language or making predictions or complex recommendations.

AI has seen increasing use in all sectors, thanks to increasing data volumes, greater computing power, and new algorithm architectures. In healthcare, there has been an exponential increase in AI research, as evidenced by an increase in publications and university funding. Most interestingly, in healthcare, the introduction of AI-based technologies have reduced costs, improved analysis and treatment outcomes, accelerated drug discovery and, consequently, increment the efficiency of the entire healthcare business [5]. AI assists healthcare professionals, particularly to help them identify the patients most at risk. It also plays an important role in preventing and detecting diseases in real-time to save people's lives. AI focuses on the analysis of big data aggregated in healthcare settings to improve and develop clinical decision

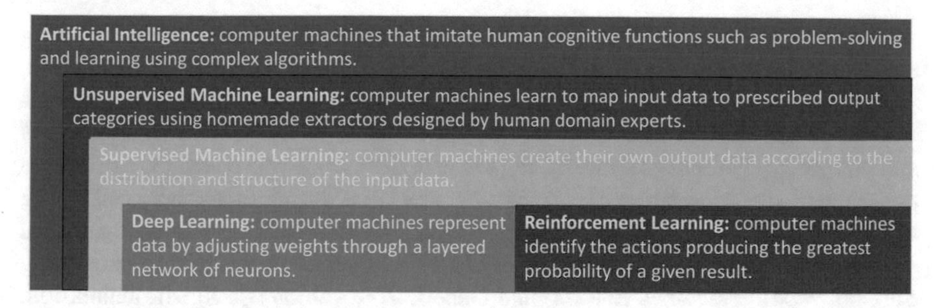

Fig. 1 Summary of AI techniques

support systems or to evaluate medical data for both quality assurance and accessibility of health services [6]. While AI can provide substantial improvements over traditional healthcare for different tasks, many researchers and scientists remain skeptical of their use when medical applications are involved. These skepticisms arise because the AI theory has not yet provided complete solutions and many issues remain unanswered. However, this chapter presents a review of research using AI in healthcare, providing a critical analysis of the relative merit and potential pitfalls of the technique as well as its prospects. To summarize, the goal thus is to outline recent advancements in AI technologies and their healthcare applications, identify the challenges for further improvement in healthcare AI models, and summarize the ethical involvements and economic cost of AI in healthcare.

The rest of this chapter is organized as follows. Section 2 provides a review of the literature on AI in healthcare. The types of AI relevance to healthcare are presented in Sect. 3. Section 4 summarizes some diagnosis and treatment applications. Section 5 presents some discussion and future trends.

2 A Review of the Literature on AI in Healthcare

AI plays a more important role in healthcare. It is currently under development or implementation for use in many targeted health applications, including patient monitoring, medical diagnostics, clinical decision support, and health systems learning [7, 8]. Many AI algorithms and software are under development or developed to support clinical decision-making and the development of public health policies. These AI algorithms usually use computerized predictive analytics algorithms to organize, filter, and search for models in large data sets from different sources and render a probability analysis that healthcare providers can take on quick and informed decisions [9]. To date, the majority of jurisdictions do not permit these AI algorithms to be the final decision-maker in healthcare. Instead, they are mainly used as a screening way or as a diagnostic aid. There are many papers published on the subject of applied AI in healthcare. In this section, we summarize published important works related to AI in healthcare.

Wu *et al.* [10] presented a system that uses logistic regression to classify triple-negative (TN) breast cancer on ultrasound images. Grayscale and color Doppler images are used to calculate Ultrasound images characteristics. The authors analyzed Ultrasonic and clinical data of 140 surgically confirmed cases for the diagnosis of TN and non-TN (NTN) breast cancer. Diagnostic performance was measured by the area under the receiver operating characteristic curve (ROC) with a sensitivity of 86.96% and a specificity of 82.91% on the dataset used. Subasi *et al.* [11] proposed a hybrid algorithm to detect epileptic seizures in Electroencephalogram (EEG) records using Support Vector Machines (SVMs) and Genetic Algorithms (GAs) ML techniques. The obtained results have shown that the proposed algorithm can achieve a classification accuracy of up to 99.38% for the EEG datasets used and is a powerful model for neuroscientists to detect epileptic seizures in EEG. Authors in [12] modeled the diabetes diagnosis using a backpropagation neural network (BPNN) and probabilistic neural network (PNN). A comparative analysis concluded that the PNN model is better than the BPNN model in terms of accuracy. For the PNN model and by using 75% of training data and 25% of testing data, they obtained a diagnostic accuracy of 97.9%. Based on the obtained results, the proposed model can effectively save doctors time and enhance diagnosis compared with the traditional diagnosis process.

Alfaras *et al.* [13] proposed an automatic and fast electrocardiograph (ECG) arrhythmia classifier based ML approach known as Echo State Networks. The obtained results achieved a sensitivity of 92.7% and a positive predictive value of 86.1% for ventricular ectopic beats, using lead II, and a sensitivity of 95.7% and positive predictive value of 75.1% when using the lead V1. A new statistical learning method, namely least absolute shrinkage and selection operator (LASSO), is proposed in [14] for removing redundant and irrelevant features from the ECG big data set. The features obtained from LASSO are trained with popular ML algorithms such as multi-class one-against-all support vector machine, artificial neural networks, and K-nearest neighbor (K-NN). The experimental results show that the proposed method of LASSO with a K-NN classifier is effective with a significant improvement in recognition accuracy of 99.1379% compared to some other existing techniques. Abdeldayem *et al.* [15] proposed an approach that exploits the spectro-temporal changes of the ECG signal to establish a personal recognition system using both short-time Fourier transform (STFT) and generalized Morse wavelets (CWT). The SFTF system achieved an average accuracy of 97.85% whereas the CWT achieves an average accuracy of 97.5%, over the eight studied databases. Prashanth *et al.* [16] use Support Vector Machine (SVM), Naïve Bayes, Boosted Trees, and Random Forests classifiers to predict and detect early Parkinson's disease (PD). The authors use important biomarkers such as cerebrospinal fluid (CSF) measurements and dopaminergic imaging markers from 183 healthy normal and 401 early PD subjects obtained from the Parkinson's Progression Markers Initiative (PPMI) database. The obtained results showed that the SVM classifier gave the best performance (96.40% accuracy, 97.03% sensitivity, 95.01% specificity). Shrivastava *et al.* [17] proposed a neural network to predict PD with a feature selection technique. For the experiments, the authors use a real-life dataset of 166 persons including both healthy controls and affected persons. The experimental results showed that the Binary Bat Algorithm outperforms traditional Algorithms like Genetic Algorithm, Particle Swarm Optimization (PSO), and Modified Cuckoo Search with an accuracy of 93.60%.

Authors in [18] used an ML-based system that uses the features in Computed tomography (CT) images taken from the parenchyma around lung nodules to identify cancerous nodules. The results showed a sensitivity of 100% and a specificity of 96%. Yang *et al.* [19] presented an ML-based system to predict the grade of Glioblastoma using radiomic features of Multiparametric-magnetic resonance imaging. The obtained results show that the grade of Glioblastoma could be predicted with an accuracy of 92% using the proposed approach. Guncar *et al.* [20] used ML algorithms to diagnose hematological disorders based solely on laboratory results. In the first method, they used all the blood tests available. In a second method, they used only a limited set more habitually measured during patient intake. They obtained a prediction accuracy of 88% and 86% for the two methods used respectively by considering the list of the five most probable diseases and 59% and 57% by considering only the most probable disease.

The comparative study of the related literature has been done in this section. The important related AI methods applied for healthcare used in the literature review are shown in Table 1.

3 Types of AI of Relevance to Healthcare

Artificial intelligence presents many techniques that can be applied in Industry 4.0. Most of these techniques have a great impact on the services of the healthcare field. Its performance in the Healthcare field depends on its parameters and the quality of data. Figure 2 illustrates the major ML Methods in Healthcare. MLis not only one method with learning but also rather a scientific discipline of them that focuses on how computers learn from data. ML is used to determine complex models, and extract knowledge, presenting novel ideas to users. ML combines two major fields such as mathematics and computer science. Many models of ML can be constructed from these two scientific disciplines and massive data. We can classify the ML methods according to their learning mode. In general, learning can be divided into many classes like supervised learning, unsupervised learning, reinforcement learning, Semi-Supervised learning, etc. In this section, we detailed the most used learning.

Supervised Learning: The supervised learning technique is the most popular mode of learning used for predictive analysis when the dataset is labeled. In the preprocessing phase, we define a couple (input, output, or target) for each input, this operation named labeling of the dataset. Supervised learning focuses on two classes of problems, classification, and regression. Both classes share the same concept of utilizing the training datasets to estimate the output parameters. Many algorithms like Support Vector Machines, Naïve Bayes, Nearest Neighbor, Linear Regression, Decision Trees, Neural Networks, etc. are categorized under SML. The main difference between the regression and classification is that the output parameter in the

Table 1 Summary of some important papers of the AI techniques in healthcare

Paper	Purpose	Domain	Technique	Topic	Best performance
Wu *et al.* [10]	Evaluate the scope of ML together with quantitative ultrasound image features for triple-negative breast cancer diagnosis	Radiology	Logistic regression	Breast cancer	Sensitivity = 86.96% Specificity = 82.91%
Subasi *et al.* [11]	Establish a hybrid algorithm for epileptic seizure detection to determine the optimum parameters of SVMs for EEG data classification	Neurology	ML	Seizure detection	Accuracy = 99.38%
Li *et al.* [12]	Use BPNN and PNN models to the diabetes diagnosis	Ophthalmology	AI neural networks	Diabetes diagnosis	Accuracy = 97.9%
Alfaras *et al.* [13]	Present a fast ECG arrhythmia classifier based on an ML approach known as Echo State Networks	Cardiology	ML	Arrhythmia	Sensitivity = 95.7% Positive predictive = 86.1%
Patro *et al.* [14]	Classify and identify the ECG data	Cardiology	Statistical learning	Biometric recognition	Accuracy = 99.1379%
Abdeldayem *et al.* [15]	Establish a personal recognition system using the deep convolutional neural network and the spectro-temporal changes of the ECG signal	Cardiology	Deep Convolutional neural network	Biometric recognition	Accuracy = 97.85%
Prashanth *et al.* [16]	Propose ML algorithms to give good accuracy detection of Parkinson's Disease using multimodal features	Neurology	ML	Parkinson's Disease	Accuracy = 96.40% Sensitivity = 97.03% Specificity = 95.01%
Shrivastava *et al.* [17]	Predict PD with feature selection technique	Neurology	Neural network	Parkinson's Disease	Accuracy = 93.60%
Uthoff *et al.* [18]	Use an ML-based system to distinguish malignant and benign lung nodules	Radiology	ML	Lungs	Sensitivity = 100% Specificity = 96%
Yang *et al.* [19]	Predict the grade of Glioblastoma	Radiology	ML	Brain	Accuracy = 92%
Guncar *et al.* [20]	Diagnose hematological disorders	Hematology	ML	Biological hematology	Accuracy = 88% Specificity = 96%

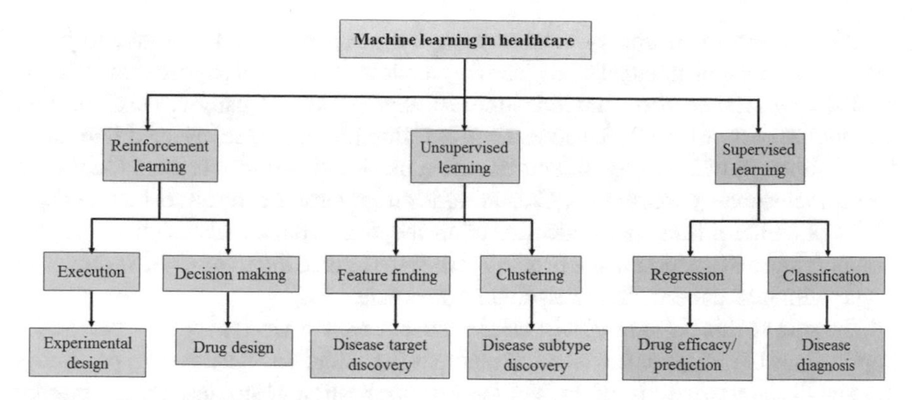

Fig. 2 Machine learning in healthcare

regression is numerical (or continuous) while that for classification is categorical (or discrete). Classification is the process permits to find or discover many parameters of the model, which assist to categorize the data into many classes or in discrete values. The classification process is applied to the data where it can be divided into binary or multiple discrete labels. In classification, data is labeled under different labels according to some parameters given in input and then the labels are predicted for the data. Regression is the process permits to find many parameters of the model to distingue the data into numerical continuous values. In regression, the model identifies the distribution movement depending on the historical data.

Many supervised learning methods have been applied in the healthcare field. The main objective is to improve the performance of the disease diagnosis system. In this paragraph, we describe the interesting contributions based on classification (disease diagnosis) and prediction (drug efficacy). For the first type of applications, many methods based on Neural networks have been used for the diagnosis of diabetes [21], a hybrid approach based on Support Vector Machine and Multilayer Perceptron with a Backpropagation (BP) algorithm is used in the diagnosis of chronic renal failure on the Internet of Medical Things (IoMT) platform [22], others researchers proposed a deep learning model that integrates S-Mask R-CNN and Inception-v3 in the ultrasound image-aided diagnosis of prostate cancer [23], a Naïve Bayes classification model integrated with temporal association rules (TARs) for coronary heart disease diagnosis [24]. For the second application, a modeling approach based on sequential dependencies to improve the prediction of the drug efficacy using recurrent neural networks [25], a hybrid approach for predicting drug efficacy for cancer treatment based on comparative analysis of chemosensitivity and gene expression data. This approach combines supervised and unsupervised learning algorithms [26], an interesting method permits to Predict novel drugs for SARS-CoV-2 using ML from a > 10 million chemical space [27].

Unsupervised Learning: In unsupervised learning, the dataset is without labeled responses i.e. no output variable to predict. In this mode, the model tries to create clusters or groups of individuals based on their similar attributes, or naturally occurring trends, patterns, or relationships. This is a real challenge task to judge the performance of unsupervised learning models. The popularly used algorithms are K-Means, Expectation maximization, association rules, density-based algorithm, etc. are the part of USML. Unsupervised learning problems are grouped into many types, although two main problems are often used in literature: they are Clustering and Dimensionality Reduction. Clustering is an important concept, it mainly deals to find or create clusters in a collection of uncategorized data. Dimensionality reduction or feature extraction is the process permits to reduce the number of variables of higher-dimensional data to lower-dimensional data.

The rest of this paragraph will briefly present related work about the two major applications of unsupervised learning methods in healthcare; disease subtype discovery and disease target discovery. For the first application, many researchers present a description of the treatment of Alzheimer's disease based on subtype-specific M1 allosteric agonists [28], others authors proposed an analytic framework permits to find of complex disease subgroups, this framework was able to subtype schizophrenia subjects into diverse subgroups with different prognosis and treatment response [29]. For the second application, the ML algorithms can be applied to segment images and compute object and spatial-based data before analysis [30], an approach computes the most relevant feature subset by taking advantage of feature selection and extraction techniques [31], the researchers discuss how to improve drug-target interaction (DTI) prediction using deep-learning models and other ML algorithms [32], a random forest model guided association of adverse drug reactions within Vitro target-based pharmacology [33].

Reinforcement Learning: Reinforcement learning is a mode of learning based on an agent that learns via interaction and feedback with the environment. The methods based on reinforcement learning solve a task by acting and receiving rewards for an environment. In general, computers take appropriate actions to increase the recompense to generate the best decision and maximize the reward use of reinforcement learning. Mostly used reinforcement learning algorithms are Q-learning, Temporal difference, deep adversarial networks.

We describe some methods based on reinforcement learning in the health field in particular on experimental design and drug design. Computer-aided drug design has become a key source for information, rationalization, and inventiveness. In this field, researchers have used many reinforcement algorithms. The authors provide numerous recent examples of applications for drug design [34]. A reinforcement learning (RL)-based optimal adaptive control approach is proposed for the continuous infusion of a sedative drug to maintain a required level of sedation. In this study, integral reinforcement learning (IRL) algorithm is designed to provide optimal drug dosing for a given performance measure that iteratively updates the control solution concerning the pharmacology of the patient while guaranteeing convergence to the optimal solution [35]. Several machine-learning methods based on reinforcement learning are used to predict drug design and discovery, which are efficient. An approach hybrid a metaheuristic algorithm with support vector machine and k-Nearest Neighbors is used for chemical descriptor selection and chemical compound activities [36].

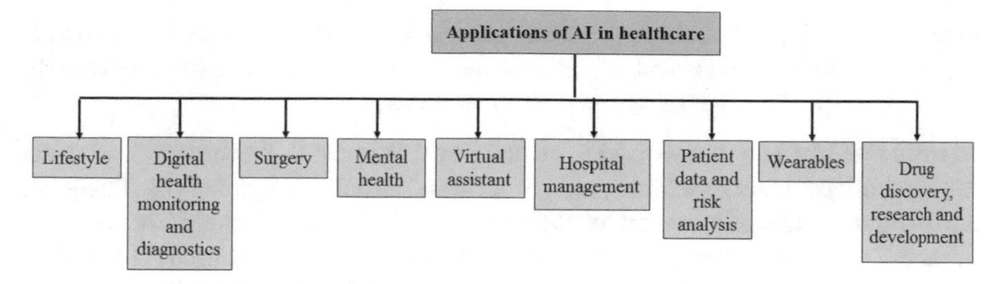

Fig. 3 Machine learning in healthcare

4 Diagnosis and Treatment Applications

AI healthcare area has been a mostly upcoming hot topic in the present years. In recent times, there has been an enormous thrust in the usage of AI for many applications, particularly in the healthcare and pharmaceutical areas. Figure 3 summarizes the important applications of AI in healthcare.

Lifestyle: By using AI, we can understand our interactions with the environment, our lifestyle improvements, and disease control with the ultimate purpose of improving our well-being. Preventing disease is more important and effective than treating disease. Therefore, if we can accurately predict our disease, we can control it properly and effectively. For instance, the authors of [37] developed prediction models based on AI to control type-2 diabetes through guided lifestyle management.

Digital Health Monitoring and Diagnostics: In the past, healthcare decision-making and diagnosis was solely based on the physician's personal experience, health knowledge, the patient's physical signs and symptoms, and laboratory diagnostic analysis. Today, in a typical healthcare monitoring system, Medical Internet of Things (MIoT) devices have emerged to play an important role in helping physicians and medical personnel to obtain accurate data and help reach a more accurate assessment, diagnosis, and decision [38]. These MIoT are designed to collect health data from the patient, which needs to be stored and analyzed or even researched in the future [39]. To analyze and take the real-time decision from the big health data collected by IoT, AI methods are needed to save patient life especially in the case of cardiac health monitoring [40].

Surgery: AI has the potential to change the manner surgery is practiced with the engagement of an optimized future for the highest quality patient care and surgeon workflow [41]. Even though robotic autonomous surgery will stay out of reach for some time, the synergy between the domains can probably expedite the competences of AI to raise surgical care. Therefore, surgeons should be engaged in the assessment of the quality and applicability of AI progress to ensure proper translation in the clinical sector. For example, AI can enhance plastic surgery practice and help plastic surgeons with the skills, knowledge, and tools to improve patient care in the future [42]. Integrating AI into surgical decision-making has the potential to change care by increasing the decision to operate, recognition and attenuation of modifiable risk factors, informed consent process, decisions about postoperative management, and shared decisions regarding the use of resources [43].

Mental Health: The future of AI in mental health is hopeful. AI can have an impact on psychological and psychiatric care in terms of information gathering, diagnosis, and treatment. Recently, mental health exercise yet has much to benefit from AI technology. Therefore, authors in [44] reviewed numerous papers that used AI-based on electronic health records, brain imaging data, mood scales, new surveillance systems, and social media platforms to predict, classify or sub-group mental health illnesses, including depression [45], schizophrenia [46] or other psychiatric illnesses [47], and suicide attempts and ideas [48].

Virtual Assistant: In eHealth, the use of virtual assistant (VA) is effective and acceptable to older adults, including those with limited health knowledge. It should be able to establish a relationship with the patient over a series of interactions. The VA presents an intelligent behavior similar to a human in the way each daily dialogue takes place and ensuring the consistency of the daily interventions towards the desirable behaviors [49]. With the application of AI, surgical robots, the diagnosis, and treatment of diseases have become smarter such as the use of VA to improve the mental health of humans. The AI accuracy diagnosis results exceed that of human doctors [50]. Overall, VA mainly acts as a bridge to communicate with doctors, patients, and medical staff. For patients, the VA can easily convert daily language into medical language thanks to the smart device, to more precisely search for the corresponding medical services. For doctors, the VA can automatically respond to relevant information based on basic patient information, helping doctors to manage patients and coordinate medical procedures more easily, so doctors can save more time. For medical staff, the application of VA can significantly save human and material resources and meet the needs of all parties more effectively.

Hospital Management: AI in healthcare is changing the obverse of hospital management [51]. The help of AI in healthcare will improve the performance of hospitals, doctors, and nurses and provide patients with more targeted and personalized services. AI makes healthcare facilities more efficient and improves the lives of healthcare providers and patients by automating tasks in the least amount of time and money. Numerous medical tasks made easier by AI in healthcare such as improving medical records and treatment solutions, automating customer relationship management, monitoring vital statistics of ICU (Intensive Care Units) patients, simplified health insurance verification, decoding of laboratory results, greater efficiency in hospital operations, facilitating patient engagement and adherence, enhancing the administrative activities and so forth.

Patient Data and Risk Analysis: AI can help us to analyze the patient data in real-time to save a life. For example, recently AI is one of the news technologies that can easily track the spread of COVID-19 virus, identify high-risk patients, and is useful in controlling this infection in real-time. It can also predict the risk of mortality by adequately analyzing past patient data [52]. AI technologies have also been widely used to analyze clinical data, including electronic health records, medical images, and physiological signals [53]. AI methodologies have been adopted also to extract information from the data collected at various stages of drug development that contains information about the mechanisms and treatments of the disease.

Wearables: Wearable devices facilitate the development of algorithms for the automated prediction, prevention and response of health events in many areas: metabolic, cardiovascular, gastrointestinal monitoring, sleep, neurology, movement disorders, mental health, maternal, pulmonary health, and environmental exposures and so forth [54]. Wearables also provide ongoing medical data to actively monitor metabolic status, diagnosis, and treatment [55]. Whereas progress in the development of wearable and AI will undoubtedly increase over the next few decades, randomized clinical studies are required to assess their real impact on patient care. In the future, the wearable will not only be diagnostic, preventive, and therapeutic methods but will also allow the uninterrupted acquisition of health data to monitor disease progression, response to drugs, and assess the effectiveness of clinical trials.

Drug Discovery, Research, and Development: At present, exceptional improvements in computing power married with advances in AI technology, could be used to revolutionize the drug discovery and development process [56]. The drug development process follows an inductive-deductive cycle, which ultimately leads to optimized hit and leads compounds. The first step is the identification of new chemical compounds with biological activity. The biological activity can result from the interaction of the compound with a specific enzyme or with an entire organism [57]. The second step is the identification of a lead molecule, which is a chemical compound that can lead to the development of a new drug to treat a disease. Using AI to automate the identification of new chemical compounds or the identification of a lead molecule can reduce errors and improve the efficiency of drug discovery and development [58]. There are other uses of AI in drug discovery and development, including pharmacological properties [59], predicting workable synthetic routes for drug-like molecules [60], drug activity against cancer cells [61], protein structural features [62], drug combination [63], drug research [64], drug reuse [65] and many others [66]. Besides, the identification of new pathways and targets using omics analysis becomes possible through the generation of new biomarkers and therapeutic targets, personalized medicine based on omics markers, and the discovery of links between drugs and diseases [67].

5 Conclusions and Future Trends

Most of the past studies applied the AI algorithms on healthcare are either focusing on unsupervised or supervised methods. Unsupervised techniques use clustering or more complex means to identify structure in data to present the data visualizations or summaries. Supervised techniques, on the other hand, focus on finding very specific and predefined type models or constructing predictive models. No matter which of these techniques is used, the families of underlying models or the similarity metric push a strong bias in the analytical process. Therefore, current AI algorithms on healthcare applications mainly focus on answering rather well-asked questions. One could support that this type of healthcare problem-solving was suitable before when healthcare data resources were significantly little and could be hoped to make sense of it using such limited techniques. To date, health care data has greatly exceeded our capacity to analyze it, and new powerful and flexible AI techniques are needed that allow the unexpected to be discovered, allowing medical personnel and the physician users to formulate new hypotheses interactively and help them discover real perspectives of understanding of the disease. Therefore, we need to develop discovery support AI algorithms rather than automated discovery AI algorithms. Such an AI algorithm should not try to do the discovery work for medical personnel and the physician users- it should rather support them by giving associative and intuitive access to everything the AI algorithm has access to unstructured and semi-structured data up to elements of expert knowledge humanly annotated.

There are various problems with ML algorithms to take into consideration when we apply these clinically [68]. The first one is its issue with overfitting which is a real issue along with every prediction method [69]. It will perform exceptionally too well on its training data, but very poorly on new data. This means that the false signal or noise in the training dataset is picked up and learned as concepts by the model. Overfitting takes place when the training algorithm acknowledges the false signal or noise in the dataset as a signal and carries out the prediction to the test dataset, showing in bad performance on the new health data or an incapability to validate the model externally. There are divers' methods to solve and recognize the problem of overfitting. First, if the precision changes drastically, for example, the 99% precision on the training set drops to 50% when the algorithm is applied to the new dataset. Second, if the precision on the training dataset increases, but the precision on the validation dataset stays the same or decreases, we need to stop training. Thirdly, the goodness of fit test can measure how well the predicted values of the algorithm match the observed true values. Finally, when we have various comparable algorithms, we can use the simpler ones so that the added benefit of any complexity can be given, which favors a simpler model among others. Besides, the issue of overfitting can be further reduced by a sampling technique including cross-validation that repeatedly partitions the sample data without method or conscious decision into training and validation sets to validate model predictions internally.

One more issue is that most clinicians and perhaps some researchers are unaware of what an ML algorithm does to produce its output. A popular example is a project to explore the outcomes of pneumonia-related hospitalization in the 1990s [70], in which asthma has appeared as a protective element against pneumonia in the project. For that, the clinicians should know from their experience that co-morbid asthma is not a protective element. In a complex machining operation generally, process parameters bear very complex mathematical relationships among them. However, as the process is costly it is not possible to generate a large dataset. In this case, we can use ML techniques with such a low number of data for supervised learning to predict and optimize the machining process parameters, as often ML employed for the problem with a large dataset.

References

1. Garbuio, M., Lin, N.: Artificial intelligence as a growth engine for health care startups: emerging business models. Calif. Manage. Rev. **61**(2), 59–83 (2019)
2. Rong, G., Mendez, A., Assi, E.B., Zhao, B., Sawan, M.: Artificial intelligence in healthcare: review and prediction case studies. Engineering **6**(3), 291–301 (2020)
3. Chang, Z., Du, Z., Zhang, F., Huang, F., Chen, J., Li, W., Guo, Z.: Landslide susceptibility prediction based on remote sensing images and GIS: comparisons of supervised and unsupervised machine learning models. Remote Sensing **12**(3), 502 (2020)
4. Ravì, D., Wong, C., Deligianni, F., Berthelot, M., Andreu-Perez, J., Lo, B., Yang, G.Z.: Deep learning for health informatics. IEEE J. Biomed. Health Inform. **21**(1), 4–21 (2016)
5. Yu, K.H., Beam, A.L., Kohane, I.S.: Artificial intelligence in healthcare. Nat. Biomed. Eng. **2**(10), 719–731 (2018)
6. Lysaght, T., Lim, H.Y., Xafis, V., Ngiam, K.Y.: AI-assisted decision-making in healthcare. Asian Bioethics Rev. **11**(3), 299–314 (2019)
7. Reddy, S., Allan, S., Coghlan, S., Cooper, P.: A governance model for the application of AI in health care. J. Am. Med. Inform. Assoc. **27**(3), 491–497 (2020)
8. Magrabi, F., Ammenwerth, E., McNair, J.B., De Keizer, N.F., Hyppönen, H., Nykönen, P., Rigby, M., Scott, P.J., Vehko, T., Wong, Z.S.Y., Georgiou, A.: Artificial Intelligence in clinical decision support: challenges for evaluating AI and practical implications. Yearbook Med. Informa. **28**(01), 128–134 (2019)
9. Persson, A., Kavathatzopoulos, I.: How to make decisions with algorithms: ethical decision-making using algorithms within predictive analytics. ACM SIGCAS Comput. Soc. **47**(4), 122–133 (2018)
10. Wu, T., Sultan, L.R., Tian, J., Cary, T.W., Sehgal, C.M.: Machine learning for diagnostic ultrasound of triple-negative breast cancer. Breast Cancer Res. Treat. **173**(2), 365–373 (2019)
11. Subasi, A., Kevric, J., Canbaz, M.A.: Epileptic seizure detection using hybrid machine learning methods. Neural Comput. Appl. **31**(1), 317–325 (2019)
12. Li, X.: Artificial intelligence neural network based on intelligent diagnosis. J. Ambient Intell. Hum. Comput. (2020). https://doi.org/10.1007/s12652-020-02108-6
13. Alfaras, M., Soriano, M.C., Ortín, S.: A fast machine learning model for ECG-based heartbeat classification and arrhythmia detection. Front. Phys. **7**, 103 (2019)
14. Patro, K.K., Reddi, S.P.R., Khalelulla, S.E., Kumar, P.R., Shankar, K.: ECG data optimization for biometric human recognition using statistical distributed machine learning algorithm. J. Supercomput. **76**(2), 858–875 (2020)
15. Abdeldayem, S.S., Bourlai, T.: ECG-based human authentication using high-level spectro-temporal signal features. In: 2018 IEEE International Conference on Big Data (Big Data), pp. 4984–4993. IEEE Press (2018)
16. Prashanth, R., Roy, S.D., Mandal, P.K., Ghosh, S.: High-accuracy detection of early Parkinson's disease through multimodal features and machine learning. Int. J. Med. Inform. **90**, 13–21 (2016)
17. Shrivastava, P., Shukla, A., Vepakomma, P., Bhansali, N., Verma, K.: A survey of nature-inspired algorithms for feature selection to identify Parkinson's disease. Comput. Methods Prog. Biomed. **139**, 171–179 (2017)

18. Uthoff, J., Stephens, M.J., Newell Jr., J.D., Hoffman, E.A., Larson, J., Koehn, N., De Stefano, F.A., Lusk, C.M., Wenzlaff, A.S., Watza, D., Neslund-Dudas, C.: Machine learning approach for distinguishing malignant and benign lung nodules utilizing standardized perinodular parenchymal features from CT. Med. Phys. **46**(7), 3207–3216 (2019)
19. Cui, G., Jeong, J.J., Lei, Y., Wang, T., Liu, T., Curran, W.J., Mao, H., Yang, X.: Machine-learning-based classification of Glioblastoma using MRI-based radiomic features. In: Proceedings of SPIE 10950, Medical Imaging 2019: Computer-Aided Diagnosis, 1095048, March 2019. https://doi.org/10.1117/12.2513110
20. Gunčar, G., Kukar, M., Notar, M., Brvar, M., Černelč, P., Notar, M., Notar, M.: An application of machine learning to haematological diagnosis. Sci. Rep. **8**(1), 1–12 (2018)
21. Erkaymaz, O., Ozer, M., Perc, M.: Performance of small-world feedforward neural networks for the diagnosis of diabetes. Appl. Math. Comput. **311**, 22–28 (2017)
22. Ma, F., Sun, T., Liu, L., Jing, H.: Detection and diagnosis of chronic kidney disease using deep learning-based heterogeneous modified artificial neural network. Fut. Gener. Comput. Syst. **111**, 17–26 (2020)
23. Liu, Z., Yang, C., Huang, J., Liu, S., Zhuo, Y., Lu, X.: Deep learning framework based on integration of S-Mask R-CNN and Inception-v3 for ultrasound image-aided diagnosis of prostate cancer. Fut. Gener. Comput. Syst. **114**, 358–367 (2020)
24. Orphanou, K., Dagliati, A., Sacchi, L., Stassopoulou, A., Keravnou, E., Bellazzi, R.: Incorporating repeating temporal association rules in Naïve Bayes classifiers for coronary heart disease diagnosis. J. Biomed. Inform. **81**, 74–82 (2018)
25. Kang, S.: Personalized prediction of drug efficacy for diabetes treatment via patient-level sequential modeling with neural networks. Artif. Intell. Med. **85**, 1–6 (2018)
26. Wan, P., Li, Q., Larsen, J.E.P., Eklund, A.C., Parlesak, A., Rigina, O., Nielsen, S.J., Bjöorkling, F., Jónsdóttir, S.Ó.: Prediction of drug efficacy for cancer treatment based on comparative analysis of chemosensitivity and gene expression data. Bioorg. Med. Chem. **20**(1), 167–176 (2012)
27. Kowalewski, J., Ray, A.: Predicting novel drugs for SARS-CoV-2 using machine learning from a 10 million chemical space. Heliyon **6**(8), e04639 (2020)
28. Lebois, E.P., et al.: Towards the treatment of Alzheimer's disease: discovery and development of novel subtype-specific M1 allosteric agonists. Alzheimers Dement. **5**(4), P335 (2009)
29. Yin, L., Chau, C.K., Sham, P.C., So, H.C.: Integrating clinical data and imputed transcriptome from GWAS to uncover complex disease subtypes: applications in psychiatry and cardiology. Am. J. Hum. Genet. **105**(6), 1193–1212 (2019)
30. Caie, P.D., Dimitriou, N., Arandjelović, O.: Precision medicine in digital pathology via image analysis and machine learning. In: Artificial Intelligence and Deep Learning in Pathology, pp. 149–173. Elsevier (2021)
31. Shah, S.M.S., Shah, F.A., Hussain, S.A., Batool, S.: Support vector machines-based heart disease diagnosis using feature subset, wrapping selection and extraction methods. Comput. Electr. Eng. **84**, 106628 (2020)
32. D'Souza, S., Prema, K.V., Balaji, S.: Machine learning models for drug-target interactions: current knowledge and future directions. Drug Disc. Today **25**(4), 748–756 (2020)
33. Ietswaart, R., Arat, S., Chen, A.X., Farahmand, S., Kim, B., DuMouchel, W., Armstrong, D., Fekete, A., Sutherland, J.J., Urban, L.: Machine learning guided association of adverse drug reactions with in vitro target-based pharmacology **57**, 102837 (2020)
34. Daina, A., Röhrig, U.F., Zoete, V.: Computer-aided drug design for cancer therapy. In: Wolkenhauer, O. (ed.) Systems Medicine, pp. 386–401. Academic Press, Oxford (2021)
35. Padmanabhan, R., Meskin, N., Haddad, W.M.: Optimal adaptive control of drug dosing using integral reinforcement learning. Math. Biosci. **309**, 131–142 (2019)
36. Houssein, E.H., Hosney, M.E., Oliva, D., Mohamed, W.M., Hassaballah, M.: A novel hybrid Harris hawks optimization and support vector machines for drug design and discovery. Comput. Chem. Eng. **133**, 106656 (2020)
37. Hsu, G.C.: Using math-physical medicine and artificial intelligence technology to manage lifestyle and control metabolic conditions of T2D. Int. J. Diab. Complications **2**(3), 1–7 (2018)

38. El Kafhali, S., Salah, K.: Performance modelling and analysis of Internet of Things enabled healthcare monitoring systems. IET Netw. **8**(1), 48–58 (2018)
39. El Kafhali, S., Salah, K., Alla, S.B.: Performance evaluation of IoT-fog-cloud deployment for healthcare services. In: 2018 4th International Conference on Cloud Computing Technologies and Applications (Cloudtech), pp. 1–6. IEEE, November 2018
40. Zhang, Q.: Artificial intelligence-enabled ECG big data mining for pervasive heart health monitoring. In: Biomedical Signal Processing, pp. 273–290. Springer, Singapore (2020)
41. Hashimoto, D.A., Rosman, G., Rus, D., Meireles, O.R.: Artificial intelligence in surgery: promises and perils. Ann. surg. **268**(1), 70 (2018)
42. Liang, X., Yang, X., Yin, S., Malay, S., Chung, K.C., Ma, J., Wang, K.: Artificial intelligence in plastic surgery: applications and challenges. Aesth. Plast Surg. 1–7 (2020)
43. Loftus, T.J., Tighe, P.J., Filiberto, A.C., Efron, P.A., Brakenridge, S.C., Mohr, A.M., Rashidi, P., Upchurch, G.R., Bihorac, A.: Artificial intelligence and surgical decision-making. JAMA Surg. **155**(2), 148–158 (2020)
44. Graham, S., Depp, C., Lee, E.E., Nebeker, C., Tu, X., Kim, H.C., Jeste, D.V.: Artificial intelligence for mental health and mental illnesses: an overview. Curr. Psychiatry Rep. **21**(11), 116 (2019)
45. Anis, K., Zakia, H., Mohamed, D., Jeffrey, C.: Detecting depression severity by interpretable representations of motion dynamics. In: 2018 13th IEEE International Conference on Automatic Face & Gesture Recognition (FG 2018), pp. 739–745. IEEE (2018)
46. Kalmady, S.V., Greiner, R., Agrawal, R., Shivakumar, V., Narayanaswamy, J.C., Brown, M.R., Greenshaw, A.J., Dursun, S.M., Venkatasubramanian, G.: Towards artificial intelligence in mental health by improving schizophrenia prediction with multiple brain parcellation ensemble-learning. npj Schizophrenia **5**(1), 1–11 (2019)
47. Thakur, A., Alam, M.S., Abir, M.R.H., Kushal, M.A.A., Rahman, R.M.: A fuzzy approach for the diagnosis of depression. In: Modern Approaches for Intelligent Information and Database Systems, pp. 199–211. Springer, Cham (2018)
48. Cook, B.L., Progovac, A.M., Chen, P., Mullin, B., Hou, S., Baca Garcia, E.: Novel use of natural language processing (NLP) to predict suicidal ideation and psychiatric symptoms in a text-based mental health intervention in Madrid. Comput. Math. Methods Med. (2016)
49. Balsa, J., Neves, P., Félix, I., Guerreiro, M.P., Alves, P., Carmo, M.B., Marques, D., Dias, A., Henriques, A., Cláudio, A.P.: Intelligent virtual assistant for promoting behaviour change in older people with T2D. In: EPIA Conference on Artificial Intelligence, pp. 372–383. Springer, Cham (2019)
50. Tian, S., Yang, W., Le Grange, J.M., Wang, P., Huang, W., Ye, Z.: Smart healthcare: making medical care more intelligent. Global Health J. **3**(3), 62–65 (2019)
51. Gu, D., Li, J., Bichindaritz, I., Deng, S., Liang, C.: The mechanism of influence of a case-based health knowledge system on hospital management systems. In: International Conference on Case-Based Reasoning, pp. 139–153. Springer, Cham (2017)
52. Vaishya, R., Javaid, M., Khan, I.H., Haleem, A.: Artificial Intelligence (AI) applications for COVID-19 pandemic. Diabet. Metab. Syndr.: Clin. Res. Revi. **14**(4), 337–339 (2020)
53. Wang, F., Preininger, A.: AI in health: state of the art, challenges, and future directions. Yearbook Med. Inform. **28**(1), 16–26 (2019)
54. Dunn, J., Runge, R., Snyder, M.: Wearables and the medical revolution. Personalized Med. **15**(5), 429–448 (2018)
55. Yetisen, A.K., Martinez-Hurtado, J.L., Ünal, B., Khademhosseini, A., Butt, H.: Wearables in medicine. Adv. Mater. **30**(33), 1706910 (2018)
56. Díaz, Ó., Dalton, J.A., Giraldo, J.: Artificial intelligence: a novel approach for drug discovery. Trends Pharmacol. Sci. **40**(8), 550–551 (2019)
57. Correia, J., Resende, T., Baptista, D., Rocha, M.: Artificial intelligence in biological activity prediction. In: International Conference on Practical Applications of Computational Biology & Bioinformatics, pp. 164–172. Springer, Cham (2019)
58. Mak, K.K., Pichika, M.R.: Artificial intelligence in drug development: present status and future prospects. Drug Disc. Today **24**(3), 773–780 (2019)

59. Klopman, G., Chakravarti, S.K., Zhu, H., Ivanov, J.M., Saiakhov, R.D.: ESP: a method to predict toxicity and pharmacological properties of chemicals using multiple MCASE databases. J. Chem. Inf. Comput. Sci. **44**(2), 704–715 (2004)
60. Merk, D., Friedrich, L., Grisoni, F., Schneider, G.: De novo design of bioactive small molecules by artificial intelligence. Mol. Inform. **37**(1–2), 1700153 (2018)
61. Lind, A.P., Anderson, P.C.: Predicting drug activity against cancer cells by random forest models based on minimal genomic information and chemical properties. PloS One **14**(7), e0219774 (2019)
62. Klausen, M.S., Jespersen, M.C., Nielsen, H., Jensen, K.K., Jurtz, V.I., Sønderby, C.K., Sommer, M.O.A., Winther, O., Nielsen, M., Petersen, B., Marcatili, P.: NetSurfP-2.0: improved prediction of protein structural features by integrated deep learning. Proteins: Struct. Funct. Bioinform. **87**(6), 520–527 (2019)
63. Tsigelny, I.F.: Artificial intelligence in drug combination therapy. Briefings Bioinform. **20**(4), 1434–1448 (2019)
64. Tu, H., Lin, Z., Lee, K.: Automation with intelligence in drug research. Clin. Ther. **41**(11), 2436–2444 (2019)
65. Hui, T.K., Mohammed, B., Donyai, P., McCrindle, R., Sherratt, R.S.: Enhancing pharmaceutical packaging through a technology ecosystem to facilitate the reuse of medicines and reduce medicinal waste. Pharmacy **8**(2), 58 (2020)
66. Chan, H.S., Shan, H., Dahoun, T., Vogel, H., Yuan, S.: Advancing drug discovery via artificial intelligence. Trends Pharmacol. Sci. **40**(8), 592–604 (2019)
67. Villalta, F., Rachakonda, G.: Advances in preclinical approaches to Chagas disease drug discovery. Exp. Opinion Drug Disc. **14**(11), 1161–1174 (2019)
68. Lee, S., Mohr, N.M., Street, W.N., Nadkarni, P.: Machine learning in relation to emergency medicine clinical and operational scenarios: an overview. West. J. Emerg. Med. **20**(2), 219–227 (2019)
69. Hawkins, D.M.: The problem of overfitting. J. Chem. Inf. Comput. Sci. **44**(1), 1–12 (2004)
70. Craven, D.E., Steger, K.A., Barber, T.W.: Preventing nosocomial pneumonia: state of the art and perspectives for the 1990s. Am. J. Med. **91**(3), S44–S53 (1991)

Intelligent System for the Protection of People

Jamal Mabrouki, Mourade Azrour, Amina Boubekraoui, and Souad El Hajjaji

Abstract The indecencies of assault, capturing, dealing, rape, and so on are unbiased in this day and age. In this article, we might want to propose a gadget that is a combination of a few gadgets and procedures that, when actuated, gives a compact contraption that can help individuals in peril to scare their friends and family and the specialists can act the hero. The gadget comprises of a switch, a Raspberry pi, an Arduino UNO, a Raspberry pi camera module, a Global Positioning System (GPS) module, a heartbeat sensor, a bell and a vibration engine. The whole design and arrangement works around the Arduino and Raspberry pi. The sensors incorporated in Arduino read information persistently and if there should arise an occurrence of deviation from the predefined limit estimation of the sensor, the data (GPS directions and photograph/video) is sent to the applicable specialists utilizing Raspberry pi. The USP of this undertaking is that it sidesteps manual impedance at a crucial time that can represent the deciding moment the situation when the casualty is in an unfortunate circumstance. It is modest, convenient and easy to understand.

Keywords Internet of Things · Smart devices · Technology · Applications · GPS · Sensors

1 Introduction

Invisible victims, unheard, unprotected, uncared for and rarely punished predators, the report draws the contours of collective abandonment and a pattern of violence

J. Mabrouki (✉) · S. El Hajjaji
Laboratory of Spectroscopy, Molecular Modeling, Materials, Nanomaterial, Water and Environment, CERN2D, Faculty of Science, Mohammed V University in Rabat, Avenue Ibn Battouta, BP1014, Agdal, Rabat, Morocco

M. Azrour
Faculty of Sciences and Techniques, Department of Computer Science, IDMS Team, Moulay Ismail University, Errachidia, Morocco

A. Boubekraoui
Laboratory of Condensed Matter and Interdisciplinary Sciences, Faculty of Science, Mohammed V University in Rabat, Avenue Ibn Battouta, BP1014, Agdal, Rabat, Morocco

N. Gherabi and J. Kacprzyk (eds.), *Intelligent Systems in Big Data, Semantic Web and Machine Learning*, Advances in Intelligent Systems and Computing 1344,
https://doi.org/10.1007/978-3-030-72588-4_11

that is repeated over and over again. At the international level, UNICEF [1], the United Nations agency entirely dedicated to children and adolescents, fights violence, including sexual violence, against children. We fight against early marriage and pregnancy, against female genital mutilation. In short, against practices that make the bodies of girls and adolescents the expression of a social order that has become unacceptable. Unacceptable in terms of equal rights and costly to society as a whole in terms of health and economic development. By strengthening statistical data in order to document the phenomenon and measure its extent, national and international awareness-raising campaigns, advocacy with States and communities, programmers to provide girls with access to education, and strengthening care and protection systems, we are trying to promote an integrated response to a multifaceted problem. UNICEF's latest report on violence against children, published at the end of 2014, reveals that 120 million girls worldwide have been subjected to sexual intercourse or other forced sexual acts. In France, every year 15% to 20% of an age group is reported to be subjected to sexual violence; more than 120,000 girls and 32,000 boys under 18 years of age have been raped or attempted rape [1].

In 2014, 76% of all trafficked persons in India were women and children. Unfortunately, children account for about 40% of prostitutes. It is estimated that more than 2 million women and children are trafficked for sexual purposes and slavery. The real figures are far from the truth and do not show the disastrous state in which our country finds itself. A woman is kidnapped every 44 min, raped every 47 min every day [2].

Therefore, it is high time that we use technological advances and design a device that can help people, regardless of gender, age or social status, to stop the increase in numbers. A security device is the need of the hour and various organizations are quite frequently offering innovative security products [3].

Artificial intelligence, the term can be frightening because we tend to think of an intrusive technology that will replace the natural. But connected objects with artificial intelligence are designed to respect the privacy of users. In the case of seniors, they are intended to provide help to compensate for the lack of caregivers and especially to promote home support. The use of artificial intelligence is becoming more and more democratic in today's society because it provides significant assistance and support to the elderly in a society with an ageing population. Connected objects, robotics, or home automation…the technology sector has a major role in the Silver Economy [4]. Artificial intelligence is designed to simplify the life of users. Home automation, for example, is able to perform household tasks or act as a personal assistant (personal diary, time to sleep and to get up, opening and closing doors, reminding people to take their medication, processing phone calls, e-mails or digital data…). Artificial intelligence also ensures the safety of seniors by giving an alert in case of unusual situations such as a break-in in the house [5].

In the following section, we have discussed the work that has already been done and this has been given the title Literature Review. After that is the section of the system description where the methodology of the project and a brief description of the present components is made. The outcome, conclusion and future work is discussed in the last half of the document. The device consists of a piezovibrator which emits a beep after one minute of actual activation of the device [6]. If someone tries to remove the bracelet and throw it, the force sensor starts and the buzzer is activated. The device then starts sending the current contact information to the registered contacts and to the police control room. On the top of the band there are two nodes that emit electric current as soon as it comes into contact with a surface after the device is activated. The current generated here is the leakage current. The device and the smartphone are connected via Bluetooth, which is responsible for global data sharing.

2 Literature Review

The model designed in this article is as follows: When a fingerprint reader module detects the user's fingerprint, the microcontroller gives high logic on the output pins [7]. The integrated electric shock system activates and dissipates a low high voltage current at the fine output. The electric shock remains active for 15 s. The shock circuit is deactivated after the period. At the same time as the fingerprint recognition is performed, the SIM 808 module starts tracking the user's location via the GPS module. This extracted location is sent to the user's family members with an alert message via the GSM module. The work in this article used a system that used a heart rate sensor, a motion sensor, GPS (global positioning system), GSM modules and an alarm system to eliminate the threat. Once the threshold value of the heart rate reading is exceeded, an alarm is triggered and the GSM module sends the victim's coordinates to the relevant authority. In the meantime, the motion detector detects the movement of the victim [8].

This article proposed a portable device with a pressure switch integrated in the system. When the user detects a danger from a stranger or a person, he can press the switch by compressing it. The pressure sensor instantly detects the pressure applied and a text message, with the victim's GPS coordinates, is sent to the parents/guardians mobile phone numbers stored in the device, followed by a call. If the call is not answered, it is redirected to the police with the SMS. In addition, if the person is in an isolated area, a message indicating the location is sent to the phone via SMS.

The system in this article consists of a piezoelectric sensor that detects vibrations produced by tapping the feet or when the emergency button is pressed [9]. The system continues to send messages every 30 s until the unit is turned off. Heartbeat Sensor is also used to detect unusual heartbeat readings that are displayed on the mobile application called "Women Security" developed on the Android platform. The simultaneous audio recording function is integrated into the application which records the incident which can later be used as evidence. At the same time, to alert people around a buzzer is heard so that they can get immediate help. A knife attached to the DC motor of our device comes out of the shoes and can be used for self-defense.

The author of this article proposed an application connected to a smart band via Bluetooth Low Energy (BLE) [10]. The system sets up a network with a smart phone via an application that acts as an interface between the system and the smart phone. Data such as pulse rate, body temperature and body movement are closely monitored by the application. In case of abuse, the application instructs the phone to send the coordinates using GPS and GSM modules. The help message is sent to family members and the nearest police station via the GSM installation integrated into the phone.

The design of this article known as SMARISA is a portable device based on Raspberry Pi that aims to help women in distress [11]. It is integrated with an application that uses GPS tracking to trace location, a camera module for recoding that can later be used as evidence, and a messaging service to help alert emergency contacts. All of this is done after pulling a trigger. The user logs in to the application that allows selection of contacts from a list. Once the desired contact is selected, the image link is retrieved from the server and a help message and the victim's current GPS location is sent to the emergency contact and the police. The buzzer connected to Raspberry-Pi is activated to collect attention in this neighbourhood.

The work done here describes a mechanism where after pressing a push button, the GPS comes into play and then finds latitude and longitude information [12]. The GPS modem then establishes a communication mode with the microcontroller and stores the data in the UART microcontroller. The GSM module interfaces with the microcontroller to send and receive messages. The UART then sends the AT + CMGS AT command to the GSM module with the recipient's number and a message that is "Google map URL". This article proposed to differentiate the problems a woman might face in different categories using existing applications [13]. Some of these problems may be specific to emergency calls when they are at risk, such as molestation, physical abuse, etc. while some applications contain comprehensive resources for victims of disease. In short, the proposed application will provide information on domestic violence prevention laws, health advice for women and an urgent appeal for women seeking help against street violence in a single system. The user must first register when all relevant data is stored in the cloud and the data is accessible by the relevant authorities.

3 Design and Implementation

3.1 System Design

The tape (system) when activated will continuously take the user's pulse using the pulse sensor built into the Arduino UNO and once the preset pulse reading threshold is exceeded, the vibration motor attached to the Arduino UNO will buzz for approximately 5 s. The vibration motor plays an important role here because it helps to bypass the trigger functionality that is seen in various systems present and helps to reduce human interference. If the user turns off the vibration engine using the button before this interval, the GPS module and the Raspberry pi camera module attached with the Raspberry pi will not come into play, otherwise the information

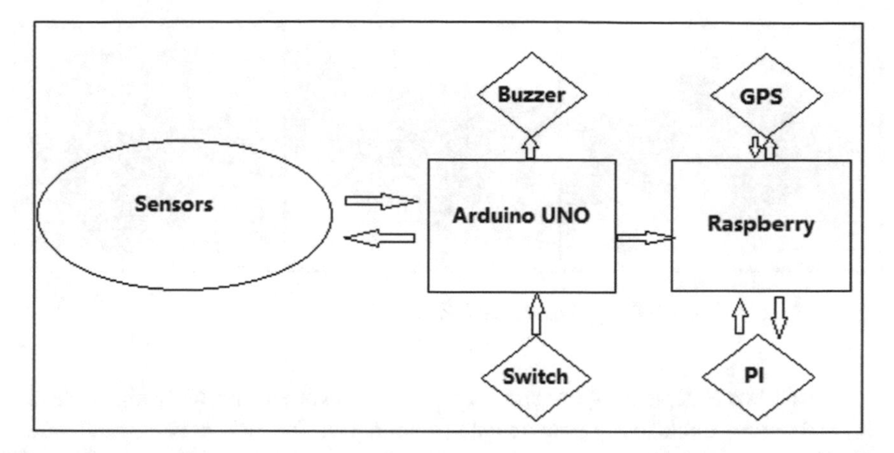

Fig. 1 Block diagram

(a time-stamped video consisting of the coordinate) will be sent to the addresses registered using the Wi-Fi Raspberry pi function via email. Figure 1 shows the flowchart of the process.

3.2 System Hardware

a. Arduino UNO

Arduino is a brand that covers hardware free electronic boards on which a microcontroller is located (of Atmel AVR architecture like the Atmega328p, and of ARM architecture like the Cortex-M3 for the Arduino Due). The schematics of these electronic boards are published under a free license. However, some components, such as the microcontroller for example, are not under free license [14].

The microcontroller can be programmed to analyze and produce electrical signals, in order to perform various tasks such as home automation (control of domestic appliances - lighting, heating…), robot control, embedded computing, etc. [15]. It is a platform based on a simple input/output interface. It was originally intended mainly but not exclusively for interactive multimedia programming for shows or artistic animations, which explains in part the descent of its Processing development environment, itself inspired by the Wiring programming environment (one designed for the production of applications involving graphics and the other for the control of theatres) [15] (Fig. 2).

b. Camera module

High resolution camera module for Raspberry Pi. It is connected to the Pi via one of the two surface mounted connectors on the Pi board. The camera is connected using the dedicated CSI interface, specially designed for this purpose.

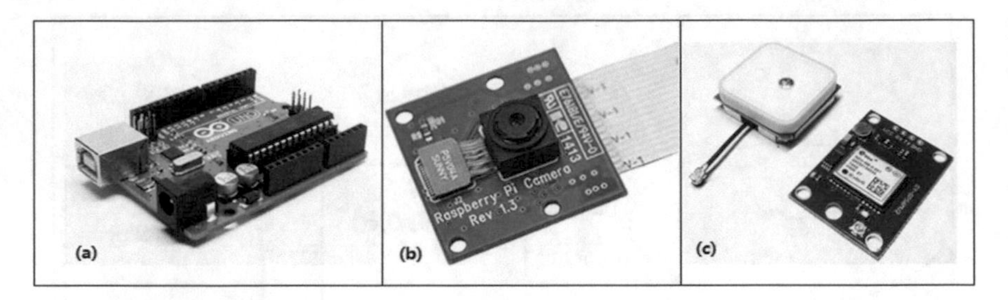

Fig. 2 **a** Arduino UNO, **b** *Camera module*, **c** *GPS module*

A CSI bus is capable of transmitting information at an extremely high data rate and carries exclusively graphic data (pixel data). The card is very small, about 25 mm × 23 mm × 9 mm (one hair larger than version 1 which was 25 × 20 × 9 mm). The weight is around 3gr, which makes it an ideal companion for your mobile applications where weight and size are important factors [16]. The camera connects to the Raspberry Pi via the ribbon, so it is connected directly to the processor via the PI's CSI bus, a higher bandwidth link to transport the graphics data from the camera to the processor. The sensor has 8 megapixel resolution and has a focus lens attached directly to the board. In terms of image, the camera is capable of capturing static images of 3280 × 2464 pixels/dot (compared to 2592 × 1944 for version 1) and also supports video in 1080p30, 720p60 and 640 × 480p90 resolution.

c. Global Positioning System (GPS) module
Low cost and high performance GPS module with ceramic patch antenna, integrated memory chip and battery backup that can be integrated into a wide range of microcontrollers, GY-NEO6MV2 NEO-6 M GPS Module with Flight Control EEPROM RS232 TTL Large Antenna For Arduino GPS APM2 APM2.5 NEO6MV2 3–5 V, Voltage: 3.3 V–5 V, Size: 36 mm * 24 mm. The module (Ublox NEO-6 M) used in this project is very popular. Gives latitude and longitude readings when connected. The patch antenna is popular in this module because they are flat [17].

d. Pulse sensor
The sensor measures heart rate, respiration, ambient temperature and even skin temperature. In particular, it allows you to distinguish precisely between the four phases of sleep by recording the heart rate on waking every morning. It is a simple sensor that has three pins: ground, Vdc and input signal. It is designed to give a digital output of the heart rate in contact with the surface of the human body. The digital output connected to the microcontroller calculates the beats per minute (BPM). However, cardiac sensors are becoming more and more efficient and are becoming more and more popular on the market, arousing the curiosity of the general public, interested in measuring cardiac exercise. In addition, these devices allow other sensors of connected objects to be more accurate, offering an appreciable synergy to these small objects carrying state-of-the-art technology.

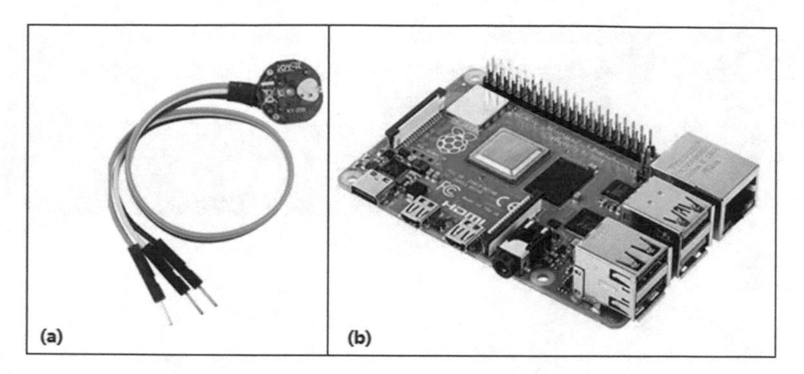

Fig. 3 **a** Pulse sensor, **b** Raspberry Pi

It works on the principle of light modulation during blood flow through the finger at each pulse. The output range of the pulse sensor operating at 3.3 V is only 0 to 675. The pulse sensor signal must be scaled so that it is dimensioned for the largest expected range [18] (Fig. 3).

e. Le Raspberry Pi
The Raspberry Pi is a single-board ARM-processor nanocomputer designed by professors from the Department of Computer Science at the University of Cambridge as part of the Raspberry Pi Foundation [19].

This credit card-sized computer is designed to encourage learning computer programming and allows several variants of the free GNU/Linux operating system, including Debian, and compatible software to run. But it also works with the Microsoft Windows operating system: Windows 10 IoT Core, Windows 10 on ARM (currently relatively unstable), Google Android Pi5, and even a version of IBM's OS/MVT with APL/360 [20]. It is supplied naked, i.e. the motherboard alone, without case, power supply, keyboard, mouse or monitor, in order to reduce costs and allow the use of recovery hardware. Nevertheless, "kits" grouping the "all-in-one" are available on the web for only a few dozen euros [21].

4 Results

During the usage of the proposed framework, we had the option to separate the beat readings to drive the entire procedure referenced in the flowchart above. The two microcontrollers functioned admirably together during combination and the transmission of data utilizing Raspberry pi was effective with no deferrals. The directions of the supported pictures were effectively sent to the pre-enrolled email addresses (Fig. 4).

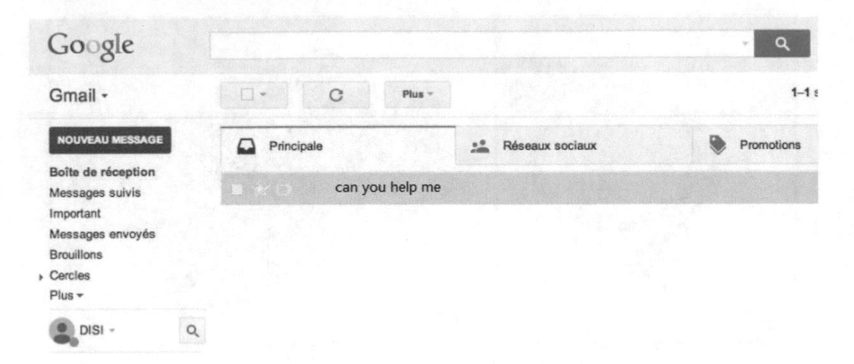

Fig. 4 Email received

5 Conclusion and Future Works

The advancement that ladies' security frameworks have made in the course of recent years is tremendous and the model we propose can at present improve existing frameworks. The majority of the work is equipment arranged with some application or programming direction. What we have done another way is that we have attempted to wipe out the utilization of a trigger at a crucial time by introducing a vibration engine and in this manner improve the security of the framework. The Raspberry pi utilized for the transmission of stepped picture directions to pre-enlisted email identifiers and furthermore gives us numerous choices for additional diminishing should the need emerge later on. In this manner, the ideal blend of the two micro-controllers to give the ideal yield gives the framework a gigantic lift. Most definitely, the utilization better camera to catch pictures in dim conditions would be an extra preferred position. Besides, AI could be incorporated into this venture to contemplate a client's pulse or heartbeat and separate the various readings for better and progressively exact outcomes. The framework could be made increasingly secure by including a biometric detail, for example, unique mark, voice, and so on for approved use. Subsequently, there are numerous chances and zones where further enhancements could be made and, in this way, we should keep on investigating such roads for the advancement of society.

References

1. United Nations Children's Fund: Hidden in plain sight: a statistical analysis of violence against children. UNICEF, New York (2014)
2. Bankar, S.A., Basatwar, K., Divekar, P., Sinha, P., Gupta, H.: Foot device for women security. In: 2018 Second International Conference on Intelligent Computing and Control Systems (ICICCS). IEEE Xplorer (2018)
3. Springer, C.: Humanités numériques et enseignement des langues en ligne: le train du progrès et de la science poursuit sa route, je voudrais descender (2016)

4. d'Ascoli, S.: Comprendre la révolution de l'intelligence artificielle. First (2020)
5. Gallissot, M.: Modéliser le concept de confort dans un habitat intelligent: du multisensoriel au comportement. Diss., Grenoble (2012)
6. Ahir, S., Kapadia, S., Chauhan, J., Sanghavi, N.: The personal stun-a smart device for women's safety. In: 2018 International Conference on Smart City and Emerging Technology (ICSCET) (2018). https://doi.org/10.1109/icscet.2018.8537376
7. Khanam, S., Shah, T.: Self defence device with GSM alert and GPS tracking with fingerprint verification for women safety. In: 3rd International Conference on Electronics, Communication and Aerospace Technology (ICECA). IEEE Xplorer (2019)
8. Helen, A., Fathila, M.F., Rijwana, R., Kalaiselvi, V.K.G.: A smart watch for women security based on IoT concept "watch me". In: 2nd International Conference on Computing and Communications Technologies (ICCCT). IEEE Xplorer (2017)
9. Punjabi, S.K., Chaure, S., Ravale, U., Reddy, D.: Smart intelligent system for women and child security. In: IEEE 9th Annual Information Technology, Electronics and Mobile Communication Conference (IEMCON). IEEE Xplorer (2018)
10. Harikiran, G.C., Menasinkai, K., Shirol, S.: Smart security solution for women based on Internet of Things (IOT). In: 2016 International Conference on Electrical, Electronics, and Optimization Techniques (ICEEOT) (2016). https://doi.org/10.1109/iceeot.2016.7755365
11. Sogi, N.R., Chatterjee, P., Nethra, U., Suma, V.: SMARISA: a Raspberry Pi based smart ring for women safety using IoT. In: Proceedings of the International Conference on Inventive Research in Computing Applications (ICIRCA 2018). IEEE Xplore (2018). Compliant Part Number: CFP18N67-ART, ISBN: 978-1-5386-2456-2
12. Velayutham, R., Sabari, M., Rajeswari, M.S.: An innovative approach for women and children's security-based location tracking system. In: 2016 International Conference on Circuit, Power and Computing Technologies (ICCPCT) (2016). https://doi.org/10.1109/iccpct.2016.7530325
13. Mahmud, S.R., Maowa, J., Wibowo, F.W.: Women empowerment: one stop solution for women. In: 2017 2nd International Conferences on Information Technology, Information Systems and Electrical Engineering (ICITISEE) (2017). https://doi.org/10.1109/icitisee.2017.8285555
14. Arduino, Store Arduino. "Arduino." Arduino LLC (2015)
15. Arduino Uno (2019). https://store.arduino.cc/usa/arduino-uno-rev3, diakses 4
16. Hinz, A., Heier, H.: The Z/I imaging digital camera system. Photogram. Rec. **16**(96), 929–936 (2000)
17. Doerflinger, E.: Les applications météorologiques du système de positionnement satellitaire GPS. La météorologie (2001)
18. Anderson, K.M., Moore, A.A.: Sensors in pacing: capteurs en stimulation cardiaque. Pacing Clin. Electrophysiol. **9**(6), 954–959 (1986)
19. Raspberry Pi: Raspberry Pi 3 model b (2015). https://www.raspberrypi.org
20. Richardson, M., Wallace, S.: Getting started with Raspberry PI. O'Reilly Media, Inc. (2012)
21. Vujović, V., Maksimović, M.: Raspberry Pi as a wireless sensor node: performances and constraints. In: 2014 37th International Convention on Information and Communication Technology, Electronics and Microelectronics (MIPRO). IEEE (2014)

Heterogeneous Integration of Big Data Using Semantic Web Technologies

Sajida Mhammedi and Noreddine Gherabi

Abstract Semantic web offers information for both individuals and computers to preserve large data scale semantically and provide a meaningful content of unstructured data. It offers new benefits for big-data research and applications. Big Data and Semantic Web are the epitome of computer sciences latest trend study subjects. Big data is a new tendency relates to a huge set of datasets including structured, semi-structured and unstructured data collected from different sources. Their integration faces many issues, as it is difficult to process this information using traditional databases and software methods. Recent works on the incorporation of both these technologies have provided a scalable approach in Data Analytics. This article attempts to give a comparative study of methods in integrating Big Data with Semantic Web, describing how Semantic Web makes Big Data smarter, revisits the difficulties and possibilities of Big Data and Semantic Web, and lastly summarizes the future direction of this inclusion.

Keywords Big data · Semantics web · Semantic heterogeneity · Structural heterogeneity · Ontology · Data integration

1 Introduction

For analytics purposes, scientists are generally spending more time extracting, reformatting and integrating data than the analytic act itself. Normally, each analytical tool expects data in a specific structure requiring a succession of procedures. The handling of the data at the terabyte level and beyond is further exasperating this process. Therefore, in order to attain the goal of determining maximum value from collections of Big Data, one must acquire the means to rapidly define strategies and execute them over extremely large collections of data.

While various definitions exist, there is no worldwide accepted understanding of Big Data. However, it has been characterized by a number of V's: volume, velocity, variety and veracity, the volume dimension has been highly stressed. Variety has

S. Mhammedi (✉) · N. Gherabi
Sultan Moulay Slimane University, ENSA Khouribga, Lasti Laboratory, Khouribga, Morocco

N. Gherabi and J. Kacprzyk (eds.), *Intelligent Systems in Big Data, Semantic Web and Machine Learning*, Advances in Intelligent Systems and Computing 1344,
https://doi.org/10.1007/978-3-030-72588-4_12

occurred in Big Data as a common feature because of data heterogeneity. in that regard, variety's Big Data concept is a generalization of semantic heterogeneity as studied in the field of databases, artificial intelligence, semantic web, and cognitive science. Variety is even more important for a number of reasons, including data heterogeneity, schema and ontology isolation, contextual dependency between knowledge collections and knowledge provenance.

Big data includes a huge collection of data from multiple sources, structured, semi-structured and unstructured. Although unstructured and semi-structured data are at least "85%" of the whole data, both types are difficult to be processed using the traditional tools because most of these data are inaccessible to users [1, 2]. Furthermore, Big-data faces many technical challenges such as (1) Acquire and record data, (2) extract and clean information, (3) integrate, aggregate, and represent data, (4) query processing and analysis, (5) interpretation, and (6) privacy and security [3].

The integration of big data means large volumes of linkage of heterogeneous data from different sources [4], with dynamic and heterogeneous data sources in their structure. Schema mapping, record linkage, and data fusion are the main challenges it faces [5]. While data integration is the process that allows users to access and use data in the enterprise while preserving its integrity and quality. In addition, it allows changes to the information stored in one source to be continuously reflected in different sources [6]. Hence to manage and use large and complex data, many companies have poorly integrated large amounts of disordered data into many different information systems in a centralized information system [7].

Semantic Web (SW) is a data structure representing meanings through connectivity, expressing various viewpoints, and using logical rules to share information across apps. It provides information for semantically manipulating large-scale data. Therefore, it overcomes the problems of formatting data in a suitable format to profit from the procedures of information recovery. Subsequently, the Semantic Web is regarded as an integrator crosswise over different content, data apps and frameworks. It is used to integrate information from distinct sources, such as web services, relational databases, and spreadsheets, etc. as a result, semantic heterogeneity and structural heterogeneity issues often involved in the data sources Due to the requirement of multiple data sources in Big Data applications.

Ontologies are a base part of SW technologies and construction in which its metadata schemes provide a controlled vocabulary of concepts [7]. The ontology describes semantics data that represents the background knowledge on the semantic level [8]. The semantic level is a collection of semantic entities, including their concepts and relations, rather than simple words applied in the thesaurus. It specifies the relations between entities and holds the facts and rules on the problem's scope. The use of ontology is essential for the integration of intelligent thinking and retrieval of big data, thus introduces new insights into the related fields and applications.

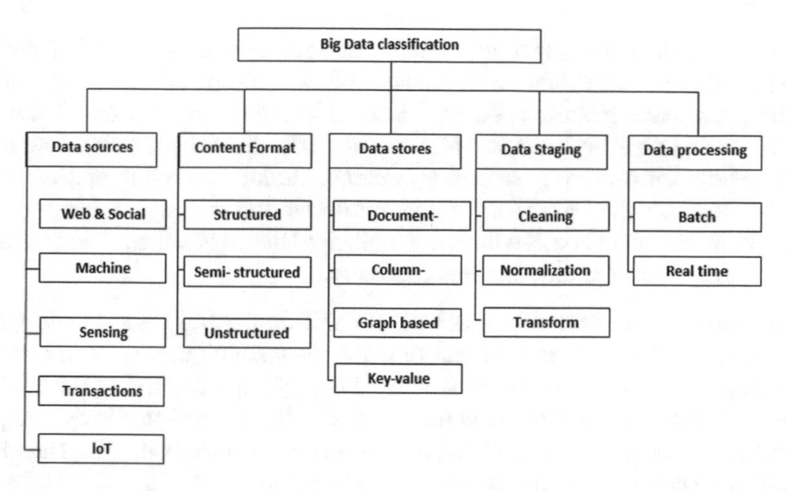

Fig. 1 Schema of big data classification

This paper is organized as follows: first we provide a general overview of big data classification and his research challenges. Section 3 provides the main concepts of the semantic web and data integration. Section 4 describes the research methodology in the Integration of big data using semantic web technologies. Section 5 concludes the overview by giving a summarizing and highlighting some ideas on future work.

2 Big Data

2.1 Classification of Big Data

Big data is classified into different categories: data sources, content format, data stores, data staging and data processing. Each category is special in complexity or characteristics. Data source include structured data, unstructured data and semi-structured data which may include inconsistencies, noise, outliers and redundant data Fig. 1 [9]. Due to this the enormous complex data cannot be readily understood.

2.2 Research Challenges

For a long-time companies have been using Business Intelligence to analyze their own data and optimize their decision-making, in order to describe and explain facts. For example, the use of statistics to better act in the optimization of business processes. However, what makes it possible, in order to have better results and grows in interest is the predictive analysis that allows us to explain the hidden causes of different actions, and to predict future transactions, purchases, or operations of a system.

The quantitative explosion of digital data has forced researchers to find new ways to process and analyze large volumes of data. It's about discovering new orders of magnitude for capturing, searching, sharing, storing, analyzing and presenting data. So was born the "Big Data", a concept for storing an unspeakable amount of information on a digital basis with many challenges like Scalability, Transformation, Data Integrity, Data Quality, and Heterogeneity.

- Scalability: Data scalability implies a fast shift in the Big Data storage system. It should be able to handle in real-time the fluctuating quantity of information. Scaling up and down according to demand is very important in Big Data. It is not very easy to scale up and down the process in Big Data. Sometimes it requires time to decide allocation and de-allocation of resources. If the Big Data is not correctly scaled, its efficiency will drop substantially.
- Transformation: The data collected from different sources is not appropriate for analyzing data because it has a variety of data formats. Before placing the data in the relational database, pre-processing should be carried out to meet schema constraints. Data Transformation is also a challenge due to complexity and scalability of the data captured.
- Data Quality: The importance of the quality of the data collected is well established with the exponential growth of the number of data generated, collected, stored, analyzed and exchanged, this question is always more sensitive. If it is technically possible to store and process immeasurable masses of data, it is likely to risk making decisions based on erroneous data. In other words, if you enter the wrong data to the system you will not be surprised at the wrong results. Data quality has become a serious problem for many cloud service providers, as data is often gathered from various sources. In the past, data processing was usually performed from well-known and limited sources on clean datasets. Therefore, the outcomes were accurate. However, with the emergence of big data, data comes from many different sources, consequently, it is a challenge to obtain high quality information from vast collections of information sources.
- Data Integration: Data integration is the process of combining data from different Multi-source into a unified view: from import to cleanup to mapping and transformation in a target field, to ultimately make the data more usable and more useful for users. Companies are implementing data integration initiatives to analyze and exploit them more effectively, especially in the face of the explosion of incoming data and the arrival of new technologies such as cloud and big data. Data integration is a necessity for innovative companies that want to improve their strategic decision-making and increase their competitive advantage. In most data integration processes, the client sends a request for data to the master server, then the master server imports the necessary datasets from internal and external sources, the required data is extracted from these sources and then combined into a coherent and unified form. The result is delivered to the customer in a consistent and usable form.

- Heterogeneity: Variety, one of the main characteristics of big data, is the consequence of the development of nearly infinite various information sources. This development leads to the heterogeneous nature of big data. Multi-source data are generally of various types and forms of representation and interconnected significantly, allowing users to store data in structured, semi-structured or unstructured formats. For today's database systems, structured data formats are adequate, while semi-structured data formats are only appropriate to certain extent. Reducing heterogeneity between data models is an important prerequisite for precise integration. Reducing heterogeneity between data models is an important prerequisite for precise integration. Many various types of heterogeneity were defined and analyzed. The most conspicuous types of heterogeneity are as follows.

  - Syntactic heterogeneity: represents the situation when two data sources are expressed in different representational language. In the case of ontologies, this situation happens when ontologies are modeled in different representation formalisms, e.g., OWL3 and KIF4.
  - Terminological heterogeneity: stands for different names of the same entity in different data models. An example, could be a usage of different natural language or usage of synonyms.
  - Semantic heterogeneity: Data that is inconsistent with each other, unable to reflect the link between the data sets and could not understand each other is referred as Semantic heterogeneous data [10] represents differences in modeling the same domain of interest. This logical mismatch arises due to a utilization of different axioms for defining the same elements from data sources. Two different mismatches may be distinguished:

    the conceptualization mismatch: differences between modeled concepts.
    And the explicitness mismatch: differences how the concepts are expressed.

    Moreover, it identifies and describes three essential reasons for conceptual differences:

    Difference in coverage: two data models describe different parts of the world at the same level of detail and from the same perspective.
    Difference in granularity: two data models describe the same part of the world from the same perspective but with different levels of detail.
    Difference in scope: two data models describe the same part of the world with the same level of detail but from a different perspective.

  - Semiotic heterogeneity: stands for a different interpretation of entities by various people. In other words, entities from two different data models with the same semantic interpretation ay be interpreted by an interpreter (human, expert system, etc.) with regard to the context. The semiotic heterogeneity is difficult to detect and solve by computer and often by a human as well.

2.3 Related Work on Big Data Integration Challenges

Han, et al. introduce that the main benefits of NoSQL database are the quick reading and writing of data, supporting enormous storage, easy expansion and cost-effective. The characteristics, data model, Classification of databases are defined according to the CAP theorem [9]. However, the significant disadvantage of NoSQL is, that it does not support industry SQL standards and cannot generate the reports.

Due to data heterogeneity, integration and interoperability play an important role in installing big data architectures because of heterogeneity of data. That's why Companies or research organizations that generate large amounts of data need to accustom themselves to new infrastructure. Kadadi et al. [11] in there paper address the challenges of data integration and data interoperability in big data which includes accommodate scope of data, data Inconsistency, query optimization, inadequate Resources, scalability, implementing support system and ETL Process in big data.as result there is no big data integration architecture that can handle all the challenges.

3 Semantic Web

Today, data frameworks need to determine the heterogeneity between information that resides in multiplc auto-ruling information sources. In particular, the use of the World Wide Web as an all-inclusive data trading medium has essentially altered our vision of access to and control of information [12]. Following the semantic web perspective, we are confident that the unambiguous introduction of informational semantics will encourage information interoperation in a variety of information control tasks.

Semantic web goes for specifically giving machine comprehensible information on the Web. It consists of two outstanding ingredients: web ontologies [13, 21, 22], and data annotation. In spite of the fact that it has a long approach for the acknowledgment of the semantic web, including the development of an ontology, inference engine and different segments, the rule of the semantic web recommends us that information semantics is to set up and keep up the correspondence from information to the subject is planned.

3.1 Semantic Data Integration

Semantic integration of information is the way to join information from divergent sources and merging it into important and significant data, however, the utilization of semantic technology is important, as organizations grow up, and so does their information. Without the right information management system, intradepartmental or application-specific, the information will storehouse rapidly emerge and impede efficiency and participation.

Semantic Data Integration provides a response that goes beyond the standard integration arrangements for applications. It uses information-driven engineering based on an institutionalized model for the distribution and exchange of data, in particular, the Resource Description Framework (RDF) [14]. In this system, the heterogeneous information of an organization, such as structured, semi-organized and unstructured is communicated [15], put away and got to the similarities. Since the information structure is communicated through the links within the information itself, it is not compelled to a structure formed by the database and does not end up outdated with the information advancement. Below are some important steps to carry out the integration of semantic data.

- Creating an Application Profile (RDF Shape) representing the desired the coveted type of the last dataset.
- Reusing existing ontologies and building new ontologies as required.
- Leveraging completely the accessible Linked Open Datasets in the area.
- Designing a basic, sensible and practical URL methodology.
- Using the assortment of accessible transformation and ETL tools to play out the integration.

To readily go through a complete lifecycle of semantic information integration, organizations need an easy arrangement to use semantic integration devices.

4 Integration of Big Data Using Semantic Web Technologies

4.1 Semantic Web for Improving ETL Process

Many research papers depended on SW and metadata to handle data integration and representation issues either in the Extract-Transform-Load (ETL) process, which is deemed one of the common and prevalent data integration solutions, or by suggesting a new method or tool. Bergamaschi et al. [16] Proposed a tool that improved ETL definitions by enabling semi-automatic, inter-attribute, semantic transforming via recognizing schemas of data sources then grouping attribute values semantically. Jiang et al. [17] Integrated heterogeneous information from database sources semantically after mapping it to a domain ontology data warehouse. Their model presumed that the type of data sources is a relational database only.

Bansal and Kagemann [13] and Bansal [2] integrated and published data from structured data sources as linked open data through adding semantic data model and semantic data instance to the transformation layer using Web Ontology Language (OWL), Resource Description Framework (RDF), and Simple Protocol and RDF Query Language (SPARQL) technologies. concerning Semantic ETL frame for Big

Data Integration [2] utilizes semantic technologies in the Transform phase of an ETL process to build a semantic data model and generate semantically linked data to be stored in a data mart or warehouse. The transformation stage will proceed to carry out other operations such as data normalization and cleanup. Extracting and loading phases of the ETL process stays the same. As limitation: The transformation stage involves a manual method of evaluating the datasets, the schema and their purpose. The schema will need to be mapped to an existing domain-specific ontology based on the findings, or an ontology will need to be created from scratch, which is a hard method that takes a lot of time, especially in real-time data, where the project is not implemented.

4.2 Integration of Big Data Using Semantic Web Technologies

Integration of data includes two main tasks. The first, at the level of the schema, includes homogenizing contrasts in the diagrams and classification used to communicate with information. The second, data-level integration, involves the recognition of documents in different data collections that allude to a comparable certifiable substance.

Li Kang, Li Yi, LIU Dong (2014) [1] analyzes the problem in the big data integration process interns of structural heterogeneity and semantic heterogeneity. To solve some research problems in integration process, they proposed a big semantic data model based on ontology. To solve the problem of data unintelligible they build ontology between the semantic models. As well as they constructed a Key/Value storage model to solve the problem of heterogeneous data storage.

A new system called Karma has been proposed by [15] in a case study, that trying to solve the variety challenge by using the semantic technology to integrate several types of Big Data sources. Starting from the importation of different types of information sources, passing by the cleaning process, to the modelling and its problems, and concluding with the integration process. Detailed information of different steps has been given. Several important problems not yet solved in this system first linking the data at the record level. Second, it did not address the issues of volume and velocity, which are the other key dimensions of big data.

Ostrowski et al. In their paper published at IEEE ICSC 2016 [18], introduced a framework for recognizing the assimilation of "Big Data" from heterogeneous sources, overcoming the shortening of the scalability of traditional ETL (Extraction-Transformation-Loading) activities. They also submitted their technical report using a case study of risk detection in the automotive sector's supply chain management. In addition, they posed several challenges in their application area to achieve the complete potential of "Big Data" applications requesting for required support for Parallelization, Real-Time Streaming, need for more domain specific Ontologies and data incompatibility.

Table 1 Comparative table of existiong methods

Methods	Data preparation	Data integration	Data processing	Data analysis & modeling	Data interpretation	Data	Metadata	Process representation
[20]	No	Yes	No	No	No	No	Yes	Yes
[21]	Yes	Yes	Yes	Yes	Yes	No	No	No
[22]	No	Yes	No	No	No	Yes	No	No
[23]	No	Yes	No	No	No	Yes	No	No

Boury-Brisset [19] designed a global architecture for intelligence data integration. Its main elements are ingestion process, ontology support, semantic enrichment, and interactions with other reasoning modules with transforming various data, which are acquired from different sources, into useful, valuable, quickly actionable intelligence. However, this work is still immature, and the experimentations showed that the incremental development and testing stages performance are still needed to be improved.

Accordingly, it is extremely difficult to extract value from large data volumes if there is no standardization or a unified language for interpreting data into information and converting information into knowledge. here the need to a unified representation by ontologies as a crucial component of SW, to give us the semantic world view, and provide possible solutions to mine the information hidden in unseen data to share a common understanding of information structures across software applications which facilitates the maximum interoperability and eliminates the need to provide code of diverse knowledge representations. Table 1, list some recent works mapped to the most pertinent stages of the Big Data pipeline. In order to clarify the role of the semantic web in the management of big data assets, and to justify the need of an intelligent system based on semantic representation, in order for us to evolve current computational systems into aware intelligent systems.

Semantic data integration can offer a solution for the heterogeneity issue and guarantee interoperability between applications exploiting the big data environment. Also, it can be used to improve big data management and analyses for decision-makers. For all these reasons the semanticization of data appears as a promising solution to overcome many challenges in data integration of enormous data both are linked to certain real problems. Furthermore, it can be seen that the main focus is knowledge of applications of both techniques, whether for greater advantage or to reveal more learning, as well as by giving significance to enormous, rapid, unstructured and uncertain information that surrounds us. Moreover, the integration of Big Data using semantic web techniques to address the variety issues is in the early stages and still a more study challenge that need future work, particularly in the healthcare, company and financial domain, creating a chance to make a higher contribution to science. We highlighted some of the challenges faced during the process of big data integration. And discuss the role of the semantic web to address those issues using semantic data integration. As a future work of this paper we will be proposing a new model to integrate the disparate big data using semantic web technologies.

5 Conclusion and Future Work

Major businesses corporations have shown interest in integrating semantic web technology with big data for added value over the past few years. Both Big Data and Semantic Web Technologies can be seen as two special research technologies, both are linked to certain real problems. Furthermore, it can be seen that the main focus is knowledge of applications of both techniques, whether for greater advantage or to reveal more learning, as well as by giving significance to enormous, rapid, unstructured and uncertain information that surrounds us. Moreover, the integration of Big Data using semantic web techniques to address the variety issues is in the early stages and still a more study challenge that need future work, particularly in the healthcare, company and financial domain, creating a chance to make a higher contribution to science. This paper highlights some of the challenges faced during the process of big data integration. And discuss the role of the semantic web to address those issues using semantic data integration. Future work of this paper will be proposing a new model to integrate the disparate big data using semantic web technologies.

References

1. Kang, L., Yi, L., Dong, L.: Research on construction methods of big data semantic model, 6 (2014)
2. Bansal, S.K.: Towards a semantic extract-transform-load (ETL) framework for big data integration. In: 2014 IEEE International Congress on Big Data, Anchorage, AK, USA, pp. 522–529. IEEE (2014)
3. Bertino, E.: Big data – opportunities and challenges panel position paper. In: 2013 IEEE 37th Annual Computer Software and Applications Conference, Kyoto, Japan, pp. 479480. IEEE (2013)
4. Bizer, C., Boncz, P., Brodie, M.L., Erling, O.: The meaningful use of big data: four perspectives – four challenges. SIGMOD Rec. **40**, 56 (2012). https://doi.org/10.1145/2094114.2094129
5. Data Integration Tools for Overcoming Integration Challenges in 2017 - DZone Integration. https://dzone.com/articles/data-integration-tools-for-overcoming-integration
6. Siva Rama Rao, A.V.S., Dhana Lakshmi, R.: A survey on challenges in integrating big data. In: Deiva Sundari, P., Dash, S.S., Das, S., et Panigrahi, B.K. (éds.) Proceedings of 2nd International Conference on Intelligent Computing and Applications, pp. 571–581. Springer, Singapore (2017)
7. Merelli, I., Pérez-Sánchez, H., Gesing, S., D'Agostino, D.: Managing, analyzing, and integrating big data in medical bioinformatics: open problems and future perspectives. Biomed. Res. Int. **2014**, 1–13 (2014). https://doi.org/10.1155/2014/134023
8. Kadadi, A., Agrawal, R., Nyamful, C., Atiq, R.: Challenges of data integration and interoperability in big data. In: 2014 IEEE International Conference on Big Data (Big Data), Washington, DC, USA, pp. 38–40. IEEE (2014)

9. Bansal, S.K., Kagemann, S.: Semantic extract-transform-load framework for big data integration. Computer **48**, 42–50 (2015)
10. Kumar, S., Singh, V., Saini, B.: A survey on ontology matching techniques. In: 2014 International Conference on Computer and Communication Technology (ICCCT), Allahabad, India, pp. 13–15. IEEE (2014)
11. Cuadra, A., Cutanda, M.M., Fuentes-Lorenzo, D., Sanchez, L.: A semantic web-based integration framework. In: 2011 7th International Conference on Next Generation Web Services Practices, Salamanca, Spain, pp. 93–98. IEEE (2011)
12. Knoblock, C.A., Szekely, P.: Exploiting semantics for big data integration. AIMag. **36**, 25 (2015). https://doi.org/10.1609/aimag.v36i1.2565
13. Bergamaschi, S., Guerra, F., Orsini, M., Sartori, C., Vincini, M.: A semantic approach to ETL technologies. Data Knowl. Eng. **70**, 717–731 (2011). https://doi.org/10.1016/j.datak.2011.03.003
14. Jiang, L., Cai, H., Xu, B.: A domain ontology approach in the ETL process of data warehousing. In: 2010 IEEE 7th International Conference on E-Business Engineering, Shanghai, China, pp. 30–35. IEEE (2010)
15. Ostrowski, D., Rychtyckyj, N., MacNeille, P., Kim, M.: Integration of big data using semantic web technologies. In: 2016 IEEE Tenth International Conference on Semantic Computing (ICSC), Laguna Hills, CA, USA, pp. 382-385. IEEE (2016)
16. Boury-Brisset, A.-C.: Managing semantic big data for intelligence, pp. 41–47 (2013)
17. Soylu, A., Giese, M., Jimenez-Ruiz, E., Kharlamov, E., Zheleznyakov, D., Horrocks, I.: OptiqueVQS: towards an ontology-based visual query system for big data. In: Proceedings of the Fifth International Conference on Management of Emergent Digital EcoSystems - MEDES 2013, Luxembourg, Luxembourg, pp. 119–126. ACM Press (2013)
18. Ardagna, C.A., Bellandi, V., Bezzi, M., Ceravolo, P., Damiani, E., Hebert, C.: Model-based big data analytics-as-a-service: take big data to the next level. IEEE Trans. Serv. Comput. 1 (2018). https://doi.org/10.1109/TSC.2018.2816941
19. Duggan, J., Kepner, J., Elmore, A.J., Madden, S.: The BigDAWG polystore system. SIGMOD Rec. **44**, 6 (2015)
20. Bondiombouy, C., Kolev, B., Levchenko, O., Valduriez, P.: Multistore big data integration with CloudMdsQL. In: Hameurlain, A., Küng, J., Wagner, R., Chen, Q. (éds.) Transactions on Large-Scale Data- and Knowledge-Centered Systems XXVIII, pp. 48–74. Springer, Heidelberg (2016)
21. Daoui, A., Gherabi, N., Marzouk, A.: A new approach for measuring semantic similarity of ontology concepts using dynamic programming. J. Theoret. Appl. Inf. Technol. **95**(17), 4132–4139 (2017)
22. Daoui, A., Gherabi, N., Marzouk, A.: An enhanced method to compute the similarity between concepts of the ontology. In: Noreddine, G., Kacprzyk, J. (eds.) International Conference on Information Technology and Communication Systems, Advances in Intelligent Systems and Computing, vol. 640, pp. 95–107. Springer, Cham (2018)
23. Bondiombouy, C., Kolev, B., Levchenko, O., Valduriez, P.: Multistore big data integration with CloudMdsQL. In: Hameurlain, A., Küng, J., Wagner, R., Chen, Q. (eds.) Transactions on Large-Scale Data- and Knowledge- Centered Systems XXVIII, vol. 9940, p. 4874. Springer, Heidelberg (2016)

Fake News Detection Approach Using Parallel Predictive Models and Spark to Avoid Misinformation Related to Covid-19 Epidemic

Youness Madani, Mohammed Erritali, and Belaid Bouikhalene

Abstract Social media and the Internet have suffered from false messages and false news since the outbreak of the COVID-19 pandemic. The intention is often to mislead readers and/or make them believe something that is not real. All that increases the need for automatic methods that can detect fake news in social media. In this paper, we proposed a classification model based on machine learning and deep learning algorithms to classify COVID-19 tweets into two classes using Apache Spark and the Python API Tweepy, the proposed idea uses the features of tweets to detect fake news. Experimental results show that the random forest algorithm gives best results with an accuracy equal to 79% and that the sentiment of tweets plays an important role in the detection of fake news. By applying the proposed model on our COVID-19 dataset, 67% of tweets are classified as REAL while 37% are classified FAKE.

Keywords Fake news · Coronavirus · COVID'19 · SARS-CoV-2 · Artificial intelligence · Machine learning · Deep learning · Spark

1 Introduction

The coronavirus (COVID-19) first appeared originated in Wuhan China in December 2019 and has rapidly spread across the continents. Until June 15, 2020, it has infected more than 6711122 and killed 395053 people worldwide. In Morocco, a total of 8132 confirmed cases, 208 deaths, and 7278 recovered cases have been reported [14]. Lastly, the invasion of this pandemic to the world is the source of a flood of fake news relayed by news sites and on social networks.

Y. Madani (✉) · M. Erritali · B. Bouikhalene
Sultan Moulay Slimane University, Beni Mellal, Morocco
e-mail: younesmadani9@gmail.com

M. Erritali
e-mail: m.erritali@usms.ma

B. Bouikhalene
e-mail: b.bouikhalene@usms.ma

N. Gherabi and J. Kacprzyk (eds.), *Intelligent Systems in Big Data, Semantic Web and Machine Learning*, Advances in Intelligent Systems and Computing 1344,
https://doi.org/10.1007/978-3-030-72588-4_13

Since March 2, 2020, date of the announcement of the first case of the coronavirus epidemic in Morocco [14], the amount of information about covid-19 circulated at great speed and came in different forms as well in social networks or in some media. This unsubstantiated information involves different sectors of Morocco. Health (detection of new cases infected with the virus, treatment of the epidemic, vaccines, etc.), training and education (school closings, white years, etc.), economic (closure of factories and businesses, cancellation of events, tourism ..) and social (confinement, remonstrance ..). Social networks allow you to publish this kind of information and share it through different devices without verifying the authentication of this information. Spreading this fake news can have serious consequences on the stability and security of the country.

To address this situation, the authorities called on citizens on March 4, 2020, to be vigilant regarding the broadcast and dissemination of false and fictitious information concerning the new coronavirus by means of new technology techniques, claiming to be attributed to official parties. The public prosecutors office said in a press release that it had given firm instructions to courts across the country to take legal action against anyone spreading fake news about the coronavirus and that all legal measures will be taken to identify those involved in publishing these allegations and lies.

We give in Table 1 the list of fake news about Covid-19 in Morocco on social networks since the beginning of this pandemic in early March 2020.

COVID-19 fake news on Moroccan social networks have been raging in the country since the beginning of the epidemic. These false news stories are spreading in a fast pace and can confuse people because of their presence on social networks. They influence public opinion and can even damage the image of the country. Thus, it is necessary to protect ourselves from misinformation and fake news.

In this work, we propose a technique based on artificial intelligence methods for a crucial classification of the fake news on twitter using the spark framework. Our idea is to classify tweets related to covid-19 as real or fake based on classification and natural language processing techniques. The growth of this type of information requires advanced processing and automatic learning based on machine learning and deep learning algorithms.

This paper is organized as follows, in the first section after this introduction we give related works. The second section deals with the proposed model and algorithms. The last two sections are devoted to results and discussion and a conclusion.

2 Related Works

Due to its easy access, social networks have become increasingly an important aspect of our lives where anyone can express their idea and publish news without hindrance or control over the authentication of what he/she published. As a result, it allows the propagation of "Fake News, news with intentionally false information. Due to the number of users in the social media platforms and the numbers of followers of what published, Fake news on social media can have significant negative societal effects.

Table 1 Covid-19 news in Morocco on social networks

News	Type
Wearing a mask is a good way to protect yourself	True, according to the World Health Organization WHO
It is possible to catch the coronavirus by receiving a package from China	False, outdoor survival of coronaviruses is 3 h or less on dry surfaces
Coronavirus is more dangerous than 2002 SARS	False, it is more contagious but less dangerous
Airport temperature controls are ineffective	True, infected patients may not show symptoms
Criminal networks usurp the identity of doctors and nurses to commit criminal acts in the homes of citizens on the pretext that they are delegated by the medical services specializing in the fight against the new coronavirus	False, according to national security
Coronavirus can travel up to 8 m away when sneezing or coughing	False, according to the World Health Organization WHO
The new coronavirus can be transmitted by mosquito bites	False, according to the World Health Organization WHO
Pneumonia vaccines protect against new coronavirus	False, according to the World Health Organization WHO
The Moroccan Association of Health Officers Operating at the Crossing Points affirmed to be mobilized for the implementation of the national plan to fight against the new coronavirus	True
A German oncologist discovered a drug against the Coronavirus	False
Morocco to move to stage 3 by Sunday or Monday	False
Coronavirus: probable drop in Morocco of 39% of tourists	True
Planes prepare to pour disinfectant on the city of Casablanca	False
Only people with symptoms are contagious	False
The Interior Ministry announces the ban, until further notice, of all public gatherings in which more than 50 people take part as well as the cancellation of all sporting, cultural and artistic events, and meetings	True
In addition to China, Italy, Spain, Algeria and France, Morocco decides to suspend, until further notice, flights to and from Germany, the Netherlands, from Belgium and Portugal	True
The masks can be sterilized and reused	False
Export license is now imposed for medical masks and disinfectant gels	True
The "Wiqaytna" application accesses the user contact database	False
The Covid-19 exposure notification mobile app will be automatically installed or activated on all phones	False
The mortality rate in Morocco is 2.6%, thus remaining stable and significantly lower than the world average (less than half)	True
The Ministry of Health has strengthened the existing contact tracking system by launching a mobile application for reporting exposure to Covid-19, dubbed "Wiqaytna"	True
The National Health Insurance Agency (ANAM) has set up and activated the electronic order office service	True

Consequently, detecting fake news on social media has attracted a lot of attention in the scientific research field.

M. Aldwairi and A. Alwahedi [1] published an article that is concerned with identifying a solution that could be used to detect and filter out sites containing fake news to help users to avoid being lured by click baits. The proposed idea consists of identifying and removing fake sites from the results provided to a user by a search engine or social media news feed. Authors use some news features to detect false information such as keywords, number of exclamation marks, number of question marks...etc.

Due to the exponential growth of information online, it is becoming impossible to distinguish which is true and which is false. Kelly Stahl in his article presented in [2] proposed a method which is a combination of Naïve Bayes classifier, Support Vector Machines, and semantic analysis.

Authors of [3] proposed a two-step method for identifying fake news on social media. The first step consists of pre-processing methods applied to the data set to convert unstructured data sets into a structured data set, such as TF weighting method and Document-Term Matrix. And the second step consists of implementing twenty-three supervised artificial intelligence algorithms on the dataset prepared in the first step. For evaluating their works, the authors use used four metrics (accuracy, precision, recall, and F1-measure). The experimental results show showed that the J48 algorithm gives the best accuracy with 65.5% for the BuzzFeed Political News Data set.

In [4], researchers tried to improve the detection of Fake-News by including online data mining. They used deep learning models based on FNN and LSTM in combination with different word vector representations. The models were combined with a live data mining section which is used to collect auxiliary information from the content/title of the news article (domain names, author details, etc.). The proposed model gives good results on the precision, recall, accuracy, and F1-measure when LSTM, in combination with word2vec representation was used.

N.X. Nyow and H.N. Chua [5] derived and transformed social media Twitter's data to identify additional significant attributes that influence the accuracy of machine learning methods to classify if the news is real or fake using the data mining approach. Authors present some new Twitter's attributes that can improve the process of classifying tweets as fake or real. They add to the FakeNewsNet dataset [6] new attributes such as t_retweet(number of retweets), t_fav(number of favorites). For the classification, researchers used the random forest and decision tree algorithms.

In the work of BalaAnand et al. [7], researchers propose an enhanced graph-based semisupervised learning algorithm (EGSLA) to detect fake users from a large volume of Twitter data. For the construction of the dataset, they used Scrapy for collecting tweets, and after they extract some useful features such as retweet, length, the fraction of URLs, and Average time between tweets.

In [8], the authors published an article that describes a deep learning model that predicts the nature of an article when given as an input. Researchers have discussed and experimented using word embedding (GloVe) for text pre-processing to construct a vector space of words and establish a lingual relationship. And the dataset of work is a Kaggle fake news dataset. For the classification, the authors proposed a new model based on the blend of the convolutional neural network and recurrent neural network architecture.

The research of [9] analyzes the mechanisms for publishing and distributing fake news according to the classification, structure, and algorithm of the construction. Authors in this article present how we can use artificial intelligence and machine learning algorithms to classify news as fake or real with high accuracy.

Atodiresei et al. [10] present a system for identifying fake users and fake news in the Twitter social network. The presented system began with crawling data from

Twitter to construct a dataset. And for the detection of fake news or fake user, authors in this works based on several modules such as hashtag sentiment, emoji sentiment, text sentiment, and named entity recognition.

Fake news has an important role in spreading misinformation and false news. The work of [11] provides a novel text analytics-driven approach to fake news detection for reducing the risks posed by fake news consumption. The proposed framework utilizes a double-layered approach for classification. The first layer performs fake topic detection and the second layer performs fake event detection.

The paper of Hosnia et al. [12] investigates the problem of minimizing the influence of malicious rumors that emerge during breaking news, which are characterized by the dissemination of a large number of malicious information over a short period. The authors design a multi-rumor propagation model named the HISBMmodel that captures the propagation process of multi-rumors in online social networks (OSNs).

The spread of fake news (false information) in the domain of health affect negatively the lives of people. Pulido et al. [13] proposed a new application to analyze social media (Reddit, Facebook, and Twitter content. The results indicate that messages focused on fake health information are mostly aggressive, those based on evidence of social impact are respectful and transformative, and finally, deliberation contexts promoted in social media overcome false information about health. For the construction of the dataset of work, researchers selected the word health as a general topic and the specific keywords vaccines, nutrition and Ebola.

3 Research Methodology

As presented earlier, our work consists of developing a new model for classifying new tweets related to Covid-19 as real or fake. To classify a tweet (that is to analyze its content) a lot of researchers use its text content by using the natural language processing (NLP) techniques and the machine learning algorithms for classifying the tweets text. But, because the tweets text is expressed using a natural language it is hard to detect and extract what the author of the tweet means, which is due to the vagueness and the imprecision of the written text, which implies low accuracy.

In this article to avoid these problems, we propose to work not only with the tweets text, but we add new useful tweets features such as favorite count, retweet count, source, length, verified, etc. The idea is that to decide if a tweet is fake or not, we will be based on 11 features instead of one (tweets text). For the classification, we used machine learning and deep learning algorithms.

3.1 Construction of Our Datasets

For the dataset of training and that of the test, we have based on the work of Sheryl Mathias and Namrata Jagadeesh [15], in which they collected tweets for our dataset based on events such as the Las Vegas shooting which took place on October 1, 2017, and Hurricane Harvey that occurred in Houston, Texas from August 17, 2017, to September 3, 2017 using Twitter's streaming API. For creating the dataset, authors used search parameters such as 'Las Vegas shooting', 'Mandalay Bay', 'Stephen Paddock', and 'Hurricane Harvey' related to these events to extract the tweets. Authors treat the tweets manually to classify them as real or fake news by referring to websites such as YouTube and Snopes.com as well as the Internet to get facts. There were two annotators for labeling the tweets. Both of them labeled the tweets separately and then cross-verified their labels and kept only those tweets for which their labeling matched. The final dataset has a total of 250 tweets (94 fake and 156 real tweets).

Each tweet in this dataset has many features (variables) such as:

- **text:** contains the text content of the tweet
- **source:** the source from where the tweet is published
- **favorite_count:** how often the tweet is liked
- **retweet_count:** how often the tweet is retweeted
- **length:** this variable contains the length of the tweet (number of characters)
- **user_name:** name the tweets author
- **user_friend_count:** number of friends of the tweets author
- **followers_count:** number of followers of the tweets author
- **user_statuses_count:** number of statuses that the tweets author published in his account
- **user_verified:** it is a boolean variable that demonstrates if the account of tweets author is verified or not.

All these variables can be collected easily from twitter using a Twitter API.

In addition to these variables and to improve the structure of the database, we added 4 new variables which are:

- **user_has_url:** it is a boolean variable that verifies if the tweet contains an URL or not.
- **sentiment:** this variable contains 3 possible values: 1, 0 or −1. the value 1 is equivalent to positive, 0 is equivalent to neutral and −1 is equivalent to negative. This variable consists in classifying the tweet into three classes (positive, negative, or neutral). For extracting the sentiment from the tweet, we used the TextBlob[1] python library, which goes through the tweet word by word to extract the degree of polarity that will give at the end the final sentiment.
- **friends/followers_count:** this variable is the ratio between the number of friends of the tweets author and the number of followers of the tweets author.
- **statuses/followers_count:** this variable is the ratio between the number of statuses that the tweets author published in his account and the number of followers of the tweets author

[1] TextBlob is a Python (2 and 3) library for processing textual data. It provides a simple API for diving into common natural language processing (NLP) tasks such as part-of-speech tagging, noun phrase extraction, sentiment analysis, classification, translation, and more.

Fig. 1 Steps to construct our unlabeled covid-19 dataset

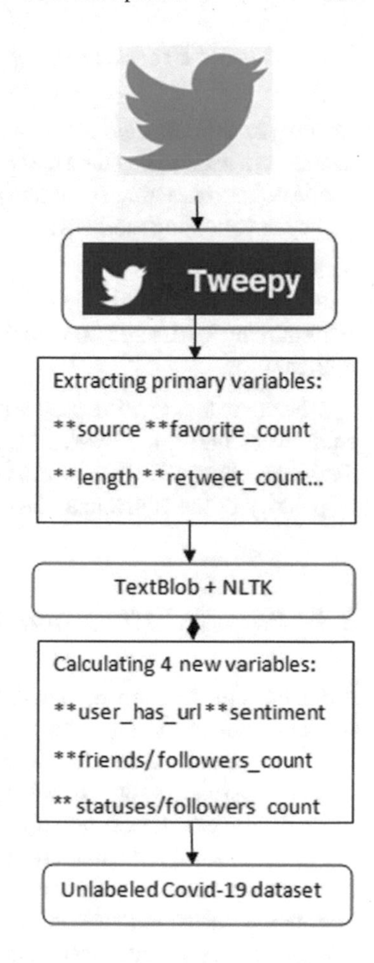

The dataset contains other variables that we will not use them in the process of classification such as: user_id, user_screen_name, user_created_at, user_description.

After all these steps our labeled dataset of training and test is ready for the phase of development of our algorithm of classification.

The second step in the construction of the dataset is the collection of Covid-19 tweets to classify into two classes fake or real. For extracting tweets, we used a twitter API called tweepy. Tweepy gives us the possibility to extract tweets related to a specific subject, for example extracting tweets related to covid-19 and coronavirus. After the collection of tweets, we use Tweepy to retrieve for each tweet the different variables described earlier (length, source, favorite_count, etc.) and we calculate the others (user_has_url, sentiment, friends/followers_count, statuses/followers_count). All that gives us our unlabeled covid-19 dataset.

Figure 1 shows the different steps to construct our unlabeled covid-19 dataset.

3.2 Tweets Preprocessing

To analyze the collected tweets either for the training/test dataset or for the covid-19 dataset, we need to do some text preprocessing methods to prepare the tweets for the analysis (for extracting sentiment). In this step, we used the python NLTK library to make the following methods:

- Tokenization
- Removing stop words
- Removing URLs and punctuation
- Stemming

The next step after the pre-treatment of tweets is the phase of analysis to classify each tweet into three classes (1: positive, 0: neutral, and −1: negative) using the TextBlob library. TextBlob works on a text and retrieves in the output either a degree of polarity or the equivalent class (−1, 0, or 1).

3.3 Proposed Model and Algorithms

After the construction of our data sets, the next step consists in developing our model to detect the fake news in the covid-19 data set. In our work, we used the Spark framework to develop the classification algorithms in a parallel way.

For constructing our model, we begin with the upload of the dataset of training/test(labeled dataset) into a spark job to construct a spark data frame, after that, we delete the columns that are unnecessary for the classification such as: user_id, user_screen_name, user_created_at, user_description. The next step is to designate the feature columns which are: retweet_count, favorite_count, source, length, user_followers_count, user_statuses_count, user_verified, user_has_url, friend/followers_count, statuses/followers_count, sentiment, and the label column which is: Final Label (Fake or Real). And Finally, it is the sampling step in which we divide our dataset into a training data set and test dataset. In this work, we used random sampling and we give to the training set 75% and 25% for the test set. Figure 2 presents all these steps.

The next step is the use of a classification algorithm to construct a model that will help us to classify a new covid-19 tweet as fake or real. In this paper, we developed 5 machine learning algorithms and one deep learning algorithm in a parallel way using Apache Spark-based on spark data-frame(Spark SQL). The next subsections present an overview of spark and the algorithms used for the classification.

3.3.1 Apache Spark

Apache Spark is a fast data processing engine dedicated to Big Data. It allows the processing of large volumes of data in a distributed manner (cluster computing).

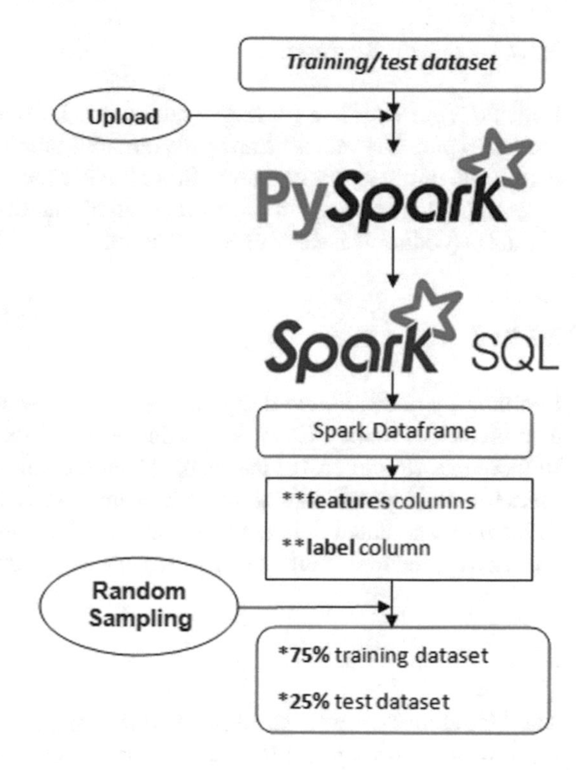

Fig. 2 Construction of training and test dataset

In recent years, it has gained a lot of popularity, this Framework is about to replace Hadoop. Its main advantages are its speed, ease of use, and versatility. Its main selling point is its speed as it makes it possible to launch programs 100 times faster than Hadoop MapReduce in memory, and 10 times faster on disk. We have based in this paper on PySpark which is a Python API for Spark.

The Spark ecosystem thus includes several tools today, in our work we have used two tools:

- Spark SQL: a SQL-like implementation for querying data. Spark SQL allows you to execute queries in SQL languages to load and transform data. The SQL language comes from relational databases, but in Spark, it can be used to process any data, regardless of its original format.
- Spark MLlib: for machine learning. It is a library of learning methods optimized for Spark. It is a machine learning library, appeared in version 1.2 of Spark, which contains all the classic learning algorithms and utilities, such as classification, regression, clustering, collaborative filtering, and reduction of dimensions.

3.3.2 Logistic Regression

Logistic regression is a predictive technique. It aims to build a model allowing to predict/explain the values taken by a qualitative target variable (most often binary, we then speak of binary logistic regression; if it has more than 2 modalities, we speak of polyatomic logistic regression) from a set of quantitative or qualitative explanatory variables (coding is necessary in this case).

3.3.3 Decision Tree

Learning by decision tree designates a method based on the use of a decision tree as a predictive model. It is used particularly in data mining and machine learning. In these tree structures, the leaves represent the values of the target variable and the branches correspond to combinations of input variables which lead to these values. In learning and data mining, a decision tree describes the data but not the decisions themselves, the tree would be used as a starting point for the decision process.

3.3.4 Random Forest

The "Random Forest" algorithm (or Random Forest sometimes also translated as a forest of decision trees) is a classification algorithm that reduces the variance of forecasts from a decision tree alone, thereby improving their performance. For this, it combines many decision trees in a bagging type approach. In its most classic formula, it performs parallel learning on multiple decision trees randomly constructed and trained on different subsets of data. The ideal number of trees, which can go up to several hundred or more, is an important parameter: it is very variable and depends on the problem.

3.3.5 Naive Bayes

Naive Bayesian classification is a type of simple probabilistic Bayesian classification based on Bayes' theorem with strong independence of hypotheses. It implements a naive Bayes classifier, belonging to the family of linear classifiers. Naive Bayes Classifier is a popular Machine Learning algorithm. It is a Supervised Learning algorithm used for classification. It is particularly useful for text classification issues.

3.3.6 Gradient Boosting

Gradient boosting is a machine learning technique for regression and classification problems, which produces a prediction model in the form of a set of weak prediction models, usually decision trees. It builds the model in stages as do the other boosting methods and generalizes them by allowing the optimization of an arbitrarily differentiable loss function.

3.3.7 SVM

Support vector machines or large margin separators (are a set of supervised learning techniques intended to solve problems of discrimination and regression. SVMs are a generalization of linear classifiers. SVM is a solution to classification problems. The SVM belongs to the category of linear classifiers (which use a linear separation of the data), and which has its own method to find the border between the categories.

3.3.8 Multilayer Perceptron MLP

The multilayer perceptron MLP is a type of artificial neural network organized in several layers within which information circulates from the input layer to the output layer only; it is therefore a direct propagation network (feedforward). Each layer is made up of a variable number of neurons, the neurons of the last layer (called "output") being the outputs of the global system. MLP is a supervised learning algorithm that learns a function f by training on a dataset. It is used for Tabular Datasets (contain data in a columnar format as in a database table) and for the Classification/Regression/prediction problems. For the development of this deep learning algorithm we used the TensorFlow[2] open-source tool.

By applying one of these algorithms on the training dataset (by specifying each time the feature columns, the label column and the necessary parameters for the launching of the algorithm) we train the algorithm which leads us to build a classification model that will be used after for classifying a new covid-19 tweet.

Figure 3 shows the different steps followed to classify new tweets related to Covid-19 into two classes: Real or Fake.

4 Results and Discussion

In this section, we will present the experiments done in our work to find the best model that will help us to easily classify new tweets related to Covid-19 into two classes (Fake or Real). For the construction of a model, we based on the training dataset described earlier that represents 75% of our training/test dataset. All our experiments are done in a parallel way using the Apache spark (pyspark, spark SQL, and spark MLIB). The idea is that we will calculate for each algorithm 4 metrics (accuracy, precision, recall, and F1-measure), and the goal is to find the best model that will be used after in the classification of our Covid-19 tweets.

[2]https://www.tensorflow.org/.

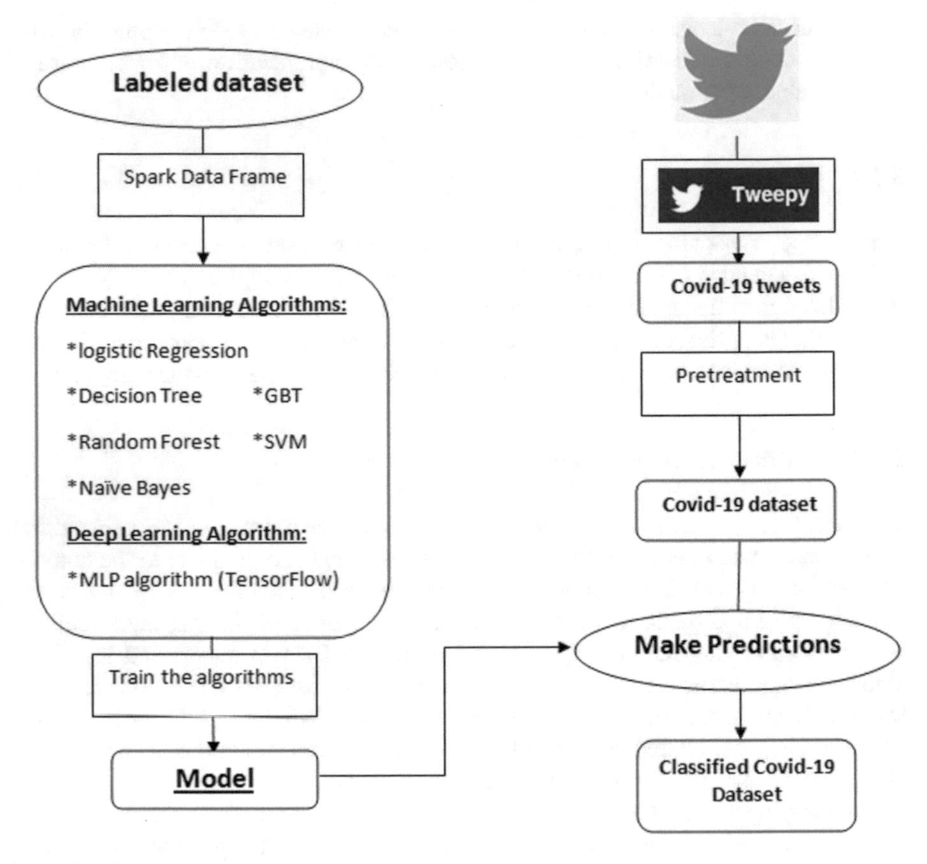

Fig. 3 Classification steps

Table 2 Confusion matrix

Confusion matrix	Predicted REAL	
Actual **REAL**	True positive	False negative
Actual **FAKE**	False positive	True negative

As presented earlier, we used in this paper 7 machine learning algorithms(Logistic Regression:**LR**, Decision Tree:**DT**, Random Forest:**RF**, Naive Bayes:**NB**, Gradient-Boosted Tree:**GBT**, Support Vector Machines:**SVM** and the neural network Multi Linear Perceptron:**MLP**), and based on the test dataset that represents 25% of our training/test dataset, we going to test the performance of these 6 models based on the confusion matrix (True positives, True negatives, False positives, False Negatives) presented in Table 2, and 4 metrics. Our final goal is to choose the best model that gives the best results which will be used after to detect fake tweets in real-time based on Tweepy and Spark.

Where:

- **TP:** True Positive: Predicted tweets correctly predicted as actual REAL
- **FP:** Predicted tweets incorrectly predicted an actual REAL. i.e., FAKE values predicted as REAL
- **FN:** False Negative: REAL Tweets predicted as FAKE
- **TN:** True Negative: Predicted tweets correctly predicted as an actual FAKE

Based on the confusion matrix we calculate for each algorithm 4 metrics:

Accuracy: the portion of all true predicted tweets against all predicted tweets.

$$Accuracy = \frac{TP + TN}{TP + TN + FP + FN} \quad (1)$$

Recall: also called sensitivity or true positive rate. It computes the ratio of REAL classes correctly detected. This metric gives how good the model is to recognize a REAL class.

$$Recall = \frac{TP}{TP + FN} \quad (2)$$

Precision: The precision metric shows the accuracy of the REAL class. It measures how likely the prediction of the REAL class is correct.

$$Precision = \frac{TP}{TP + FP} \quad (3)$$

F1-score: a harmonic average of precision and recall.

$$F1\text{-}score = \frac{2 * Precision * Recall}{Precision + Recall} \quad (4)$$

Figure 4 shows the results obtained for the 4 metrics by applying the 7 models of machine learning on the test dataset.

According to the figure, we remark that random forest algorithm outperforms all other models for the 4 metrics. RF model achieved an accuracy of 79% in comparison with SVM (72%). The degree of recall reached up to 100% using RF while that of DT for example is equal to 94%. RF have also 85% for the precision and 83% for the F1-score. From all these results, we choose the RF algorithm for classifying new covid-19 tweets.

After we find our best classifier, our next step is the collection of the covid-19 tweets for detecting fake tweets in real-time based on Tweepy and Spark. For that, we collected 2000 tweets using the keyword covid-19, and for each tweet collected after the pretreatment step and the collection of the necessary variables (favorite_count, user_friend_count, length, etc.), we pass it through the RF model to predict its class(FAKE or REAL). All the prediction steps are done using Spark SQL, Tweepy, and the RF model.

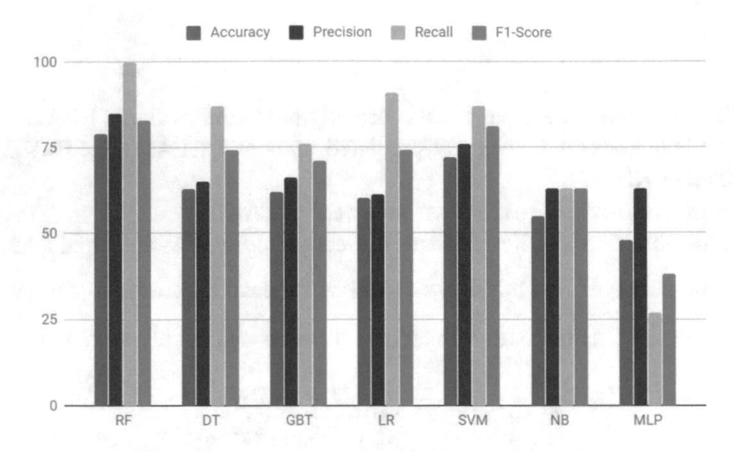

Fig. 4 Classification results

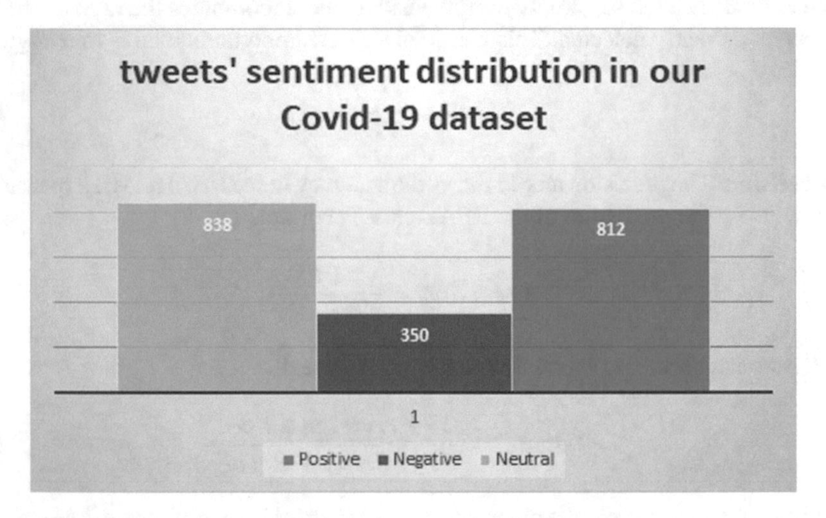

Fig. 5 Distribution of positive, negative, and neutral tweets in our covid-19 datasets

Figure 5 shows the distribution of positive, negative, and neutral tweets in our covid-19 datasets.

After we apply our random forest model on our unlabeled covid-19 dataset, each tweet is classified as FAKE or REAL based on the feature columns(length, favorite_count, retweet_count, user_friend_count, etc.). Figure 6 presents the distribution of REAL and FAKE covid-19 tweets after the prediction using RF model.

From this figure, we remark that from the 2000 COVID-19 tweets collected using Tweepy in real-time; 63% of them are REAL tweets and 37% are FAKE.

The last experience we made in this work, is to show the role of tweets sentiment to detect fake tweets. Figure 7 shows that for example for 812 positive tweets, 543 are REAL while 269 are FAKE.

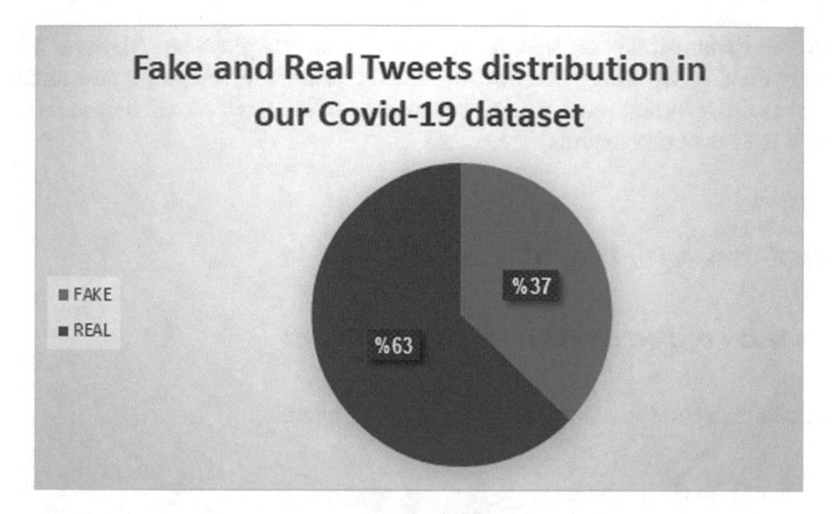

Fig. 6 Distribution of REAL and FAKE covid-19 tweets

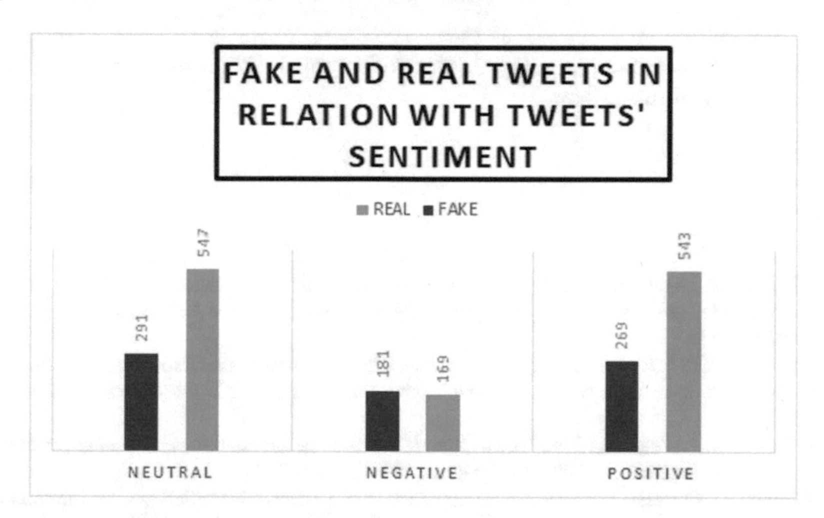

Fig. 7 Effect of the sentiments on classification

5 Conclusion

With the advent of social networks, everyone can publish news regardless of place and time. For instance, Twitter users can publish new content without hindrance, which helps in the spread of fake and misleading news. In this article, we have presented a model based on Spark, Tweepy, and machine learning algorithms to detect fake news in tweets related to the pandemic of coronavirus using the keyword covid-19. Experimental results show that the random forest model gives the best results, and by applying it on 2000 new covid-19 tweets, we detect 37% of fake news. We demonstrate also that the sentiment of tweets plays an important role in the classification of tweets. Our next work will consist in developing a new model to extract useful information from tweets related to covid-19 which will help authorities to fight the novel coronavirus.

Compliance with Ethical Standards

Disclosure of Potential Conflicts of Interest:

The authors declare that they have no conflict of interest.

Research Involving Human Participants and/or Animals:

This article does not contain any studies with human participants or animals performed by any of the authors.

References

1. Aldwairi, M., Alwahedi, A.: Detecting fake news in social media networks. In: The 9th International Conference on Emerging Ubiquitous Systems and Pervasive Networks (EUSPN 2018) (2018)
2. Stahl, K.: Fake news detection in social media, Department of Mathematics and Department of Computer Sciences, California State University Stanislaus, 1 University Circle, Turlock, CA 95382
3. Ozbay, F.A., Alatas, B.: Fake news detection within online social media using supervised artificial intelligence algorithms. Physica A **540**, 123174 (2020)
4. Deepak, S., Chitturi, B.: Deep neural approach to Fake-News identification. In: International Conference on Computational Intelligence and Data Science (ICCIDS 2019), Procedia Computer Science, vol. 167, pp. 2236–2243 (2020)
5. Nyow, N.X., Chua, H.N.: Detecting fake news with tweets' properties. In: 2019 IEEE Conference on the Application, Information and Network Security (AINS) (2019)
6. Shu, K., Mahudeswaran, D., Wang, S., Lee, D., Liu, H.: FakeNewsNet: A Data Repository with News Content, Social Context and Spatiotemporal Information for Studying Fake News on Social Media. arXiv:1809.01286v3 [cs.SI], 27 March 2019
7. BalaAnand, M., Karthikeyan, N., Karthik, S., et al.: An enhanced graph-based semi-supervised learning algorithm to detect fake users on Twitter. J. Supercomput. **75**, 60856105 (2019). https://doi.org/10.1007/s11227-019-02948-w
8. Agarwal, A., Mittal, M., Pathak, A., et al.: Fake news detection using a blend of neural networks: an application of deep learning. SN Comput. Sci. **1**, 143 (2020). https://doi.org/10.1007/s42979-020-00165-4

9. Zhuk, D., Tretiakov, A., Gordeichuk, A., Puchkovskaia, A.: Methods to identify fake news in social media using artificial intelligence technologies. In: Alexandrov, D., Boukhanovsky, A., Chugunov, A., Kabanov, Y., Koltsova, O. (eds.) Digital Transformation and Global Society. DTGS. Communications in Computer and Information Science, vol. 858. Springer, Cham (2018)
10. Atodiresei, C.-S., Tănăselea, A., Iftene, A.: Identifying fake news and fake users on Twitter. In: International Conference on Knowledge-Based and Intelligent Information and Engineering Systems, KES 2018, 3–5 September 2018, Belgrade, Serbia (2020)
11. Zhang, C., Gupta, A., Kauten, C., Deokar, A.V., Qin, X.: Detecting fake news for reducing misinformation risks using analytics approaches. Eur. J. Oper. Res. **279**(3), 1036–1052. ISSN 0377-2217 (2019). https://doi.org/10.1016/j.ejor.2019.06.022
12. Hosni, A.I.E., Li, K.: Minimizing the influence of rumors during breaking news events in online social networks. Knowl.-Based Syst. **193**, 105452. ISSN 0950-7051 (2020). https://doi.org/10.1016/j.knosys.2019.105452
13. Pulido, C.M., Ruiz-Eugenio, L., Redondo-Sama, G., Villarejo-Carballido, B.: A new application of social impact in social media for overcoming fake news in health. Int. J. Environ. Res. Public Health. **17**(7), 2430 (2020). https://doi.org/10.3390/ijerph17072430.
14. Wikipedia. https://en.wikipedia.org/wiki/COVID-19_pandemic_in_Morocco
15. Mathias, S., Jagadeesh, N.: Detecting Fake News Tweets from Twitter. College of Information Studies, University of Maryland, College Park, USA

Detection, Analysis and Classification of Skin Lesions: Challenges and Opportunities

My Abdelouahed Sabri, Youssef Filali, Soumia Fathi, and Abdellah Aarab

Abstract Health authorities consider skin cancer to be one of the deadliest cancers in the world. Computer-Aided Diagnosis (CAD) systems are a widely used solution for the detection and the classification of skin cancers. Such systems reduce significantly doctors' effort and time with a very high classification accuracy. Several challenges are encountered in setting up such systems. In the literature, two categories of approaches are proposed and which depend directly on the size of the skin lesion images Dataset. Thus, for small datasets, Machine Learning based approaches are the most commonly used, starting with the identification of the lesion region, as the analysed areas contain a lot of noise. A good lesion identification will allow the extraction of relevant features and at the end an excellent classification accuracy. For large datasets, Deep learning based approaches are the most widely used and the most efficient where several architectures have been proposed. Very promising hybrid ideas have recently been proposed combining the power of ML-based and DL-based approaches. This chapter presents the different challenges and opportunities encountered in each skin lesion classification implementations and steps.

1 Introduction

The reduction of the ozone layer that protects the human body from the radiation, the abusive body exposition to the sun and the use of the tanning are the reasons of the increasing rate of skin cancer incidence in the last decades. If diagnosed earlier, skin cancer can cure and many lives can be saved. So, it is very important to invest in the development of techniques that can be used in the early prevention of this cancer. Two main categories of skin lesion can be diagnosed; Melanoma and Non-Melanoma and the difficulty is how to distinguish between them. The medical community is investing money, time and effort in raising awareness of the risk of this type of cancer. Computer-Aided Diagnosis (CAD) systems can be used to help dermatologist in skin lesion detection and the classification.

M. A. Sabri (✉) · Y. Filali · S. Fathi · A. Aarab
LISAC Laboratory, Faculty of Sciences Dhar-Mahraz, University Sidi Mohamed Ben Abdellah, Fez, Morocco

N. Gherabi and J. Kacprzyk (eds.), *Intelligent Systems in Big Data, Semantic Web and Machine Learning*, Advances in Intelligent Systems and Computing 1344,
https://doi.org/10.1007/978-3-030-72588-4_14

Machine learning (ML) and deep learning (DL) approaches are used to propose efficient systems for skin lesion classification. The performance of each type is related to the size of the data set. So, ML based approaches are less powerful than DL based approaches when learning from very large datasets and DL based approaches are less powerful than ML when learning from very small datasets. Both categories of approaches based on ML and DL represent enormous challenges and opportunities, and especially those based on ML.

DL and ML based approaches use images of lesions acquired from either dermoscopic or macroscopic devices but does not the follow the same workflow. ML based approaches follow three main steps that are: ii) Segmentation; ii) features engineering; ii) classification. Each of these three stages represents a challenge and may present opportunities for scientific research. On the other hand, deep learning is based on a convolutional neural network (CNN) that includes (convolution, pooling, and fully connected layers). Each stage represents a challenge and may present opportunities for scientific research.

In skin lesion classification, Image segmentation aims to distinguish between the skin and the lesion. In this processing case, we begin by identifying for each image the lesion and the healthy skin that surrounds it using a segmentation algorithm. Authors can propose their personal algorithm or use one of the many algorithms that have been proposed for skin lesion images segmentation [45, 48]. Segmentation is not always easy in image processing and especially in the case of skin images since there is always noise, i.e. hairs, artifacts, ... This is why segmenting an image to identify lesion represents a great challenge in the thematic and especially since the quality of the segmentation directly influences the following steps in the classification process. Many research works have been proposed in this sense and we find that the best ones are those using a pre-processing allowing a good segmentation [16]. Pre-processing can be used to enhance image, remove noise, or to separate objects from texture.

Features engineering is the second step after segmentation. Skin lesion classification accuracy depends on the features relevance which depends also on the segmentation quality. In this part four steps are used: features extraction, features normalization, dimensionality reduction, and features selection [14]. Extracted features from skin lesion are based essentially on the ABCD (Asymmetry, Border, Colour, and Diameter) rules used by dermatologists. Authors can use already proposed features from literature or propose new ones. Shape, texture, colour are the most used features. The extracted features are usually in different ranges and, for good classification accuracy, the features must be normalized so that the values of each features have a mean of zero and a unit variance. Classification algorithms are in the majority multidimensional projection algorithms. Thus, too many characteristics imply high dimensionality and can lead to classification errors. Some extracted features can be correlated and can be combined to reduce the number of features and thus the number of dimensions for the classification algorithm. Dimensionality reduction therefore refers to techniques to reduce the number of characteristics. Several algorithms are proposed and the most widely used is PCA (Principal Components Analysis). Features selection can be seen as a dimensionality reduction technique. It consists in selection the best and relevant features that can efficiency represent the lesion. Many algorithms have been proposed in the literature and there is still work in progress to propose new selection approaches.

The last step in such a system is classification which allows us to classify the lesion as Melanoma or Non-Melanoma. In the literature, several classification algorithms are proposed and used, namely; Artificial Neural Network (ANN), Decision Tree (DT), Support Vector Machine (SVM) and ensemble methods [24, 34].

Each of the steps in a skin lesion classification system offers scientific research challenges and opportunities. So, in this chapter we will present in detail and for each step; pre-treatment, segmentation, feature engineering and classification the different problems encountered and the solutions proposed in the literature.

2 AI for Skin Lesion Classification

Artificial Intelligence is widely used to provide valuable assistance to Doctors in their diagnosis by relying on image processing and machine learning and/or deep learning algorithms. Several studies have focused on this combination to provide robust tools to assist Doctors in their diagnosis and decisions. Recently, there are even devices which combine image processing and artificial intelligence.

In recent years, hundreds (if not thousands) of scientific publications have emerged involving image processing and advanced machine learning in medicine [18]. These applications generally concern:

- Assist in the diagnosis of neurological diseases. For example, various studies have shown that AI is able to predict Alzheimer's disease years before it manifests itself [47],
- Measuring and predicting the risk of cardiovascular disease and identifying abnormalities in common medical tests such as chest X-rays [5, 41],
- Early diagnosis of cancer. A common example is skin cancer where image processing combined with AI algorithms can predict and classify skin lesions [15, 16],
- ...

Machine learning (ML) and deep learning (DL) approaches are used to propose efficient systems for skin lesion classification. ML based approaches are less powerful than DL based approaches when learning from very large datasets and DL based approaches are less powerful than ML when learning from very small datasets. ML uses a complex process comparing DL it consists in 5 major steps: pre-processing, segmentation, feature engineering and classification. The challenge in ML is to extract and select the relevant features from the image dataset.

Deep learning is based on a convolutional neural network (CNN) that includes (convolution, pooling, and fully connected layers). Authors can build their own architecture or implement an existing architecture. A comparative study of Deep learning architectures used in skin lesion classification has been presented in [26]. The architectures differ in (i) the number and type of connection in the convolutional layer, (ii) the pooling layer can be max, min or mean, (iii) same architectures use unpooling and deconvolution layers.

On the other hand, Machine learning based approaches follow three main steps that are: ii) Segmentation; ii) features engineering; ii) classification in addition to the prepossessing step to enhance the segmentation. In the segmentation. The aim of pre-treatment is to remove the artifact that the lesion contains and to improve the segmentation performance. Many solutions have been proposed in the literature such as filtering, colour space transformation, contrast enhancement artefact removal, and multi-scale decomposition [12–14]. In the skin lesion segmentation step, several algorithms are proposed in the literature. The thresholding algorithms, segmentation by region growing, the active contour-based algorithm and algorithms based on Machine learning (clustering) have been widely used [11, 43]. Also, depth learning with unsupervised learning has been used to segment skin lesions [6].

Features extraction, features normalization, dimensionality reduction and features selection are parts of features engineering. The main objective is to extract and to identify relevant features to be used in the classification.

Features extraction: is a fundamental step to obtain a good classification accuracy. Extract an appropriate feature is a big challenge that a lot of research has confronted [32]. Handcrafted features are essentially based on the rules called ABCD (Asymmetry, Border, Color and Diameter) used clinically by dermatolog to well describe the skin lesion [8, 42]. Many works proposed to used also textural chrematistics of the skin and the lesion [2, 25, 40]. Recently lesion skeleton has been proposed as a powerful and relevant feature [15]. Deep learning was also use as features extractor that uses a convolutional layer to describe well and learn from the images [28, 29].

A normalization step is mandatory to make all features value in the same scale and to improve the classification result. Many algorithms have been used the best one is the Zscore transformation is the best [36].

Too many features imply high dimensionality and can imply to classification errors. Some of the extracted features are correlated and, thus, can be combined to reduce the number of features and thus the number of dimensions for the classification algorithm. Dimensionality reduction therefore refers to techniques to reduce the number of characteristics. Several algorithms are proposed such as Principal Components Analysis (PCA), Linear Discriminant Analysis (LDA), Neighbourhood Component Analysis (NCA) and the most widely used is PCA. Features selection phase consist in selecting relevant features that can efficiency represent the lesion. Many algorithms have been proposed in the literature and there is still work in progress to propose new selection approaches. Relief, Correlation-based Feature Selection (CFS), Chi2, Recursive Feature Elimination (RFE), and genetic algorithm which belong to the evolutionary algorithms family [4, 14].

Classification step: is the last step a skin lesion classification system. The final objective is to be able to classify a new skin lesion into melanoma and non-melanoma with a high accuracy. Many classification algorithms are used for this field such as; Artificial Neural Network (ANN), Decision Tree (DT), Support Vector Machine (SVM), and Ensemble Methods [24, 34].

There are some works that combine Machine learning and Deep learning. Some authors combine their handcrafted features with those extracted from Deep learning architectures [19]. Others proposed Ensemble learning approaches for skin lesion classification [19, 22]. Somme used Deep learning to segment skin lesion images and they process to the rest of machine learning flow chart (features engineering and classification) [46].

The main issue is to design a robust system to help in the diagnosis and classification of skin lesion images. Despite they use Machine learning, Deep learning or a combination of ML and DL each of the steps present a challenge and opportunities for researchers.

3 Skin Lesion Classification Challenges and Opportunities

Many challenges and opportunities are encountered when working on skin lesion detection, analysis and classification. As said before, the aim is to build a model able to classify correctly a new skin lesion image into melanoma and non-melanoma. This model is created using AI (machine learning and/or deep learning). The model is built using a training dataset, validated on validation dataset and tested on a test dataset.

3.1 Dataset

Despite the technology used, the classification model is built using a training dataset, validated on validation dataset and tested on a test dataset. Many datasets have been used in the literature and many challenges have been lunched using their own datasets. The most known challenge is International Skin Imaging Collaboration (ISIC) launched in 2017, 2018 and 2019 [7]. The ISIC dataset in the most used dataset [10] which contains 2000 images: 1626 non-melanomas (254 seborrheic keratoses, 1372 atypical nevi), and 374 melanomas skin cancer. These images are RGB (red, green, blue) colour system and have a resolution of 767 * 1022 pixels. The PH2 dataset is also used as a small dataset, it contains 200 images: 160 non-melanomas (80 common nevi, 80 atypical nevi), and 40 melanomas skin cancer [31]. The two databases are classified by experts and contain the segmentation ground-truth to evaluate the segmentation accuracy. The choice of these two datasets is because Ph2 which is a small dataset is suitable to be used with Machine learning classification and ISIC which is a large dataset is suitable to be used with Deep learning architectures.

The skin lesion images in both ISCIC and Ph2 datasets contains noise and artifact and propose segmentation Ground-Truth to evaluate the segmentation (see Fig. 1).

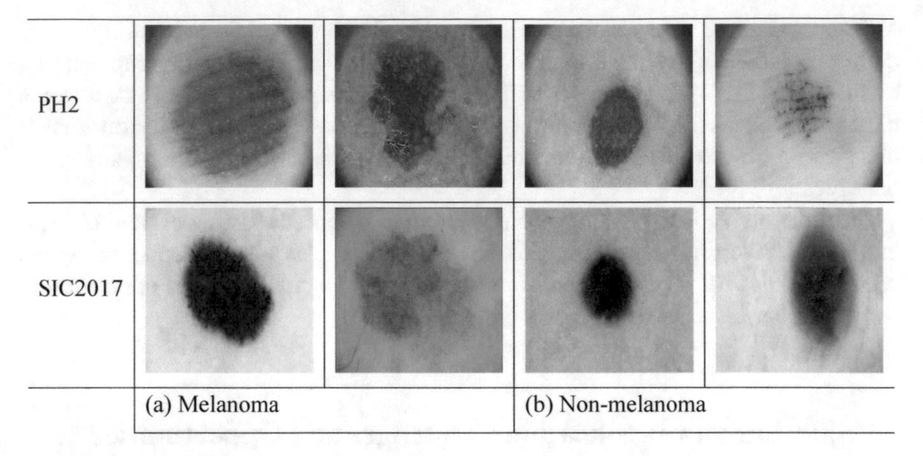

Fig. 1 Images randomly selected from the used datasets

The ISIC dataset proposes separately training dataset, validation dataset and test dataset. Meanwhile, Ph2 is proposed as a one datasets and authors should use cross-validation or k-folds cross validation. To improve the built classification model, authors using Deep learning architecture can use data augmentation.

The two datasets can be used to both evaluate the segmentation step and the classification accuracy.

3.2 *Workflows of Classification Schemes*

The workflow in classification approach differs from Machine learning to Deep learning. In a supervised machine learning approach, the workflow is as presented in Fig. 2.

The objective of a classification approach is de build a classification model able to classify correctly an input image. In skin lesion classification we use generally 2 classes; Melanoma and Non-melanoma.

The build model should be evaluated before its deployment. So, the model is first built using a labelled training dataset. And then, we use a labelled test (validation) dataset where we try to classify each image of the test dataset and we compare the predicted class with the original class.

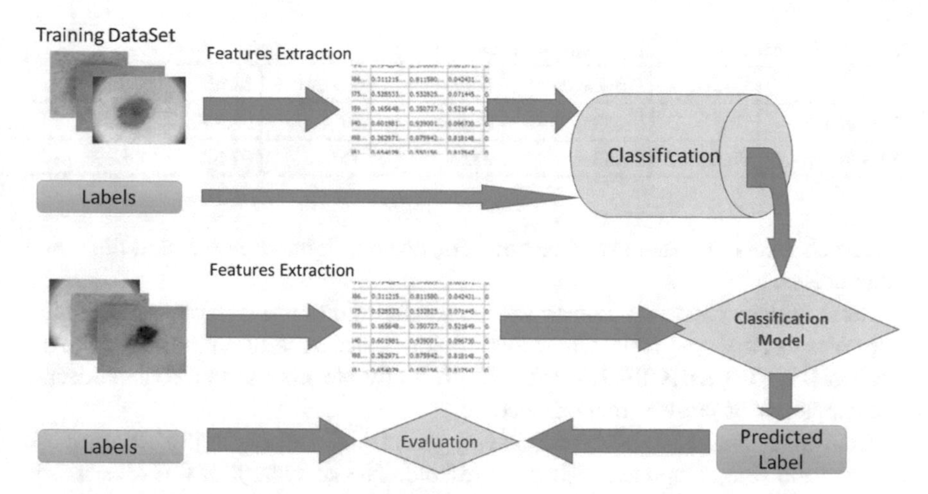

Fig. 2 Supervised machine learning workflow

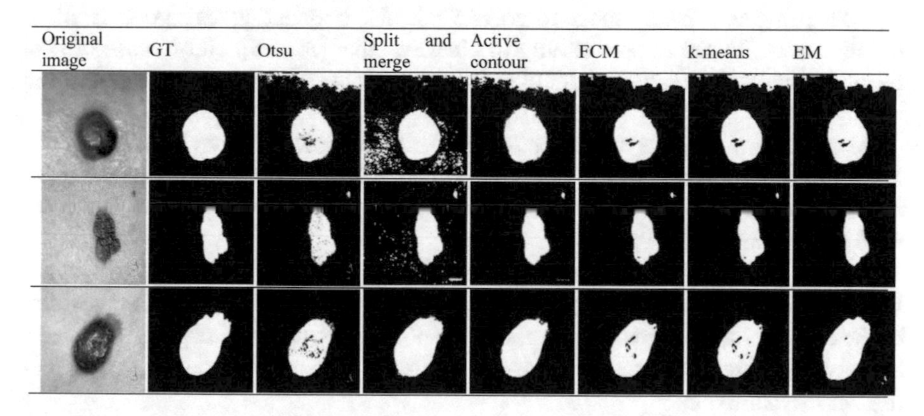

Fig. 3 Segmentation results using the well-known algorithms

3.3 Pre-processing and Segmentation

Segmentation of Skin lesion images is a challenge because they contain noise and artifacts. Many published research segment directly the images without segmentation evaluation. The classification accuracy in these cases are not very convenient.

To show the issue of segmenting directly this type of image, in Fig. 3 we present the segmentation results without pre-processing using the well-known algorithms; Otsu, Split and merge, Active contour, FCM, K-means and EM. The Ground of Truth (GT) is also presented to evaluate these algorithms.

From Fig. 3 we can see that the segmentation using the well-known algorithms is not perfect and the lesion will not be extracted correctly which will affect the relevance of extracted features and so the classification rate.

Table 1 Segmentation sensitivity and specificity average

	CMY	Grayscale	HSV	CIELAB	EDP	Without
Sensitivity	91.15	88.57	95.30	94.07	98.17	88.57
Specificity	90.55	89.33	95.89	95.75	97.32	89.33

The challenge here is to remove artefacts, texture, light reflexion, skin lines, and other noises.

Some authors choose to convert the input skin lesion images to other representations such as HSV, grayscale. Others use a colour space transformation such as LAB, CMY, YUV, CIELAB, CIELUV, and UIQ. Recently, we proposed to use a multiscale decomposition as pre-treatment procedure.

Converting image to HSV space is very promising. Authors in [26, 46] propose segment the Hue component using a threshold. The advantage of this idea is it is very easy implement but need to know in advance the hue component ranges to use and this value may change from image to another.

Converting skin lesion image to grayscale is also easy and gives a good segmentation results. It used a thresholding to binarize input [30, 35]. But in case of skin lesion images which contains a lot of details and colors and shapes grayscale method is very weak.

Many authors propose to use space Color transformation such as LAB [9], CMY, YUV, CIELAB, CIELUV [33] and YIQ [21]. Indeed, some space color transformation is very complex and the output is not natural to human. Segmentation is then used base on manual thresholding.

Recently we propose to use Multiscale decomposition based on PDE (Partial Differential Equation) to remove noises and artefacts [11, 43]. The decomposition of the original image gives two components; object and, texture images. The Object component is used to segment the lesion. While the Texture component is used later in features extraction.

To evaluate the performance of the each of the pre-processing approaches on the segmentation results, Table 1 presents the sensitivity and specificity values. The K-means is used based on a study presented in [39].

It is clear from results presented in Table 1 and Fig. 4 that the pre-treatment improves the segmentation results. The multiscale decomposition based on the EDP is the best one with a sensitivity value equal to 98.17. That's mean that there is still a research opportunity to propose a new pre-processing approach to enhance more the segmentation quality especially when we know that its success is elementary to a better classification.

Some new idea consists in using DL and ML to segment the skin lesion images. To have a good segmentation quality with DL we need a huge dataset and it is not very useful in real-time applications. Working with ML can be a good idea and we can have a powerful segmentation Model.

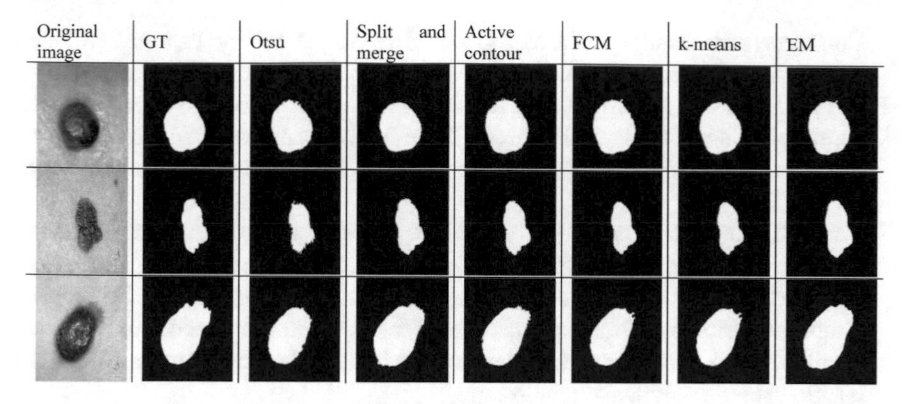

Fig. 4 Segmentation results after pre-processing

3.4 Features Engineering

Features engineering is a set of steps used to extract and select relevant features to be used in the classification used algorithm. It consists in features extraction, features normalization, dimensionality reduction and features selection are parts of features engineering.

3.4.1 Features Extraction

Characteristic or features are information extracted that can represent effectively the image. So, in any field we must choice the best characteristics that's why features extraction is considered as a big challenge and propose many opportunities [32]. Basically, handcrafted features are based on the rules used by dermatology noted ABCD (Asymmetry, Border, Color and Diameter) [8, 42]. Features can be extracted from the lesion and/or from the skin. Follow we describe the most used features:

– Colour features: usually used by dermatologists, color's information can help in skin lesion classification. Jason R. Hagerty et al. [22] extracts features using detection of specific colors. [15, 16] used the percentage of colours identified on the skin and the lesion. Colours used are white, red, light-brown, dark-brown, blue-gray, and black. They also used the lesion's colour variation. The derived characteristics are: maximum, minimum mean and variance of the pixel intensity within the R, G and B plan, a total of 12 features.
– Texture: To measure the textural information present in the lesion mathematical texture descriptors have been used. Authors is [13, 14] propose to use 10 textural features: Contrast; correlation; Energy; Homogeneity; Entropy; inverse difference Moment; Smoothness; Standard Derivation; Kurtosis; Root Mean Square. Some works propose also to use features extracted from the Gray Level

Co-Occurrence Matrix (GLCM), Gabor filte, Local binary Pattern(LBP) and Histogram of Oriented Gradient (HOG). They extract 34,828 features.

- Shape: The lesion's form plays an important role in the lesion classification as the dermatologist uses in their diagnosis. Eight features are extracted: area, greatest diameter, smallest diameter, perimeter, eccentricity, extent, and the circularity [20]
- Skeleton: Authors in paper [15] show that the skeletons of non-melanoma are symmetric and are very different to the melanoma skeleton. So they propose to use nine features extracted from the segmented lesion. The features are: Number of Endpoint, Number of Branchpoints, Size of the skeleton, Number of Subbranches, Width of the skeleton, Length of the skeleton image, Maximum length, Minimum length, skeleton width and length ratio.
- Naser Alfed in [1] uses Histogram of Gradients (HG) and the Histogram of Lines (HL) and color vector angle (CVA)
- SIFT: Scale-Invariant Feature Transform is a feature detection algorithm is used for feature extraction. This method has shown good accuracy [3].
- Deep learning is used also as features extraction tools. This is called Transfer learning when we transfer the deep learning architecture to a standard machine learning [15–18, 26]. In [19] they use the transfer learning to extract features from four different architectures of pre-trained networks: AlexNet, VggNet, GoogLeNet, and ResNet. Features are extracted from the last layer of the fully connected network in case of having multiple layers.

Table 2 presents the classification accuracy of approaches form literature using different features and for the studied datasets. The ISIC dataset is very hard to classify. In this section the presented approaches are those who did not use dimensionality reduction neither features selection. The adopted classification algorithm was the SVM.

Table 2 Classification accuracy using different extracted features using Ph2 and ISIC datasets

	Approaches	Used charsacteristics	Accuracy	
			Ph2	ISIC
ML	[15]	Skeleton, Color, Texture	90.50	80.30
	[24]	Color and texture	85.00	80.09
	[44]	Color in HSV and RGB space and texture using GLCM	88.20	80.25
	[27]	area, mean, variance and the standard deviation	89.50	80.33
	[17]	GLCM, Gabor filter, LBP, HOG, color and Asymmetry	90.30	81.30
DL	AlexNet	1000 features	87.00	80.70
	GoogLeNet	1000 features	87.5	82.00
	Vgg16	1000 features	87.00	81.6
	Resnet18	1000 features	88.00	81.5
TL	[19]	Fusion of Handcrafted features and 4 DL architectures	91.50	82.00

3.4.2 Features Normalization

A normalization step is mandatory to make all features value in the same scale and to improve the classification result. Many algorithms have been used such as Zscore, Min–Max, Mean, and the best one is the Zscore transformation is the best [36].

3.4.3 Dimensionality Reduction

Too many features imply high dimensionality and can imply to classification errors. Some of the extracted features are correlated and, thus, can be combined to reduce the number of features and thus the number of dimensions for the classification algorithm. Dimensionality reduction therefore refers to techniques to reduce the number of characteristics. Several algorithms are proposed such as Principal Components Analysis (PCA), Linear Discriminant Analysis (LDA), Neighbourhood Component Analysis (NCA) and the most widely used is PCA.

3.4.4 Features Selection

Features selection phase consist in selecting relevant features that can efficiency represent the lesion. Many algorithms have been proposed in the literature and there is still work in progress to propose new selection approaches. Relief, Correlation-based Feature Selection (CFS), Chi2, Recursive Feature Elimination (RFE), and Genetics Algorithm (GA) which belong to the evolutionary algorithms family [4, 14]. Not all approaches presented in the literature use features selection.

To conduct a comparative study on features selection algorithms we conduct our study using our features extracted based on transfer learning scheme [19].

From results presented in Table 3 we can conclude that features selection is a crucial step and should be conducted to first eliminate the non-relevant features and second to increase the classification accuracy. Genetics algorithm is the best features selection algorithm which select the best relevant features followed by the RelieF algorithm. Authors can use other algorithms (check for new algorithms) to get the best features to use in the classification.

3.5 Classification

Classification step: is the last step in the skin lesion classification system. The final objective is to be able to classify a new skin lesion into melanoma and non-melanoma with a high accuracy. Many classification algorithms are used for this field such as; Artificial Neural Network (ANN), Decision Tree (DT), Support Vector Machine (SVM), and Ensemble Methods [24, 34].

Table 3 Features selection and classification accuracy results

	Nbre of features	Accuracy	Features selection algorithm	Nbre of selected features	Accuracy
Ph2	4034	91.50	GA	406	98.00
		89.00	RFE	600	92.5
		90.35	RelieF	440	93.00
		88.78	CFS	600	92.5
		85.14	Chi2	250	91.75
ISIC	4034	82.00	GA	160	87.80
		81.30	RFE	210	82.00
		82.00	RelieF	170	83.15
		80.45	CFS	210	81.98
		80.10	Chi2	115	81.75

The classification accuracy depends on the earlier steps; pre-processing, segmentation, features extraction and features selection. When using DL or ML authors should use a classification algorithm. In this part there are not much to do instead of searching the classification algorithm that is suitable to the used data.

But it is much important to know which is the best classification algorithm in case of skin lesion images. In Table 3 we present first a comparative study between the most used classification algorithm from literature. The approach adopted in this is based on a new proposed approached based on fusion of handcrafted features and and pre-trained CNNs features (AlexNet, VggNet, GoogLeNet, and ResNet). The extracted features were normalized and selected using genetics algorithm to get at the end 406 features for Ph2 dataset and 160 features for ISIC dataset.

The Sensitivity (True Positive rate), Specificity (True Negative rate) and Accuracy measures are used to evaluate and to compare each of the classifiers:

$$\text{Recall} = \frac{\text{TP}}{\text{TP} + \text{FN}} \tag{1}$$

$$\text{Specificity} = \frac{\text{TN}}{\text{FP} + \text{TN}} \tag{2}$$

$$\text{Precision} = \frac{\text{TP}}{\text{TP} + \text{FP}} \tag{3}$$

$$\text{Accuracy} = \frac{TP + TN}{TP + TN + FP + FN} \tag{4}$$

where TP (True Positives) is be the number of melanoma correctly classified, TN (True Negatives) is the number of non-melanoma correctly classified, FP (False Positives) is the number of melanoma classified as non-melanoma and FN (False Negatives) is the number of non-melanoma classified as melanoma.

Table 4 Comparative study of classification on Ph2 and ISIC dataset using different classification algorithms

	Features extraction	Logistic regression				K-Nearest neighbors				Support vector machine				Decision tree			
		R (%)	S (%)	P (%)	A (%)	R (%)	S (%)	P (%)	A (%)	R (%)	S (%)	P (%)	A (%)	R (%)	S (%)	P (%)	A (%)
Ph2	Texture	86	46	88	78.5	87	49	88	79.5	89	**57**	**89**	**83**	**90**	48	83	79
	Shape	88	44	83	77	89	57	89	83	95	57	84	84	84	60	95	82
	Skeleton	87	49	88	79	71	62	90	85	94	62.1	88	86	89	57	89	83
	Color	91	44	79	76.5	84	60	95	82	89	52	88	81	90	48	83	79
	AlexNet	90	38	72	71	86	75	97	85	91	68	93	87	89	52	87	81
	GoogleNet	87	36	76	71.5	89	81	97	88	89	78	96	87.5	91	54	86	82
	Vgg16	85	31	47	68.5	84	77	98	86.5	89	75	95	87	86	46	88	78.5
	Resnet18	90	43	79	76	86	93	99	86.5	93	82	96	91	90	62	91	84.5
	Fusion of all	87	30	66	64.5	89	94	99	87	91	85	97	91.5	89	56	89	82.5
ISIC	Texture	66	90	32	67.7	38	86	39	77.3	16	83	34	78.5	25	83	26	72.6
	Shape	32	77	24	86.7	23	84	53	81.8	20	84	50	81.3	32.1	89	40	78.5
	Skeleton	35	90	31	64.7	24	84	44	80.2	29	85	55	82.3	23	84	45	80.3
	Color	37	90	32	66.3	37	86	33	77	40	86	36	75.7	33	80	27	71.2
	AlexNet	44	85	29	69.7	23	84	43	81.1	35	86	48	80.7	32	84	30	73.4
	GoogleNet	47	86	29	68.8	18	84	47	80.8	22	84	55	82	17	83	34	78.3
	Vgg16	41	85	27	68	13	83	35	79.3	19	84	52	81.6	17	83	33	78
	Resnet18	45	85	29	69.1	18	83	43	80.2	21	84	51	81.5	11	82	29	78.4
	Fusion of all	33	73	24	68.7	29	85	55	82.3	37	92	52	82	34	89	42	79

Table 5 The confusion matrix on Ph2 and ISIC datasets

	PH2	
	Melanoma	Non-melanoma
Melanoma	37	1
Non-melanoma	3	159

	ISIC	
	Melanoma	Non melanoma
Melanoma	206	76
Non-melanoma	168	1550

In the case of skin lesion classification, the specificity value is very important and should be used as an important factor to evaluate the classification results. It tells us about the % of the melanoma classified as non-melanoma.

From results shown in Table 4 and for the Ph2 dataset we can conclude that KNN and SVM are more suitable for classification. And as we say before, the Specificity can be taken as a referent parameter to evaluate a classifier we can say that SVM is the best in term of classification accuracy and KNN in term of Specificity. In case of ISIC dataset in addition to SVM and KNN we can use DT in case of elementary features. But in case of the fusion of handcrafted and Deep learning features KNN is the best in term of classification accuracy and SVM in term of Specificity. So as opportunity authors can use both SVM and KNN to classify skin lesion images.

To show how it is important to use the Specificity value in case of skin lesion classification we will present in Table 5 the confusion matrix for both Ph2 and ISIC datasets. The classification is based on the fusion of handcrafted and Deep learning features and using SVM classifier.

The confusion matrix using genetic algorithms as a features selection with Ph2 and ISIC databases.

For the Ph2 dataset it is clear the classification results is perfect with only one melanoma skin lesion was misclassified as Non-melanoma. For the ISIC dataset we have 76 melanoma skin lesions was misclassified as Non-melanoma which is a good value when knowing that ISIC dataset is used in a challenge and the images used are very hard to classify.

There are some works that combine Machine learning and Deep learning. Some authors combine their handcrafted features with those extracted from Deep learning architectures [19]. Others proposed Ensemble learning approaches for skin lesion classification [19, 22]. Somme used Deep learning to segment skin lesion images and they process to the rest of machine learning flow chart (features engineering and classification) [22].

4 Discussion

The objective of a classification system is to design a powerful classification model to classify successfully a new input a lesion image as melanoma or non-melanoma. Two types of approaches are proposed; Machine Learning-based approaches and Deep Learning-based approaches. ML-based skin lesion classification process is essentially composed of 3 major steps; Segmentation, Features Extraction and Classification. However, for standard DLit consists in using directly the training dataset to design a classification model. Segmentation is considered a key phase in this process. Thus, a bad segmentation will not allow the identification of the lesion. Our study has shown that the most used algorithms are inefficient due to the presence of artifacts, hair, ... for this reason it is strongly recommended to perform a pre-processing whose objective is to reduce noise to facilitate the segmentation. The multi-scale approach based on PDE is an ideal solution and has allowed a better segmentation. There is another multi-scale approach that can give very good results, namely the BEMD (Bidimensional Empirical Modal Decomposition) [37, 38]. It should be noted that DL has been used to segment skin lesions images but without much success and especially for small datasets [23].

Next comes the feature extraction step, which consists of identifying relevant descriptors that can correctly represent the lesion and the skin. The basic characteristics are those based on the rules used by dermatologists in their diagnostics; namely the ABCD rules (Asymmetry, Border, Colours, Diameter). This part is very sensitive and the relevance of the extracted features will automatically increase the classification accuracy. In this part, feature engineering is used, which consists, among other things, in homogenizing the extracted features, creating new features based on the extracted ones, reducing the number of features by merging them (dimensionality reduction), and removing the outliers and leaving only the most relevant ones (Features selection). Standard features that are based on shape, texture and colour do not provide an excellent classification rate. Thus, you have to find new features. Many authors have proposed new features that gave a very satisfactory classification rate. There is still a lot of margins and opportunity to be undertaken to find new relevant features to have a classification rate almost equal to 100%. The DL has also been used to only extract the features for merging them with the handcrafted ones and which have allowed to get a very good classification rate, and very good precision and which is useful for both small and large datasets.

The last step of the process is the classification, which consists of using the final features to design a design model capable of correctly classifying as melanoma or non-melanoma a new image containing a skin lesion. Several algorithms are used and the studies presented have shown that the classification accuracy varies from one classifier to another. The challenge here is to define beforehand the algorithm to be used. In this sense, and as a research opportunity is to use a set of classification algorithms with a voting system. The voting should not be in a fair and traditional way but use a weighting based on a preliminary study of the percentage of precision of each classification algorithm for each class.

5 Conclusion

Skin cancer is one of the deadliest cancers in the world. Computer-Aided Diagnosis (CAD) systems are used for the detection and the classification of skin cancers and which can reduce significantly effort and time. Machine learning and Deep learning approaches are used to propose efficient systems. Many challenges are encountered in setting up such systems. Machine Learning based approaches consist in three main steps; segmentation to identify the lesion's region, features extraction and classification. Each step presents a specific challenge and can present a good opportunity for research. The challenge in the segmentation step is to detect correctly the lesion knowing that the used images contain many noises. So the use of multiscale decomposition as pre-treatment can be a good solution especially that they can automatically isolate the noise in the texture component. Features engineering can be a good solution at the second step and especially features selection which can reduces the number of extracted features and keeps only relevant ones. In classification, as several algorithms are used and the classification accuracy varies from one classifier to another. The opportunity is to use a set of classification algorithms with a weighted voting system. For approaches based on Deep learning, several architectures are used and each one gives different classification accuracy for each dataset. Thus, it is possible to use several DL architectures and to use a weighted voting system to improve the classification rate. Some recent uses of DL to extract features. The disadvantage of DL-based implementations is that they are less efficient with small databases as opposed to ML-based approaches. Proposed solutions include merging the features extracted from DL with Handcrafted ones.

References

1. Alfed, N., Khelifi, F.: Bagged textural and color features for melanoma skin cancer detection in dermoscopic and standard images. Expert Syst. Appl. **90**, 101–110 (2017)
2. Almansour, E., Jaffar, M.A.: Classification of Dermoscopic skin cancer images using color and hybrid texture features. IJCSNS Int. J. Comput. Sci. Netw. Secur. **16**(4), 135–139 (2016)
3. An, F., Zhou, X.: BEMD–SIFT feature extraction algorithm for image processing application. Multimed. Tools Appl. **76**, 13153–13172 (2017). https://doi.org/10.1007/s11042-016-3746-y
4. Anirudha, R.C., Kannan, R., Patil, N.: Genetic algorithm based wrapper feature selection on hybrid prediction model for analysis of high dimensional data. In: 2014 9th International Conference on Industrial and Information Systems (ICIIS), 15 December 2014, pp. 1–6. IEEE (2014)
5. Ardila, D., Kiraly, A.P., Bharadwaj, S., et al.: End-to-end lung cancer screening with three-dimensional deep learning on low-dose chest computed tomography. Nat. Med. **25**, 954–961 (2019). https://doi.org/10.1038/s41591-019-0447-x
6. Bissoto, A., Perez, F., Ribeiro, V., Fornaciali, M., Avila, S., Valle, E.: Deep-learning ensembles for skin-lesion segmentation, analysis, classification: RECOD titans at ISIC challenge 2018 (2018). arXiv preprint arXiv:1808.08480
7. Codella, N.C., Gutman, D., Celebi, M.E., Helba, B., Marchetti, M.A., Dusza, S.W., Kalloo, A., Liopyris, K., Mishra, N., Kittler, H., Halpern, A.: Skin lesion analysis toward melanoma detection: a challenge at the 2017 international symposium on biomedical imaging (ISBI), hosted by the international skin imaging collaboration (ISIC). In: 2018 IEEE 15th International Symposium on Biomedical Imaging (ISBI 2018), 4 April 2018, pp. 168–172. IEEE (2018)
8. Correa, D.N., Paniagua, L.R., Noguera, J.L., Pinto-Roa, D.P., Toledo, L.A.: Computerized diagnosis of melanocytic lesions based on the ABCD method. In: 2015 Latin American Computing Conference (CLEI), 19 October 2015, pp. 1–12. IEEE (2015)
9. Dalila, F., Zohra, A., Reda, K., Hocine, C.: Segmentation and classification of melanoma and benign skin lesions. Optik **140**, 749–761 (2017)

10. Esteva, A., Kuprel, B., Novoa, R. A., Ko, J., Swetter, S.M., Blau, H.M., Thrun, S.: Dermatologist-level classification of skin cancer with deep neural networks. Nature **542**(7639), 115–118 (2017)
11. Fan, H., Xie, F., Li, Y., Jiang, Z., Liu, J.: Automatic segmentation of dermoscopy images using saliency combined with Otsu threshold. Comput. Biol. Med. **85**, 75–85 (2017)
12. Filali, Y., Sabri, M.A., Aarab, A.: An improved approach for skin lesion analysis based on multiscale decomposition. In: 2017 International Conference on Electrical and Information Technologies (ICEIT), 15 November 2017, pp. 1–6. IEEE (2017)
13. Filali, Y., Ennouni, A., Sabri, M.A., Aarab, A.: Multiscale approach for skin lesion analysis and classification. In: International Conference on Advanced Technologies for Signal and Image Processing (ATSIP), . Fez, Morocco, 22–24 May 2017 (2017)
14. Filali, Y., Ennouni, A., Sabri, M.A., Aarab, A.: A study of lesion skin segmentation, features selection and classification approaches. In: 2018 International Conference on Intelligent Systems and Computer Vision (ISCV), 2 April 2018, pp. 1–7. IEEE (2018)
15. Filali, Y., El Khoukhi, H., Sabri, M.A., Yahyaouy, A., Aarab, A.: New and efficient features for skin lesion classification based on skeletonization. J. Comput. Sci. **15**(9), 1225 (2019). https://doi.org/10.3844/jcssp.2019.1225.1236
16. Filali, Y., Abdelouahed, S., Aarab, A.: an improved segmentation approach for skin lesion classification. Stat. Optim. Inf. Comput. **7**(2), 456–67 (2019)
17. Filali, Y., El Khoukhi, H., Sabri, M.A., Yahyaouy, A., Aarab, A.: Texture Classification of skin lesion using convolutional neural network. In: 2019 International Conference on Wireless Technologies, Embedded and Intelligent Systems (WITS), 3 April 2019, pp. 1–5. IEEE (2019)
18. Filali, Y., El Khoukhi, H., Sabri, M.A., Yahyaouy, A., Aarab, A.: New and efficient features for skin lesion classification based on skeletonization. J. Comput. Sci. **15**(9),1225–1236 (2019)
19. Filali, Y., EL Khoukhi, H., Sabri, M.A., Aarab, A.: Efficient fusion of handcrafted and pre-trained CNNs features to classify melanoma skin cancer. Multimed. Tools Appl. (2020). https://doi.org/10.1007/s11042-020-09637-4
20. Filali, Y., Sabri, M.A., Aarab, A.: Improving skin cancer classification based on features fusion and selection. In: Embedded Systems and Artificial Intelligence, pp. 379–387. Springer, Singapore (2020)
21. Gupta, A., Issac, A., Dutta, M.K., Hsu, H.H.: Adaptive thresholding for skin lesion segmentation using statistical parameters. In: 2017 31st International Conference on Advanced Information Networking and Applications Workshops (WAINA), pp. 616–620. IEEE (2017)
22. Hagerty, J.R., Stanley, R.J., Almubarak, H.A., Lama, N., Kasmi, R., Guo, P., Drugge, R.J., Rabinovitz, H.S., Oliviero, M. and Stoecker, W.V.: Deep learning and handcrafted method fusion: higher diagnostic accuracy for melanoma dermoscopy images. IEEE J. Biomed. Health Inform. **23**(4), 1385–1391 (2019)
23. Hesamian, M.H., Jia, W., He, X., et al.: Deep learning techniques for medical image segmentation: achievements and challenges. J. Digit. Imaging **32**, 582–596 (2019). https://doi.org/10.1007/s10278-019-00227-x
24. Immagulate, I., Vijaya, M.S.: Categorization of non-melanoma skin lesion diseases using support vector machine and its variants. Int. J. Med. Imaging **3**(2), 34–40 (2015)
25. Jain, S., Pise, N.: Computer-aided melanoma skin cancer detection using image processing. Procedia Comput. Sci **48**, 735–40 (2015)
26. Kassani, S.H., Kassani, P.H.: A comparative study of deep learning architectures on melanoma detection. Tissue Cell **58**, 76–83 (2019)
27. Korotkov, K., Garcia, R.: Computerized analysis of pigmented skin lesions: a review. Artif. Intell. Med. **56**, 69–90 (2012). https://doi.org/10.1016/j.artmed.2012.08.002
28. Li, Y., Shen, L.: Skin lesion analysis towards melanoma detection using deep learning network. Sensors. **18**(2), 556 (2018)
29. Litjens, G., Kooi, T., Bejnordi, B.E., Setio, A.A., Ciompi, F., Ghafoorian, M., Van Der Laak, J.A., Van Ginneken, B., Sánchez, C.I.: A survey on deep learning in medical image analysis. Med. Image Anal. **42**, 60–88 (2017)

30. Mahmoud, K.A., Al-Jumaily, A.: Segmentation of skin cancer images based on gradient vector flow (GVF) snake. In: 2011 IEEE International Conference on Mechatronics and Automation, pp. 216–220. IEEE (2011)
31. Moura, N., Veras, R., Aires, K., Machado, V., Silva, R., Araújo, F., Claro, M.: Combining ABCD rule, texture features and transfer learning in automatic diagnosis of melanoma. In: 2018 IEEE Symposium on Computers and Communications (ISCC) 25 June 2018, pp. 00508–00513. IEEE (2018)
32. Oliveira, R.B., Mercedes Filho, E., Ma, Z., Papa, J.P., Pereira, A.S., Tavares, J.M.: Computational methods for the image segmentation of pigmented skin lesions: a review. Comput. Methods Programs Biomed. **131**, 127–41 (2016)
33. Oliveira, R.B., Papa, J.P., Pereira, A.S., Tavares, J.M.R.: Computational methods for pigmented skin lesion classification in images: review and future trends. Neural Comput. Appl. **29**(3), 613–636 (2018)
34. Ozkan, I.A., Koklu, M.: Skin lesion classification using machine learning algorithms. Int. J. Intell. Syst. Appl. Eng. **5**(4), 285–289 (2017)
35. Patel, B., Dhayal, K., Roy, S., Shah, R.: Computerized skin cancer lesion identification using the combination of clustering and entropy. In: 2017 International Conference on Big Data Analytics and Computational Intelligence (ICBDAC), pp. 46–51. IEEE (2017)
36. Patro, S., Sahu, K.K.: Normalization. A preprocessing stage (2015). arXiv preprint arXiv:1503.06462, 19 March 2015
37. Sabri, A., Karoud, M., Tairi, H., Aarab, A: An efficient image retrieval approach based on spatial correlation of the extrema points of the IMFs. Int. Rev. Comput. Softw. (I. RE. CO. S) **3**(3), 597–604 (2008)
38. Sabri, A., Senhaji, S., Aarab. A.: Accelerating the BEMD by reducing the number of extrema points to interpolate in the SP. Int. Rev. Comput. Softw. **6**(2), 264–268 (2011)
39. Sabri, M., Filali, Y., Ennouni, A., Yahyaouy, A., Aarab, A.: 2 An overview of skin lesion segmentation, features engineering, and classification. In: Intelligent Decision Support Systems, pp. 31–52. De Gruyter, Berlin. https://doi.org/10.1515/9783110621105-002
40. Schaefer, G., Krawczyk, B., Celebi, M.E., Iyatomi, H.: An ensemble classification approach for melanoma diagnosis. Memetic Comput. **6**(4), 233–240 (2014)
41. Upadhyay, R.K.: Emerging risk biomarkers in cardiovascular diseases and disorders. J Lipids. **2015**(2015), 971453 (2015). https://doi.org/10.1155/2015/971453
42. Vasconcelos, M.J., Rosado, L., Ferreira, M.: A new risk assessment methodology for dermoscopic skin lesion images. In: 2015 IEEE International Symposium on Medical Measurements and Applications (MeMeA) Proceedings, 7 May 2015, pp. 570–575. IEEE (2015)
43. Victor, A., Ghalib, M.: Automatic detection and classification of skin cancer. Int. J. Intell. Eng. Syst. **10**(3), 444–451 (2017)
44. Waheed, Z.: An efficient machine learning approach for the detection of melanoma using dermoscopic images. In: Proceedings of the International Conference on Communication, Computing and Digital Systems, Islamabad, Pakistan, 8–9 March 2017, IEEE Xplore Press, pp. 316–319 (2017). https://doi.org/10.1109/C-CODE.2017.7918949
45. Xu, L., Jackowski, M., Goshtasby, A., Roseman, D., Bines, S., Yu, C., Dhawan, A., Huntley, A.: Segmentation of skin cancer images. Image Vis. Comput. **17**(1), 65–74 (1999)
46. Yang, X., Zeng, Z., Yeo, S.Y., Tan, C., Tey, H.L., Su, Y.: A novel multi-task deep learning model for skin lesion segmentation and classification (2017). arXiv preprint arXiv:1703.01025
47. Ding, Y., et al.: A deep learning model to predict a diagnosis of alzheimer disease by using 18F-FDG PET of the brain. Radiology **2018**, 180958 (2018). https://doi.org/10.1148/radiol.2018180958
48. Zhous, H., Schaefer, G., Celebi, M.E., Iyatomi, H., Norton, K.A., Liu, T., Lin, F.: Skin lesion segmentation using snake model. In: 2010 Annual International Conference of the IEEE Engineering in Medicine and Biology, 31 August 2010, pp. 1974–1977. IEEE (2010)

Deduplication Over Big Data Integration

M. El Abassi, Med. Amnai, and A. Choukri

Abstract Entity Resolution is the process of matching records from more than one database that refer to the same entity. In case of a single database the process is called deduplication. This article proposes a method to solve deduplication problem using Scala over Spark framework. The concept of data quality is very important for good data governance in order to improve the interaction between the different collaborators of one or more organizations concerned. The presence of duplicate or similar data creates significant data quality concerns. A panorama of the methods of calculation of distance similarity between the data as well as algorithms for the elimination of similar data are presented and compared.

Keywords Entity resolution · Data matching · Spark · Scala · Data quality · Similarity

1 Introduction

Big Data phenomenon is usually defined in four dimensions (Vs): Volume, Velocity, Variety and Veracity. Volume refers to managing large volumes of data. Velocity is the time it takes to collect and process data. The variety deals with structured, semi-structured and unstructured data. Finally, veracity ensures data quality and reliability. In this article, we are interested in the latter V. Today, data is a wealth for businesses and administrations and contribute to their development. The quality of these data is an important issue. The cost of non-quality can be very high: making a decision based on bad information can harm the organization, its customers or partners. Data governance is an important issue in businesses and administrations. It allows for improved interaction between the different staff of one or more organizations concerned [13].

M. El Abassi (✉) · Med. Amnai · A. Choukri
Laboratory of Computer Sciences, Faculty of Sciences, Ibn Tofail University, Kenitra, Morocco
e-mail: merieme.elabassi@uit.ac.ma

Med. Amnai
e-mail: mohamed.amnai@uit.ac.ma

A. Choukri
S.A.R.S Group, ENSA Safi, Cadi Ayyad University, Safi, Morocco

N. Gherabi and J. Kacprzyk (eds.), *Intelligent Systems in Big Data, Semantic Web and Machine Learning*, Advances in Intelligent Systems and Computing 1344,
https://doi.org/10.1007/978-3-030-72588-4_15

There are many methods to identify measure and resolve certain data quality problems. The proposed tools do not yet address all the issues raised. They most often focus only on raw data and not on the meaning of it. The data, in order to be useful, must be interpreted in the context of its use [13]. At present, few companies have implemented a data quality management program at both the database (DB) and warehouse (DW) levels built from them. Integration projects are likely to lack sufficient support for the importance of data quality.

While the ETL (Extract1, Transform2, and Load3) have now reached a high degree of maturity and offer many components, the data integration results contain far too many anomalies and do not inspire confidence to help in decision-making.

The current work on extracting knowledge from an information environment characterized by very large amounts of heterogeneous data and distributed in data warehouses (DWs), focuses mainly on the search for potential, useful and previously unknown information, so the quality of the information collected depends on that of the data. So making decisions from bad information can harm the organization, hence a cost of non-quality that can be very high. The construction of a DW, resulting from the integration of totally heterogeneous sources of variable quality, and decision support tools derived from these masses of information requires the development of new data extraction and processing tools (ETL). The latter must take account of the heterogeneity of the data and their constraints and ensure the quality of the new data set built.

This paper is structured as follows: Sects. 2 summarizes related work on Data Integration, Sect. 3 presents the data integration by giving its complexity and the architectures of these two approaches, the Sect. 4 is dedicated to conducting a comprehensive study of the basic concepts associated with data quality with algorithms that solve the deduplication problems and the different existing methods of calculating the distance of similarity by data type. Problem formulation is described in Sect. 5. In Sect. 6 we present our proposal, and in Sect. 7 we experience our solution. Conclusions and ideas for future extensions are discussed in Sect. 8.

2 Related Work

In [1] a method is proposed to distribute the load among different computing nodes in order to solve the entity resolution problem with MapReduce using standard blocking. In [2] a method for entity resolution on Hadoop that works with semi structured data, including preprocessing tasks, indexing, comparisons and ranking results. It explains how to use different classification techniques and its results to improve future comparisons. The authors in [3] created a method called multi-sig-er, which supports structured and unstructured data type and the tasks contemplate preprocessing of data and reducing comparisons. Another algorithm derived from standard blocking is proposed in [4], however no practical results are provided. Different techniques preprocessing of large amounts of data in MapReduce have been discussed and testing in [5]. In [6] Authors proposed a simple, effective solution to efficiently distribute the calculation of similarity between two entities in entity resolution over MapReduce using two Jobs which the output of the first Job is the input of the second Job. For each Job we must define the Map, Partitioner and Reducer processes. Job 1 is responsible for preparing the information in order for Job 2 can evenly distribute comparisons avoiding the problem of data skew, therefore obtaining a method that is not sensitive to information distribution, so its performance depends only upon the size of the data sets to compare.

3 Data Integration

Data integration consists of combining data from different sources and providing the user with a unified view of the data. The problem of designing data integration systems is important in today's real world applications and is characterized by a number of problems of theoretical interest.

Data integration is the combination of technical and business processes used to combine data from disparate sources into meaningful and information. A complete data integration solution trusted data from various sources. This definition is by IBM.

3.1 Architecture of Data Integration System

Information in a data system is represented and stored in a variety of data sources with a heterogeneous way. The first approaches to integrating these data sources, to make them work together, were carried out in the framework of relational database systems, objects/relational or objects, through the establishment of a database federation. The essential need is to be able to query different data sources simultaneously and to give the user the impression that they are querying a single data source. The virtual approach and the materialized approach attempt to address this problem.

3.2 Virtual Approach

The virtual approach, or mediator, refers to a comprehensive view, through a single representation scheme, of all the different heterogeneous data sources. This global scheme can be defined automatically using tools, or extracting schemas. In this virtual approach user queries are formulated using the extracted global schema semantics. The execution of these queries requires translating them into subqueries adapted to each of the subschemas of the different data sources [12] (Fig. 1).

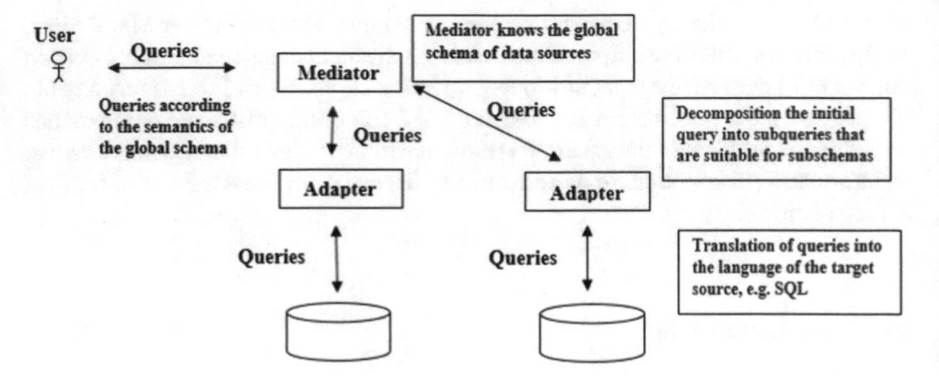

Fig. 1 Illustration of a virtual approach architecture

3.3 *Materialized Approach*

A Data Warehouse addresses the problems of overflowing and localized data on multiple heterogeneous systems, an architecture that can serve as a foundation for decision-making applications. To bc usablc, all data from distributed systems must be organized, coordinated, integrated and finally stored to give the user a comprehensive view of the information.

3.4 *Integration Steps*

We differentiate two levels in the construction of data warehouses. The first level is the construction of operational data sources, and the global data warehouse. The second tier includes all local data warehouses. The reason for this distinction is that at each level, different processing steps and technical difficulties are associated. At the first level, the construction process is broken down into four main steps, which are: (1) extraction of data from operational data sources, (2) transformation of data at the structural and semantic levels, (3) integration of data, and (4) storage of data integrated into the target system. Figure 2 summarizes the sequence of these processing steps.

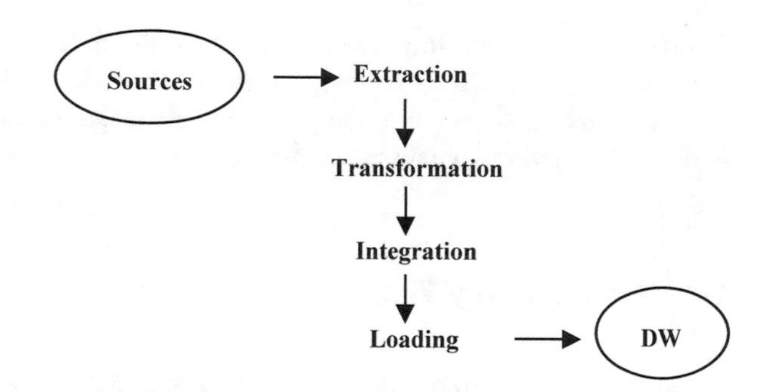

Fig. 2 Data warehouse first level processing steps

4 Data Integration Problems

Data integration is one of the most important and challenging computer science problems The first major step of the data integration is that of schema alignment, which takes up the challenge of semantic ambiguity and aims at including which attributes have the same signification and which are not it. In this step we have three major stages: creating a mediated schema, attribute matching, and schema mapping, and during all this stage we have some problems and for to provide best-effort services on dataspaces, uncertainty needs to be handled at various levels. First, when the number of data sources is large, there will be uncertainty as to how to model the domain; thus, there is uncertainty as to the creation of the mediated scheme. In the second levels, attributes can have ambiguous meanings, some attributes can overlap in their meanings, and meanings can change over time; thus, there is uncertainty as to the correspondence of the attributes [11].

The second major step of the data integration is that of record linkage, which takes up the challenge of instance representation ambiguity and aims to understand which recording represent the same entity and which are not it. In this part we have three methods for resolve this type of problems, in the first we have Pairwise matching method in which we compares a pair of records and makes a local decision of whether or not they refer to the same entity, but this step is not globally consistent, so there is an inconsistency if pairwise matching declares that record pair R1 and R2 match, record pair R2 and R3 match, but record pair R1 and R3 do not match. In such a scenario, the goal of the clustering step is to arrive at a globally consistent decision on how to partition the set of all records so that each partition refers to a separate entity and that the different partitions refer to different entities. Pairwise matching and clustering guarantee the desired semantics of the record linkage, but are possible be rather ineffectual and even unachievable for big group of records. So the Blocking method was proposed as a strategy to scale record linkage to large data sets. The last step is that of data fusion, which addresses the challenge of data quality [11].

The need for consistent access to multiple data sources is becoming increasingly acute, especially in decision-making systems that require comprehensive data analysis. With the development of Data Integration Systems (SIDs), the quality of information has become a top-level property increasingly demanded by users.

In order to assess and improve the quality of the data, it is important to carry out a continuous improvement strategy "at the source". In addition to this, tools may be needed to deal with past problems (double, inconsistencies) and to detect difficulties in entering them (misspelling, confusion).

4.1 Data Quality Tools

In most cases, existent processes based on databases draw away a deficient quality; consequently, data are not adequate for the usage for which they are intended (even in cases where laudable efforts were made to discern and to improve errors at specific instants). The targeting of the quality of data should therefore be a process of uninterrupted improvement. This process includes a cycle of working on a database in state of constant fluctuation (entries and updates), the management of variable quality requirements, waitings of the users and of changes in an environment in evolution.

Because of the size of modern databases, it is generally more possible to do it manually and it is necessary to automate. To this end, the quality tools of data give a potential support of the administrator of the database. These include a big number of functionalities (most automated possible) as well as group of integrated knowledge (for example, zip codes, street names,), having offered the user via an interface. From a total point of view, we can break down the functionality habitually given by the quality tools of data in 4 concepts:

Profiling: formal audit of data to test their suitability for corresponding metadata. Based on the results of this audit, corrective measures may be taken (adaptation of meta data or inconsistent values).

Standardization: application of rules to ensure that all data is encoded according to the same conventions, including complex types (addresses, names) in a multilingual context and with the help of address knowledge bases covering international zones (or internal knowledge bases).

Matching: comparison of records within a file or between competing databases to detect inconsistencies or duplicates and improve the quality of records.

Monitoring: time-bound monitoring of the specified quality indicators.

4.2 Data Quality Dimensions

On the surface, it is obvious that the quality of the data is to clean the incorrect data - missing, incorrect or invalid data in some way. But to ensure that data are reliable, it is important to understand the key dimensions of data quality to assess how data are "bad" first.

Completeness: The completeness is defined as the expected completeness. Data can be complete even if optional data is missing. As long as the data meets expectations, the data is considered complete.

Consistency: Consistency means that all systems' data reflect the same information and are synchronized across the enterprise.

Compliance: Compliance means that data follows the set of standard data definitions such as data type, size, and format. For example, the date of birth of the client is in the format "mm/dd/yyyy".

Accuracy: Precision is the degree to which the data correctly reflects the actual object OR an event described.

Integrity: Integrity means the validity of data in all relationships and ensures that all data in a database can be traced and connected to other data.

Timeliness: Data are available in a timely manner when they are up-to-date (current) and when information is available in a timely manner.

4.3 Data Matching

Data Matching (also known as record or data linkage, entity resolution, object identification, or field matching) involves identifying, comparing, and merging records that match the same entities from multiple databases or even within a single database. Through research in various fields, including applied statistics, health informatics, data mining, automatic learning, artificial intelligence, database management and digital libraries, significant progress has been made over the past decade in all aspects of the data linkage process, including the accuracy and scalability of data linkage for large databases [6]. Fundamentals it is the ability to identify duplicate data in large data sets. These duplicates can be people with multiple entries in one or more databases. It could also be duplicate items, of any description, in stock systems. The Data Matching allows you to identify duplicate or duplicate data, and then enables you to take measures such as merging the two identical or similar entries into one. It also allows you to identify nonduplicates, which can be just as important to identify, because you want to know that two similar things are certainly not the same.

4.3.1 Matching Field

As we have seen, the Data Matching allows you to compare two data sources, in order to identify common records (or to detect a series of records in which a crucial subset of data is equivalent). If you compare data sources and want to detect duplicate data (redundant records in a single directory). The comparison between the two sources is usually done in two phases:

In a first step, the similarity between records is defined field by field.

In a second step, field-by-field similarities are aggregated to achieve an overall result: for each possible pair of records, it is decided whether it is a match, certainly not a match, or perhaps a match and it can be indicated why [10] (Table 1).

The many methods for making field-by-field comparisons can be divided into families. We will describe some of them, but we will not want to be exhaustive. On

Table 1 Families of the comparison algorithm

Boolean	Predicates and rule	phonetics
	Equal Suffix of Stems from	Soundex metaphone
Similarity	Word-based	Token-based
	Edit distances Letters communes	Cosines Recursive

the one hand, we have the Boolean methods, which result from a comparison always results in a match or a non-match (0 or 1). These are either phonetic algorithms or methods consisting of a combination of predicates (logical) and rules. On the other hand, there are comparison methods whose result is in the form of a similarity score, allowing a more subtle match or non-match appreciation. These methods can be word-based (comparison between strings of individual words) or token-based (comparison between strings of multiple words).

4.3.2 Matching by Record

Record Linkage is used to create a frame, remove duplicate files, or combine files to study relationships on two or more data items from separate files. In the past, most of the record linkage work has been done either manually or through elementary but ad hoc rules. Simply put, record linkage is based on the search for unique identifiers. A unique identifier is a property that most likely will not change in the future. But it's not that simple. Why? Because in reality, things change over time. So the most important aspect is to identify properties that cannot change. For people, it could be the date of birth, size and sex. For objects, it can be either shape or size. There are two main methods of record linkage:

Deterministic: it is determined by the number of corresponding identifiers **Probabilistic**: it is determined by the probability that a number of identifiers match [7].

This part aims to introduce some existing ER blocking techniques. The general approach to blocking techniques is to process all data set records, to insert each record into one or more blocks, in which all pairs are considered candidate records, which may refer to the same entity. Below are the most commonly used blocking approach [8].

4.3.2.1 Sorted Neighborhood Blocking

Sorted Neighborhood (SN) is a popular blocking approach that works as follows. A K blocking key is determined for each of the n entities. Typically, concatenated prefixes of some attributes form the blocking key. Entities are then sorted by this blocking key. A fixed-size window w is then moved to the sorted records and, at each step, all the entities in the window, that is, entities located at a distance between w − 1, are compared [9].

4.3.2.2 Duplicate Count Strategy

Duplicate Count Strategy (DCS) is based on the Sorted Neighborhood Method (SNM) and varies the window size according to the number of duplicates identified. Due to the increase and decrease in the window size, all of the compared records differ from the original NMS. Adjustment of window size does not necessarily lead to additional comparisons; it can also reduce the number of comparisons.

4.3.2.3 Q-Gram Based Indexing

For data that are dirty and contain large amounts of errors and variations, standard blocking and the approach of sorted neighborhood may not be able to insert records into the same blocks, for example if the beginning of a sort key value is different for two name variations. Q-gram indexing is designed to overcome this disadvantage by generating variations of each BKV, and to use these variations as real index keys for a blocking-based approach. Each record is inserted into several blocks depending on the variations generated by its BKV.

5 Problem Formulation

Entity Resolution is also known as record de-duplication or record linkage among others, and is a well-considered problem where duplicate records are found and all records are provided. This can be a difficult problem because the same "entity can be referenced with different names, and these names can even contain typos, so ordinary string matching will not suffice. In this paper, we design a technique to address resolution of the company, using the rich API of spark (graphframes, ml, sparksql etc.…).

Given a collection of records (addresses in our case), find records that represent the same entity. This is a difficult problem because the same entity can have a different lexical (textual) representation. Therefore, direct string matching will fail to identify duplicates. See Fig. 3 below for address differences even if they represent the same entity.

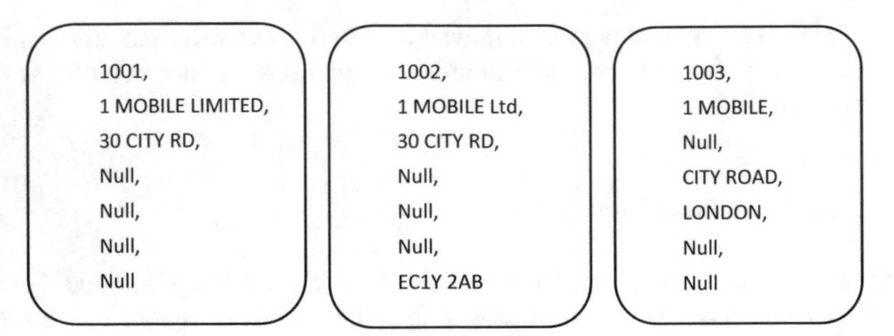

Fig. 3 Example of duplicate record

6 Proposed Solution

Consider a dataset R = {r1, r2, r3,…,rn} formed by records with attributes A = {R.A1, R.A2,..., R.Am}. Consider also a function G(ri) that maps a record ri ∈ R to a graph, and a function S(ri,rj) for measurer similarity. We say that a record a is similar to a record b if S (ra, rb) is closer to 1. In general the mapping G(ri) is a graphical approach that allows to transform relational data into graphs. In simple terms, a property graph represents data in vertices and edges. Here we reorganize our relational data into a chart so that entities share the nodes they have in common.

Property Graph Model
As with most technologies, there are few different approaches to what makes up the key components of a graph database. One such approach is the property graph model, where data is organized as nodes, relationships, and properties (data stored on the nodes or relationships). Nodes are representing the entities or instance in the graph, they can hold any information called properties. Edges, also Relationships, are the line that connect between two node entities, they always has a direction, a type, a start node, and an end node.

The main objectives of the proposed system are:

Identity potential duplicates: we use spark graphframes in this case that limiting our search space in our data.

TF-IDF Computation: apply tf-idf method on potential duplicates.

Compute Similarity: we use the cosines similarity algorithm on the TF-IDF vector (Fig. 4).

Create Graph: The first step is blocking graph where a collection of record is taken as input constructs a two types of entities, first type belong to master address, and second type to which other addresses would be matched (Fig. 5).

Find Potential Duplicates: Then comes what is probably the most important and also the most expensive calculation step is to search for records that are potential matches. The goal is to find records that look like records without necessarily being the same field by field. This step is essentially an inner join of the table with itself.

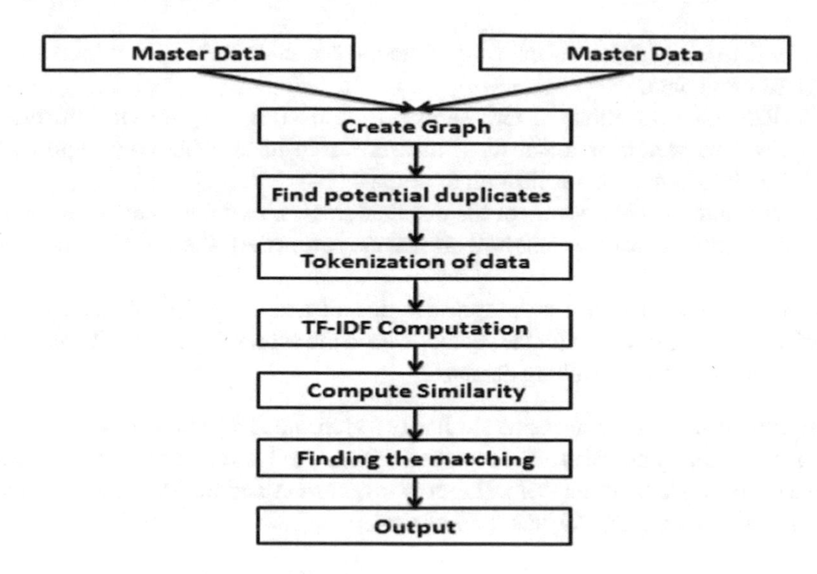

Fig. 4 Architecture of the proposed approach

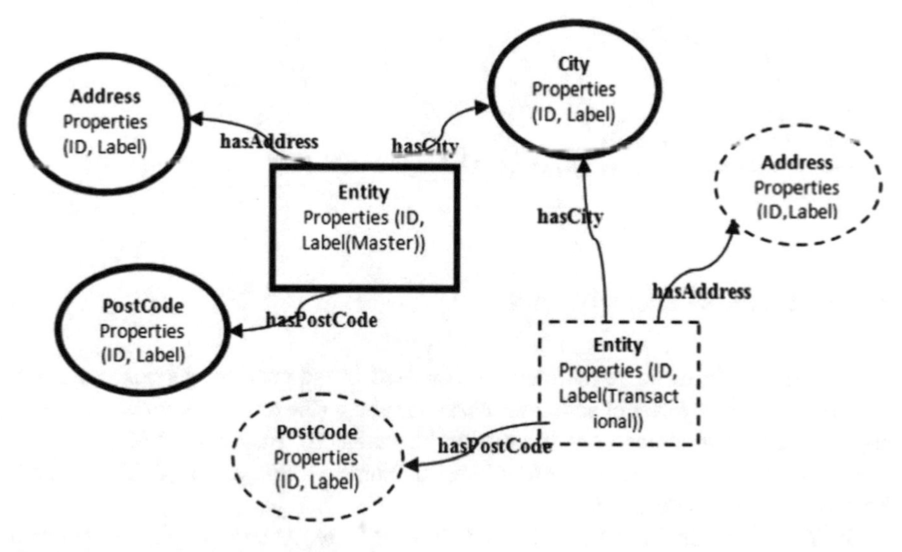

Fig. 5 Example of propriety graph

The join condition is very specific: we have chosen to use GraphFrameMotives; we remove the search space and identify potential duplicates. This is broad enough that catches differ, but selective enough to avoid a Cartesian product (Cartesian products should be avoided at all costs). Above the query (gf.find("(a)-[ae1]-> (p1); (b)-[be1]-> (p1)")), the pairs of potential duplicates are output.

Tokenization of data: Tokenization function is used to form a set of tokens from the given input data.

TF-IDF Computation: tf-idf, short for term frequency–inverse document frequency, this weight is a statistical measure used to evaluate how important a word is to a document in a collection or corpus.

We compute the tf-idf vector of the detail addresses, then compare the pair using the cosine similarity to determine how similar they are at the 0–1 scale with 1 perfectly similar.

Where we calculate a vector representation of addresses (TFIDF) and use the broadcast variable to search for the vector given to the address. Similarity is calculated on these vectors in the resulting database.

Compute Similarity: The cosine similarity (or cosine measurement) allows calculating the similarity between two documents represented as vector by determining the cosine of the angle between them. Either two vectors A and B, the angle is obtained by the scalar product and the standard of vectors:

$$\cos\theta = \frac{\mathbf{A.B}}{||\mathbf{A}|| \cdot ||\mathbf{B}||}$$

```
Algorithm cSim
Input: A,B;
Output : S;
Compute dotProuct(A,B)
Compute magnitude(A)
S=dotProduct(A,B)/magnitude(A)/magnitude(B);
```

7 Results and Experiments

In this section we evaluate the method proposed in the previous section, we use a data set of two categories of addresses Master address (the reference/correct address) and Transactional addresses (addresses to be matched). Each record of these data sets includes 7 fields, namely the id, CompanyName, AddressLine1, AddressLine2, PostTown, County, PostCode.

In Fig. 7, the addresses "adetail" and "bdetail " are potential duplicates because they share the same city. The structure of the graph may allow us to make even more sophisticated queries to limit the search space.

After finding the potential duplicate data, we use a function of tokenization. With that function we can form a set of tokens from the given input data. (In our example the input data is "aDetail" and "bDetail" in Fig. 7). Then we compute tf-idf vectors (show afeatures and bfeatures in Fig. 8) and using the defined "cSim" function, we calculate the similarity of feature vectors as shown below (Fig. 8).

Figure 9, clearly shows the similarity between the master data and transactional data, that allows us to match the records that are most closely matched, such as in

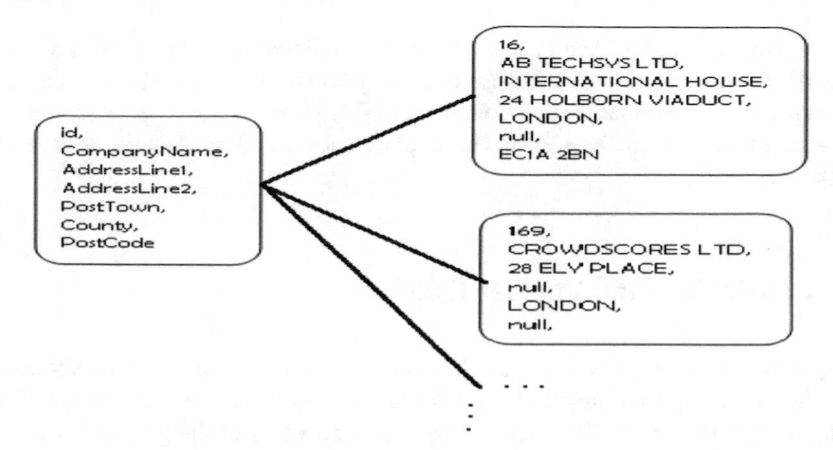

Fig. 6 Example of records

ae1	be1	aId	bId	aDetail	bDetail
[38, london, hasC...	[1006, london, ha...	38	1006	almaradix ltd 145...	1 tech 57 CHARTE...
[172, london, has...	[1006, london, ha...	172	1003	ct consultants lt...	1 mobile null CIT...
[179, ec1m6ha, ha...	[1006, london, ha...	179	1005	dandan digital lt...	1 tech ltd 57 CHA...
[200, london, has...	[1006, london, ha...	200	1004	digital goods sto...	1 tech ltd 57 CHA...
[222, london, has...	[1006, london, ha...	222	1006	electronic market...	1 tech 57 CHARTE...

Fig. 7 Example of potential duplicate pairs

aid	bid	adetail	bdetail	similarity	afeatures	bfeatures
280	1004	global technology...	global technology...	0.8684236217138475	(50000,[2758,1027...	(50000,[10277,320...
1	1002	1 mobile limited ...	1 mobile ltd. 30 ...	0.8606114041279506	(50000,[4744,8009...	(50000,[4744,8009...
2	1006	1 tech ltd 57 CHA...	1 tech 57 CHARTE...	0.8139991661912716	(50000,[4744,8595...	(50000,[4744,8595...
280	1005	global technology...	global technology...	0.7264278128943454	(50000,[2758,1027...	(50000,[2758,1027...
1	1001	1 mobile limited ...	1 mobile limited ...	0.6354935445843789	(50000,[4744,8009...	(50000,[4744,1944...

Fig. 8 Similarity calculator

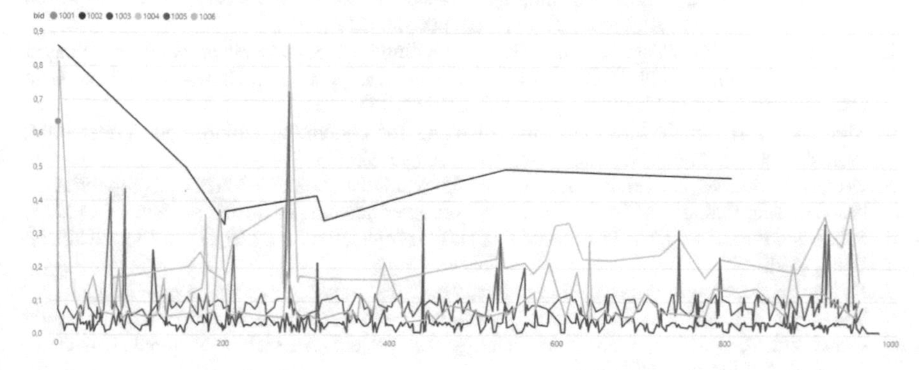

Fig. 9 Similar pairs comparison

this figure, each color identify a record in transactional data (bid), and the x-axis identify the records of master data (aid), so the power of this method is its simplicity in terms of implementation and complexity of calculus, because we structure our data in graphical form, this extended us to detect the different duplication in any database.

8 Conclusions and Further Research

In this document, we present a method for solving entity resolution and deduplication problem using Spark Frameworks proposed. We started with the discovery of the study of the data integration system; first we did a study of the general context of integration by its definition, and its architectures. Then, we treated the different data quality tools to present the different features that allow cleaning and improving the quality of the databases and especially the resolution of problems with the detection of duplicates.

This project can be extended to include more improvement to have less comparison between different records and to reduce the calculation time, for example using parallel calculation. Also, clustering analysis can be performed using the algorithms of the Apache Spark MLib libraries.

Future research could incorporate the use of non-relational and distributed databases such as HBase3 with the aim of reuse the results of comparisons in future comparisons within the same process, as the authors in [2].

References

1. Karapiperis, D., Verykios, V.S.: LoadBalancing the distance computations in record linkage. ACM SIGKDD Explor. Newslett. **17**(1), 1–7 (2015)
2. PrabhakarBennya, S., Vasavi, S., Anupriya, P.: Hadoop framework for entity resolution within high velocity streams. In: CMS 2016, vol. 85, pp. 550–557 (2016)
3. Yan, C., Song, Y., Wang, J., Guo, W.: Eliminating the redundancy in MapReduce-based entity resolution. In: 15th IEEE/ACM International Symposium on Cluster, Cloud and Grid Computing (CCGrid), Shenzhen, pp. 1233–1236 (2015)
4. Mon, A.C., Thwin, M.M.S.: Effective blocking for combining multiple entity resolution systems. Int. J. Comput. Sci. Eng. **2**(4), 126–136 (2013)
5. He, Q., Tan, Q., Ma, X., Shi, Z.: The high-activity parallel implementation of data preprocessing based on MapReduce. In: Lecture Notes in Computer Science, vol. 6401, pp. 646–654 (2015)
6. Albanese, P.A., Ale, J.M.: Data Matching and Deduplication Over Big Data Using Hadoop Framework
7. By Andrei Popescu in Data Matching. https://winpure.com/blog/what-is-record-linkage/
8. Towards a Scalable and Robust Entity Resolution—Approximate Blocking with Semantic Constraints COMP8740: Artificial Intelligence Project Australian National University Semester 1, 30 May 2014 (2014)
9. Parallel Sorted Neighborhood Blocking with MapReduce Lars Kolb, Andreas Thor, Erhard Rahm Department of Computer Science, University of Leipzig, Germany

10. Reprinted from Data Quality: Tools by (p. 35) Dries Van Dromme (2007)
11. Big Data Integration, Xin Luna Dong, Divesh Srivastava, March 2015
12. Integration donnees—approche materialisee virtuelle, 17 October 2008
13. Qualité contextuelle des données: détection et nettoyage guidés par la sémantique des données, Aïcha Ben Salem. Accessed 13 Dec 2017

Flight Arrival Delay Prediction Using Supervised Machine Learning Algorithms

Hajar Alla, Lahcen Moumoun, and Youssef Balouki

Abstract Over the last few decades, the air transport growth has shown good numbers in comparison with the other years. In fact, most of tourists, who crossed international borders last years, did so by airways. Over the next two decades, the demand for air transport is expected to double. Hence, the density of traffic is going to increase which will result in traffic delays. Delay is one of the most memorable performance indicators in the air transport system. It hurts passengers, airports and airlines. Pilots, air traffic controllers and other aviation personnel were questioned in this study and a survey was established to identify the importance of flight delay reduction. Thus, delays prediction turns out very useful. Flight delay prediction studies have been modeled in different ways. The approach of this work is based on machine learning algorithms. Our model is able to predict whether a scheduled flight will be on-time or delayed. We used relevant and filtered features that, to the best of our knowledge, some of them were not adopted in the previous studies. Holidays, seasons, day of week and the importance of the airport used were added as new features to enhance the accuracy of the prediction system. The resulting model was deployed and used as a flight delay prediction tool. The aim of the deployed application is to inform airport personnel and airlines about flight delays in advance to avoid losses and terminal crowdedness.

Keywords Flight delay prediction · Machine learning algorithms · Regression analysis

H. Alla (✉) · L. Moumoun · Y. Balouki
Laboratory of Mathematics, Computer and Engineering Sciences, Mathematics and Computer Science Department, Faculty of Science and Techniques, Hassan First University of Settat, Km 3, P.B. 577 Route of Casablanca, Settat 26000, Morocco
e-mail: h.alla@uhp.ac.ma

L. Moumoun
e-mail: lahcen.moumoun@uhp.ac.ma

Y. Balouki
e-mail: youssef.balouki@uhp.ac.ma

N. Gherabi and J. Kacprzyk (eds.), *Intelligent Systems in Big Data, Semantic Web and Machine Learning*, Advances in Intelligent Systems and Computing 1344,
https://doi.org/10.1007/978-3-030-72588-4_16

1 Introduction

Nowadays, air transport is considered as the best transportation mean. It is safer, time saver, useful for long distances, and with affordable prices. It also takes a vital role in giving medical assistance [1] which has lead to an aviation industry growth and a high demand. The ability to yield and respond to this demand will require the industry to shine in several operations, such as: boarding and disembarking facilitations, travelling comfort and passengers satisfaction, especially delay avoidance. Many factors affect flight delays including air traffic density, airports congestion, bad weather, boarding or disembarking problems, security measures, or ground operations issues such as aircraft maintenance, personnel absence, baggage loading, fuelling/refueling, aircraft cleaning, equipment congestion, etc. Flight delays not only affect passengers, airlines and airports but also the environment by increasing fuel consumption and gas emission [2]. Hence, flight delays avoidance turns out very important. To measure how useful is minimizing delays, a survey[1] was created and shared with pilots, air traffic controllers, airport personnel and passengers. 80% of the target population has considered flight delays annoying and necessary to be reduced. Most of them voted for delay avoidance as a solution for a lot of issues such as time and money saving, as many of them said Time is money. Also, being on-time reduces operational and financial risks and enhances company profit. The passengers considered delays avoidance very essential to reduce arriving late problems at destination. In fact, most of hotels check-in service do not have a 24 h reception, so they close at the night. A passenger even had to spend the night in a garden according to his survey responses. Flight delays minimizing will also help in arriving on-time in a business meeting or a doctor appointment. To solve flight delay problems, researchers have adopted different approaches. Figure 1 resumes the categories of the models used to predict flight delays.

Since artificial intelligence applications are less common but very efficient in this field, we decided to use Machine Learning to predict whether a scheduled flight will be delayed or on-time. For this, historical flight data extracted from Bureau of Transportation Statistics (BTS) were employed to help build the predictive model. Our model demonstrates various supervised machine learning algorithms, implemented in Python such as Multiple Linear Regression, Decision Tree, Random Forest, Gradient Boosting and XGBoost Regressors. It also evaluates the efficacy of each algorithm used to predict flight delays. Unlike previous studies that utilized only the features from the database, in our study, we created new features to enhance the efficacy of our work. Holidays, seasons, day of the week, airport frequenting and density are highly contributory to flight delays. The idea from selecting these new features is explained later in the paper. To strengthen the study, a Flight Delay Prediction Tool was built using our model to predict flight delays.

This paper is organized as follows: Sect. 2 shows a brief review of the previous researches related to our study. The proposed methodology used in this paper is

[1] https://forms.gle/nohRaMmAsgvphkXU7.

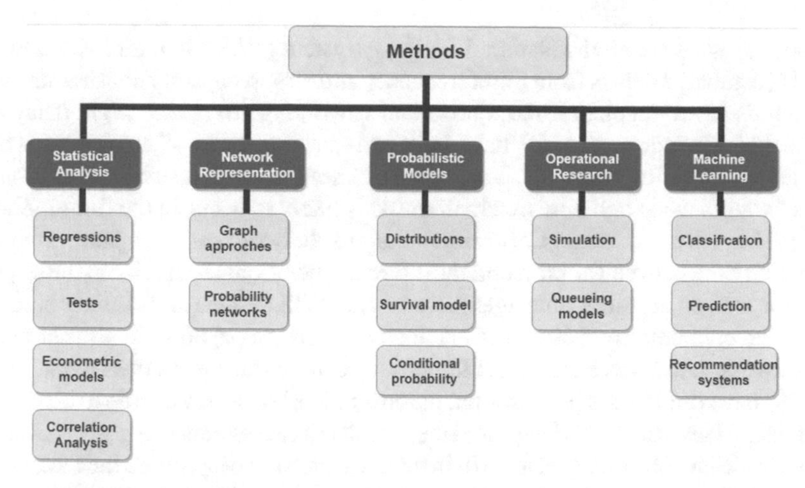

Fig. 1 Categories of methods utilized to predict flight delays

described in Sect. 3. The results of the predictive model are discussed in Sect. 4. It also visualizes and explains the Flight Delay Prediction Tool. Finally, it is concluded with future work and perspectives in Sect. 5.

2 Related Work

2.1 Traditional Methods

Flight delay has always been the major concern of airlines, airports and passengers. Hence, predicting and analyzing delays and their causes become the preoccupation of many researchers. The flight delay prediction problems were modeled in different ways according to the research goal. Nogueira et al. [3] used the ant colony methodology to reduce taxiing time for a set of aircraft at an airport by finding the optimal taxiing routes. The idea of the algorithm is captivating. However, it shows some contradictions. The aircraft will be able to minimize the taxi time by choosing the shortest path and avoiding conflicts with other traffic at the same time. In other terms, the system will be able to choose the adequate route and make its own separation with other nearby traffic, which is not practical because of many things. First, it is up to the ground traffic controller of the airport to give the instructions of push back, taxi routes, traffic avoidance, or any movement at the ground. Second, in case of bad weather conditions especially bad visibility, aircraft cannot avoid each other. Finally, if we let all the pilots or the aircraft systems detect their optimal taxiing routes, this may lead to taxiways congestion and automatically to delay problems as all the aircraft will choose the minimized path. Hao et al. [4] applied both econometric and simulation models to show how the delays of three New York airports can be propagated to all the National Airspace System (NAS). In their model, Evans et al. [5] estimated delays from flight frequency and airport capacity constraints using queuing theory to predict airline routing and scheduling. To reduce flight delay and total delay, Kotegawa et al. [6] have built and tested a series of algorithms which forecast restructuring of the US commercial airline network. The model is capable to predict where, when and how much air traffic is likely to occur in the future. Zhong et al. [7] have created Monte Carlo simulations to study runway occupancy time, the separation time between aircraft and the different operational scenarios at the airport. They focused on air traffic management to estimate flight delays. Baspinar et al. [8] applied an epidemic spreading methodology to delay propagation in air transportation network under stress using real flight-track data of Europe. In their work, Wong et al. [9] have used Cox's proportional hazards model for survival-time to clarify the relationship between flight delays and the associated causes considering a Taiwanese domestic airline. Mofokeng et al. [10] in their research have identified the factors that result in delays during a check maintenance. Interviews and maintenance documents were used in order to extract the causes of delays. Triangulation was done to validate authors' findings.

2.2 Machine Learning

Flight delay can also be studied otherwise. Recently, researchers and projects that work on flight systems recommend Machine Learning techniques. Thiagarajan et al. in their paper [11] have developed a two-stage predictive model for flight delay. The first one is a classification model to predict whether a flight will be delayed or not. The second one is a regressive model to calculate the value of the delay. The two existing models contain redundant information: predicting the occurrence of delays using classification and calculating the time using regression. However, we can use a regression algorithm to predict the flight delay time and with a simple function, we can deduce if a flight is delayed or not by comparing if it is less, equal or greater than 15 min as we will be using in our predictive model. Priyanka [12] examined delays using K Nearest Neighbor algorithm. The model was implemented with features including flights and weather conditions for both departure and arrival airports with an accuracy of 90%. The model in [13] shows that, using only top-3 features of the data, the decision tree classifier, logistic regression and neural network were able to achieve a test accuracy of approximately 91% for each one. According to Manna et al. [14] Gradient Boosted Decision Tree has shown a better accuracy in their model as compared to other methods. Alok Dand et al. in their paper [15] used binary supervised and unsupervised machine learning classification algorithms to predict airline delays. They have taken into account two other features such as airline hub and aircraft size. Chakrabarty et al. [16] made a model based on Gradient Boosting, Support Vector Machine, Random Forest and K-Nearest Neighbour to accurately predict the arrival delay of American Airlines flights. The result was 79.7% best performance of Gradient Boosting Classifier in comparison to the other algorithms.

Ebenezer et al. [17] combined flight US data and weather observations then used a sampling technique to balance the data before applying machine learning algorithms to predict the flight delays with a better accuracy.

3 Proposed Approach

The methodology proposed in this work and the process of applying Machine Learning to a real-world problem is pictured in Fig. 2.

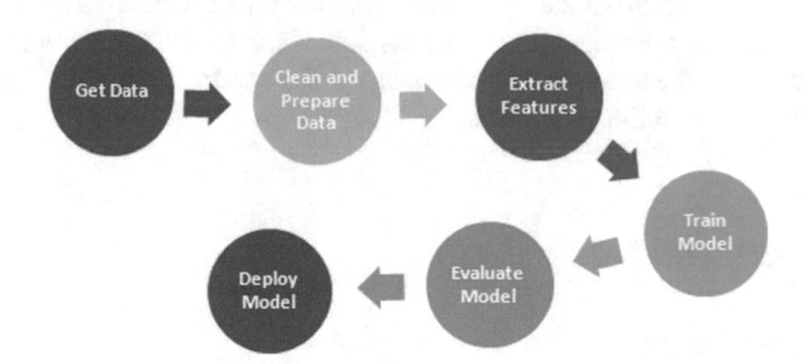

Fig. 2 Proposed approach process

3.1 Data Description

In our proposed model, we used flight data which are sourced from the Bureau of Transportation Statistics and contain about 760 thousand flights performed in the year 2018. An amount of 18 airlines and 122 airports was considered in our study. We chose US BTS mainly because of its reliability and also due to the large volume of statistical data saved since 1987 which leads to a well training of the model. Rigorous analysis, transparent data quality, and independence from political influence are the slogan of the BTS. We selected from the database different features we considered relevant and necessary for our model. Following the BTS flight delay definition, flights whose difference between scheduled and actual arrival times is 15 min or greater are considered delayed.

3.2 Data Cleaning

As we know, real-world data is often incomplete, may contain errors and contradictions. Also, unnecessary multiple data can lead to a low accuracy. Hence, data pre-processing and cleaning turn out useful to show best results. As we will be using machine learning algorithms, some of them need information in a specified format. For example, Random Forest algorithm does not support null values. First, we started by removing examples containing missing values. Second, we converted all categorical variables into numerical ones since other machine learning algorithms show better performance with numeric such as multiple linear regression. Then, we removed also the airports data that occur less than 10 times. To summarize, our dataset includes three csv files. One contains data about flights performance in 2018. The other contains the most used and frequented airports in US in 2018. The last one resumes federal US holidays in 2018.

3.3 Features Analysis

3.3.1 Dataset Features

Features were selected from the data based on a strong correlation with flight delays and due to many reasons. Some of the features are described as follows:

Flight Distance: The long-term flights are more likely to arriving late in several cases: low performance type of aircraft used to perform the flight, bad en-route meteorological conditions, density of traffic or crew stress and fatigue. According to Lowden et al. [18], in long distance flights, especially after an extra dimension of night work, most of air crew suffers from Jet Lag. It is a flight fatigue and disorder resulting from travelling across multiple time zones.

Origin and Destination Airports: An airport has a high contribution into delays depending on its capacity, its size, its personnel, its infrastructure and facilities, etc. Also, as we evocated before, if the distance between the origin and destination airports is important, the flight is likely to be late.

Carrier Code (or Airlines Code): Airlines may lead to flight delays in several ways: An airline which has lazy or inattentive personnel (push-back car drivers, mechanical engineers, fuel service staff), or a non professional crew (pilots, stewards), or an airline that uses a low performance type of aircraft or which does not have standby aircraft in case of failures. Furthermore, airlines whose personnel suffer from stress and fatigue contribute in delays. All these parameters prove that we should consider this feature in our study. We plot Carrier Code of airlines against mean arrival delay, in Fig. 3, to find out which ones got delayed more than others in order to show how involved are companies in flight delays. It revealed that JetBlue, Endeavor and Delta airlines were respectively the most delayed in 2018.

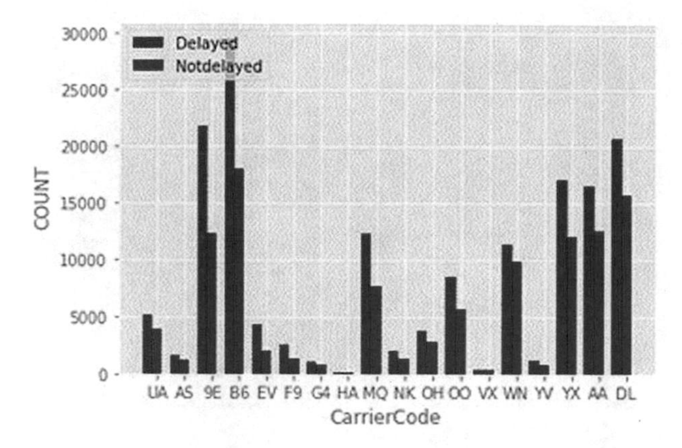

Fig. 3 Plotting Carrier Code and Arrival delay

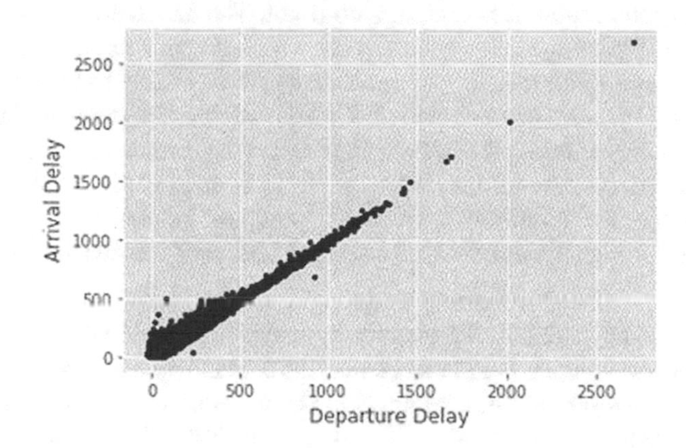

Fig. 4 Departure and arrival delay features correlation

Departure Delay: A flight delayed on departure is likely to be delayed on arrival. The correlation between Departure Delay and Arrival Delay features from our dataset are scattered and plotted in Fig. 4. From the plot results, we can see that Departure and Arrival delays are totally linked and dependant. A traffic which had a delay on departure will obviously land later than the scheduled arrival time.

3.3.2 Added Features

We created four other relevant features that, to the best of our knowledge, have not been selected so far in the previous studies.

Airport Frequenting: The first one FREQUENTED is related to the density of the airport, whether it is most frequented by airlines and passengers or not. It is equal

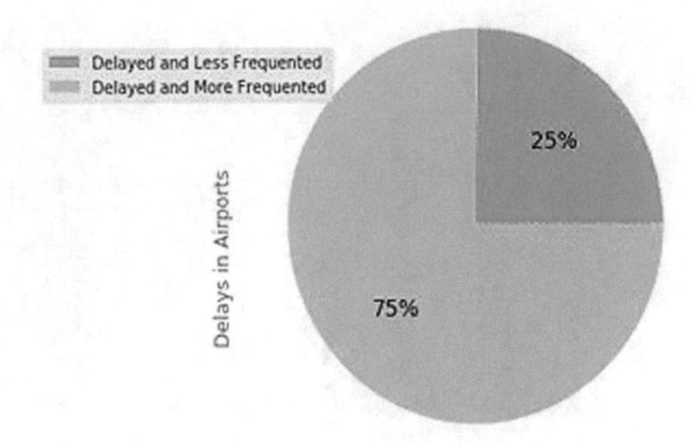

Fig. 5 Airports density and delayed flights

to 1 if the airport is more frequented, zero if not. The data were extracted from the World Airport Codes database.[2] An airport which is more frequented with a high traffic density and a huge number of movements is more likely to suffer from delays. In fact, Central Flow Management Unit (CFMU) allocates to every flight crew a slot time in which they should take-off which is related to the density of the airport and the airspace. The pie model presented in Fig. 5 compares the number of delayed flights from our data in both most frequented and less frequented airports. It means that most of the flights were delayed mainly because of the density of the airports they used.

Holiday: Obviously, in a holiday, everybody tend to travel using the safest and fastest mean of transport which is the plane. Hence, the demand increases and so does traffic density, which involves in flight delays. The second added feature HOLIDAY is true if the day of the flight is a federal holiday (Christmas's day, New Year, Thanksgiving day, etc.) and false if not. These data were selected from US Office of Personnel Management OPM.[3]

Season: A season in which there is a high demand of travelling is likely to suffer from delays. The third feature SEASON was created to monitor in which season the flight was delayed the most, SUMMER, AUTUMN, WINTER, or SPRING. In Fig. 6, arrival delays are plotted against the four seasons of the year. It shows that most of delays occurred respectively in summer, autumn and spring, mainly because of the weather in these seasons which is generally warm and in normal meteorological conditions, which makes people more interested in travelling.

Day of the Week: Not all the days of the week have the same effect on flight delays. For example, in some countries, Saturdays and Sundays affect the most air travel since it is the Weekend. Hence, we extracted the fourth feature from DATE OF FLIGHT column and called it WEEKDAY in which the flight was performed. In Fig. 7, arrival

[2]https://www.world-airport-codes.com.

[3]https://www.opm.gov.

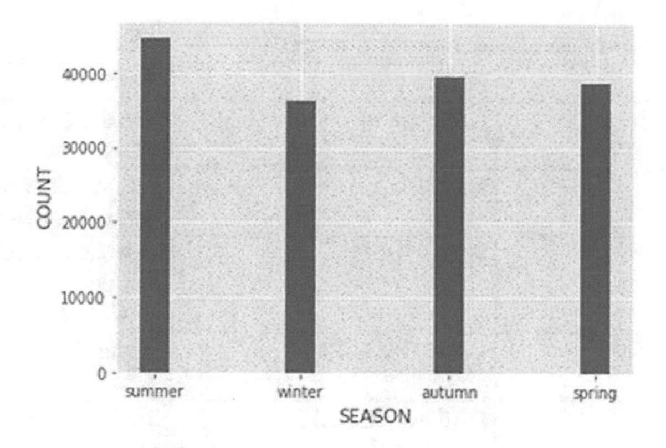

Fig. 6 Plotting Seasons and Arrival delay

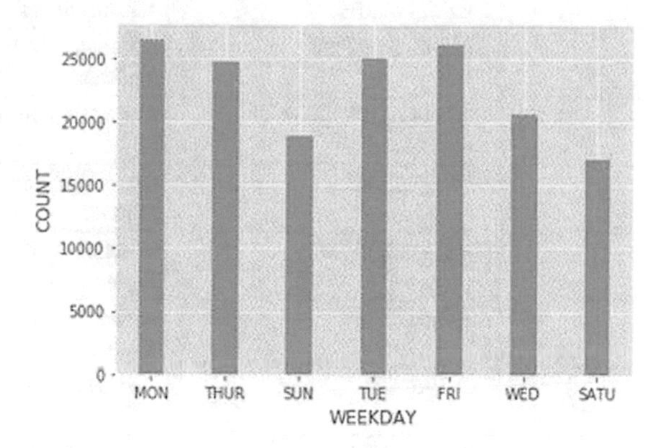

Fig. 7 Plotting Weekdays and Arrival delay

delays are represented according to days of the week to find when flights are more often delayed. It shows that in the US, most of the flights in 2018 were delayed respectively on Mondays, Fridays, Tuesdays and Thursdays. Analyzing the Holiday data revealed that Monday was the most celebrated day in 2018, United States.

In order to simplify the training, we extracted from DATE OF FLIGHT column the parameters month and day and placed it in two new columns: MONTH and DAY. Since WEEKDAY column was extracted and the year used for our predictive model is 2018, obviously, we removed the column DATE OF FLIGHT to avoid redundancy. The fifteen features utilized in our study are described in Table 1.

We removed also the airports data that occur less than 10 times. To summarize, our dataset includes three csv files. One contains data about flights performance in 2018. The other contains the most used airports in US in 2018. The last one resumes federal US holidays in 2018.

Table 1 Description of the proposed model features

Number	Feature	Description
1	DAY	The day in which the flight was performed (example: 24)
2	MONTH	The month in which the flight was performed (example: 09)
3	WEEKDAY	The day in which the flight was performed (example: FRIDAY)
4	CARRIER CODE	Airline IATA[a] code (example: AA: American Airlines)
5	ORIGIN_NAME	The airport of departure (example: JFK: John F. Kennedy International Airport)
6	DEST_NAME	The airport of arrival
7	CRS_DEP_TIME	Scheduled departure time of the flight
8	ACTUAL_DEP_TIME	Actual departure time of the flight
9	DEP_DELAY	Departure delay of the flight in minutes
10	CRS_ARR_TIME	Scheduled arrival time of the flight
11	ARR_DELAY	Arrival delay of the flight in minutes
12	DISTANCE	Distance between airports in miles
13	FREQUENTED	1 if the airport of departure or arrival is among the most frequented airports in US in 2018 0 if not
14	HOLIDAY	1 if the day of flight is a federal US holiday 0 if not
15	SEASON	Summer, Winter, Autumn or Spring

[a] International Air Transport Association

3.4 Training Algorithms

With the big increasing amount of data becoming available in the last years there is a good reason that smart data analysis becomes necessary. According to [19], the aim of machine learning, whether supervised or unsupervised is being able to perform well, based on past experiences. In unsupervised learning, the machine tries to find hidden structure in unlabeled data. As [20] said, unsupervised machine learning involves pattern recognition without the involvement of a target attribute. In supervised machine learning, we find regression-based systems and classification-based systems. Classification algorithms use features to classify an item into a category. Regression algorithms employ features for predicting an output that is a continuous variable. In our case, we are using regression algorithms since our result will be a value which is flight delay.

Multiple Linear Regression: To understand the relationship between input (independent variables) and output (dependent variable), Linear regression is often used. The Eq. 1 represents the Linear Regression formula. The goal is to create a best fit line through all of the data points to easily understand predictions by minimizing margin or residual variation [21].

$$y = \beta_0 x_0 + \beta_1 x_1 + \beta_2 x_2 + \cdots + \beta_i x_i + £ \tag{1}$$

Where y is the output (dependent variable), x is the predictor (independent variable), β_0 is the bias or the intercept of y-axis, β_1 is the slope or rate of the independent variable and £ is the residuals of the model.

Decision Tree: A decision tree is an algorithm that helps in classifying an event or predicting the output values of a variable. It starts from a root and contains internal nodes that represent features and branches (outcomes) [22]. To measure the uncertainty and the homogeneity of a sample in a node and determine how impure or disordered a data set is, decision tree algorithm uses indices such as entropy or Gini impurity measured respectively by Eqs. 2 and 3 [19].

$$H(S) = \sum_{i=1}^{C} -p(i) \times \log_2 p(i) \tag{2}$$

$$G = \sum_{i=1}^{C} p(i) \times (1 - p(i)) \tag{3}$$

Where C is the number of classes, S is the set of examples, and p(i) is the probability of randomly picking an element of class i.

Random Forest: Random forest is based on multiple trees running to produce an average prediction. Each decision tree works on a sample of the data. The output of the multiple decision trees is combined to generate the prediction. The major tuned parameters in random forest are the number and depth of trees [23].

Gradient Boosting: Gradient boosting is a technique for both regression and classification problems in which the model is trained sequentially [24]. Each decision tree predicts the error of the previous one by boosting and minimizing the loss function.

Extreme Gradient Boosting: XGBoost, or extreme gradient boosting, is a gradient boosting variation. It adds regularization and uses a distributed, multithreaded processing to speed up the operations and enhance the efficiency. More explanations of machine learning algorithms are given in [25].

4 Results and Experiments

4.1 Performance Metrics

MAE (Mean Absolute Error): is a measure of how close forecast or predictions are to the eventual outcome [26]. Represented by Eq. 4, it is the most natural and unambiguous measure of average error [27].

$$MAE = \frac{1}{n}\sum_{i=1}^{n}\left|y_i - \hat{y}_i\right| \quad (4)$$

Where $\hat{y}_i$ is the predicted value of the i-th sample, y_i is the corresponding true value and n is the number of samples.

RMSE (Root Mean Squared Error): is the square root of the mean of squares of all the errors, as given in Eq. 5. It is usually better at revealing model performance differences. The errors are unbiased and follow a normal distribution [28].

$$RMSE = \sqrt{\frac{1}{n}\sum_{i=1}^{n}(y_i - \hat{y}_i)^2} \quad (5)$$

Where $\hat{y}_i$ is the predicted value of the i-th sample, y_i is the corresponding true value and n is the number of samples.

MdAE (Median Absolute Error): is a robust measure to outliers. The loss is calculated by taking the median of all absolute differences between the target and the prediction [29]. It is calculated by Eq. 6.

$$MdAE(y_i, \hat{y}_i) = median(\left|y_1 - \hat{y}_1\right|, \ldots, \left|y_n - \hat{y}_n\right|) \quad (6)$$

Where $\hat{y}_i$ is the predicted value of the i-th sample, y_i is the corresponding true value and n is the number of samples.

4.2 Results

As discussed in Sect. 3, we used 760 thousand samples to train and test the model. We split flight data into 80% for training and 20% for testing. We trained the model

Table 2 Metrics values of applied algorithms

Algorithm	Accuracy	MAE	Median Absolute Error	RMSE
Multiple Linear Regression	93.10%	16.76	13.5	22.49
Decision Tree	91.72%	16.95	12.54	24.64
Random Forest	98.41%	5.73	2.0	10.78
Gradient Boosting	86.52%	21.81	17.46	31.43
XGBoost	95.09%	13.54	10.66	18.96

Table 3 Accuracy Benchmark using the algorithms in common

Accuracy					
Algorithm	Our work	Chakrabarty et al. [16]	Dand et al. [15]	Thiagarajan et al. [11]	Ebenezer et al. [17]
Decision Tree	91.72%	–	81.60%	93.73%	95.62%
Random Forest	98.41%	78.70%	80.90%	94.09%	96.89%
Gradient Boosting	86.52%	79.72%	–	94.35%	–

using Multiple Linear Regression, Decision Tree, Random Forest, Gradient Boosting and XGBoost Regressors that were coded in Python. To analyze and evaluate the performance of the proposed model results, accuracy, mean absolute error (MAE), median absolute error and root mean squared error (RMSE) were used and displayed in Table 2. We propose in Table 3 a Benchmark which consists of comparing the accuracy of our model to that of other studies from the literature review [11, 15–17].

From Tables 2 and 3, Random Forest was the best performing method in predicting flight arrival delays with the highest percentage of accuracy in our study 98.41% and in Ebenezer et al. [17] paper with a percentage of 96.89%. The values of the errors were weak compared to the others algorithms.

4.3 Flight Delay Prediction Tool

In order to organize the airport gates, the disembarking doors and avoid terminal crowdedness, also to minimize costs and fuel consumption, we developed a Flight Delay Prediction Tool (FDPT) that will help airport personnel and airlines to predict flight delays in advance. This application was built using Python. Figure 8 represents the interface of the FDPT. This tool can be used immediately after the take-off of the flight, so the actual departure time feature will be known. A flight delayed on

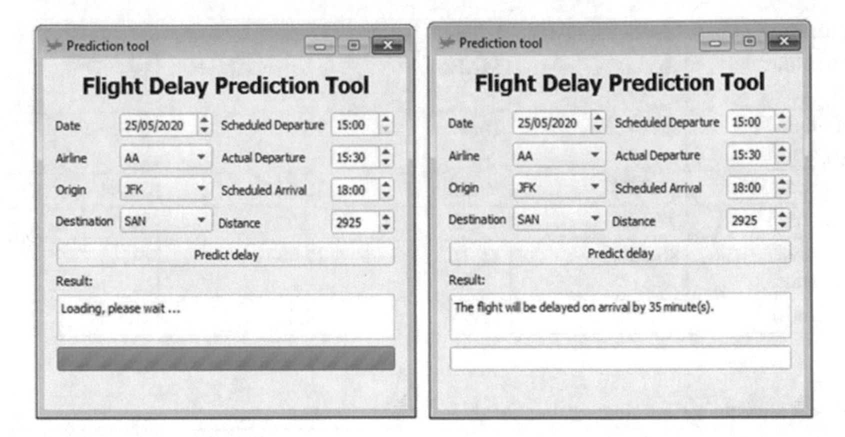

Fig. 8 Interface of the FDPT

departure will be surely delayed on arrival. But when the scheduled departure time is equal to the actual departure time (departure delay is null), the flight still may be delayed due to many other reasons such as en route bad weather, traffic density, in-flight incident or accident, destination airport density, etc. When the flight is delayed, the message "The flight will be delayed on arrival by x minute(s)." is displayed, otherwise "the flight will be on time." is visualized.

5 Conclusion and Future Work

To summarize, Multiple Linear Regression, Decision Tree, Random Forest, Gradient Boosting and XGBoost were used for predicting the arrival delay using 15 important features including created ones. In this project, we were able to achieve a high accuracy prediction such as 98.41% for Random Forest, 95.09% for XGBoost, and 93.10% for Multiple Linear Regression. To boost and strengthen our model, a tool that allows airport users to predict flight delays was developed. As perspectives, we consider, in the future, using other features such as the number of passengers of the flight, departure and destination airports size and infrastructure. Moreover, weather conditions for en route, departure and destination airports can also achieve good results for analyzing flight delays. We will also focus on improving the efficacy of the FDPT which we will try to make available two days or more before the flight. Finally, we vision to use Neural Network to predict future flight arrival delays.

References

1. Musaddi, R., Jaiswal, A., Girdonia, M., Sanjudharan, M.S.M.: Flight delay prediction using binary classification. Int. J. Emerg. Technol. Eng. Res. (IJETER) **6**, 34–38 (2018)
2. Sternberg, A., Soares, J., Carvalho, D., Ogasawara, E.: A review on flight delay prediction, arXiv preprint arXiv:1703.06118: Computers and society (2017)
3. Nogueira, K.B., Aguiar, P.H.C., Weigang, L.: Using ant algorithm to arrange taxiway sequencing in airport. Int. J. Comput. Theory Eng. **6**, 357 (2014)
4. Hao, L., Hansen, M., Zhang, Y., Post, J.: New York, New York: two ways of estimating the delay impact of New York airports. Transp. Res. Part E: Logist. Transp. Rev. **70**, 245–260 (2014)
5. Evans, A., Schafer, A., Dray, L.: Modelling airline network routing and scheduling under airport capacity constraints. In: The 26th Congress of ICAS and 8th AIAA ATIO, p. 8855 (2008)
6. Kotegawa, T., DeLaurentis, D., Noonan, K., Post, J.: Impact of commercial airline network evolution on the US air transportation system. In: Proceedings of the 9th USA/Europe Air Traffic Management Research and Development Seminar (ATM 2011) (2011)
7. Zhong, Z.W., Varun, D., Lin, Y.J.: Studies for air traffic management R&D in the ASEAN-region context. J. Air Transp. Manag. **64**, 15–20 (2017)
8. Baspinar, B., Koyuncu, E.: A data-driven air transportation delay propagation model using epidemic process models. Int. J. Aerosp. Eng. **2016**, 11 p. (2016). ID 4836260
9. Wong, J.-T., Tsai, S.-C.: A survival model for flight delay propagation. J. Air Transp. Manag. **23**, 5–11 (2012)
10. Mofokeng, T.J., Marnewick, A.: Factors contributing to delays regarding aircraft during A-check maintenance. In: IEEE Technology and Engineering Management Conference (TEMSCON), pp. 185–190 (2017)
11. Thiagarajan, B., Srinivasan, L., Sharma, A.V., Sreekanthan, D., Vijayaraghavan, V.: A machine learning approach for prediction of on-time performance of flights. In: 2017 IEEE/AIAA 36th Digital Avionics Systems Conference (DASC), vol. 6, pp. 1–6 (2017)
12. Priyanka, G.: Prediction of airline delays using K-nearest neighbor algorithm. Int. J. Emerg. Technol. Innov. Eng. **4**(5), 87–90 (2018). ISSN: 2394 – 6598
13. Kuhn, N., Jamadagni, N.: Application of machine learning algorithms to predict flight arrival delays, CS229 (2017)
14. Manna, S., Biswas, S., Kundu, R., Rakshit, S., Gupta, P., Barman, S.: A statistical approach to predict flight delay using gradient boosted decision tree. In: International Conference on Computational Intelligence in Data Science (ICCIDS), vol. 6, pp. 1–5 (2017)
15. Dand, A., Saeed, K., Yildirim, B.: Prediction of Airline Delays based on Machine Learning Algorithms. In: AMCIS (2019)
16. Chakrabarty, N., Kundu, T., Dandapat, S., Sarkar, A., Kole, D.K.: Flight arrival delay prediction using gradient boosting classifier. In: Emerging Technologies in Data Mining and Information Security, pp. 651–659 (2019)
17. Ebenezer, K., Brahmaji Rao, K.N.: Machine learning approach to predict flight delays. Int. J. Comput. Sci. Eng. **6**, 231–234 (2018)
18. Lowden, A., kerstedt, T.: Eastward long distance flights sleep and wake patterns in air crews in connection with a two-day layover. J. Sleep Res. **8**, 15–24 (1999)
19. Panesar, A.: Machine learning algorithms. In: Machine Learning and AI for Healthcare, pp. 119–188. Apress, Berkeley (2019)
20. Berry, M.W., Mohamed, A., Yap, B.W.: Supervised and Unsupervised Learning for Data Science. Springer, Heidelberg (2019)
21. Ayyadevara, V.K.: Linear regression. In: Pro Machine Learning Algorithms, pp. 17–47. Apress, Berkeley (2018)
22. Ayyadevara, V.K.: Decision tree. In: Pro Machine Learning Algorithms, pp. 71–103. Apress, Berkeley (2018)
23. Ayyadevara, V.K.: Random forest. In: Pro Machine Learning Algorithms, pp. 105–116. Apress, Berkeley (2018)

24. Ayyadevara, V.K.: Gradient boosting machine. In: Pro Machine Learning Algorithms, pp. 117–134. Apress, Berkeley (2018)
25. Ayyadevara, V.K.: Pro Machine Learning Algorithms. Apress, New York (2018)
26. Osisanwo, F.Y., Akinsola, J.E.T., Awodele, O., Hinmikaiye, J.O., Olakanmi, O., Akinjobi, J.: Supervised machine learning algorithms: classification and comparison. Int. J. Comput. Trends Technol. (IJCTT) **48**, 128–138 (2017)
27. Willmott, C.J., Matsuura, K.: Advantages of the mean absolute error (MAE) over the root mean square error (RMSE) in assessing average model performance. Climate Res. **30**(1), 79–82 (2005)
28. Chai, T., Draxler, R.R.: Root mean square error (RMSE) or mean absolute error (MAE)? Arguments against avoiding RMSE in the literature. Geosci. Model Dev. **7**(3), 1247–1250 (2014)
29. Metrics and scoring: quantifying the quality of predictions. https://www.scikit-learn.org

Arabic Machine Translation Based on the Combination of Word Embedding Techniques

Nouhaila Bensalah, Habib Ayad, Abdellah Adib, and Abdelhamid Ibn El Farouk

Abstract Automatic Machine Translation is a computer application that automatically translates one source-language sentence into the corresponding target-language sentence. With the increased volume of user-generated content on the web, textual information becomes freely available and with a gigantic quantity. Hence, it is becoming increasingly common to adopt automated analysis tools from Machine Learning (ML) to represent such kind of information. In this paper, we propose a new method called Enhanced Word Vectors (EWVs) generated using Word2vec and FastText models. These EWVs are then used for training and testing a new Deep Learning (DL) architecture based on Convolutional Neural Networks (CNNs) and Recurrent Neural Networks (RNNs). Moreover, special preprocessing of the Arabic sentences is carried out. The performance of the proposed scheme is validated and compared with Word2vec and FastText using UN dataset. From the experimental results, we find that in most of the cases, our proposed approach achieves the best results, compared to Word2vec and FastText models alone.

Keywords Arabic machine translation · Arabic preprocessing · ATB · Word embedding · Word2vec · FastText · RNN · CNN

1 Introduction

Machine Translation (MT) is one of the most sought after areas of research in the linguistics and computational community. It is a field of study under Natural Language

N. Bensalah (✉) · H. Ayad · A. Adib
Team Networks, Telecoms and Multimedia, University of Hassan II Casablanca, 20000 Casablanca, Morocco
e-mail: nouhaila.bensalah@etu.fstm.ac.ma

A. Adib
e-mail: abdellah.adib@fstm.ac.ma

A. I. El Farouk
Teaching, Languages and Cultures Laboratory Mohammedia, Casablanca, Morocco

N. Gherabi and J. Kacprzyk (eds.), *Intelligent Systems in Big Data, Semantic Web and Machine Learning*, Advances in Intelligent Systems and Computing 1344,
https://doi.org/10.1007/978-3-030-72588-4_17

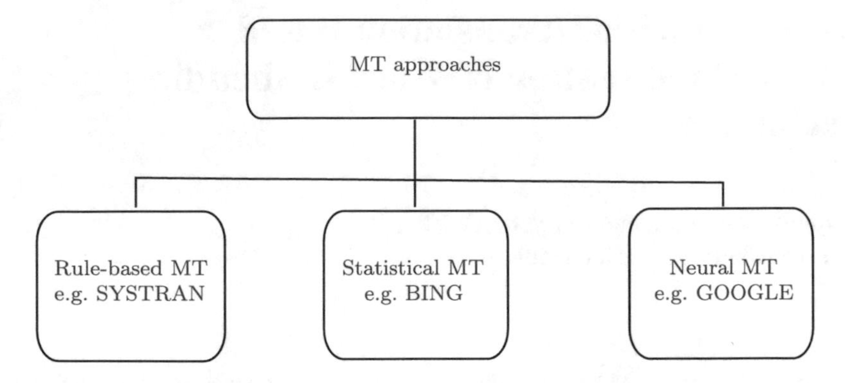

Fig. 1 General MT approaches

Processing (NLP) which deals with the automatic translation of human language, from one language to another with the help of computerized systems.

As shown in Fig. 1, MT can be categorized as rule-based, Statistical and Neural Machine Translation (NMT). The Rule-Based MT (RBMT) involves a set of linguistic rules which are usually constructed by a language expert. The downside of this approach is that it strongly relies on a large dictionary and great linguistic knowledge [20]. The latter is expensive to build and it is impossible to write rules that cover all languages. The Statistical MT (SMT) is a data-driven approach which focuses on probabilistic models during the translation process. SMT system is composed of three main stages, which are: language model, translation model and decoder model. First, the translation model is trained on bilingual corpus and used to estimate the probability that the source sentence is the translated version of the target sentence. Then, the language model is trained on a monolingual corpora so as to enhance the fluency of the output translation. Finally, in the decoding phase, the most probable sentence in the target language is determined based on the language model as well as the translation one. The strength of this approach is that it can deal with ambiguity by recording phrase-based translations with their frequency of occurrence on a phrase table [3, 5]. Hence, the generated translations are more fluent and natural than RBMT. However, as stated in [21], it gives poor results when translating sentences that are not similar to content from the training data. Recently, NMT has been expeditiously attracting the attention of the research community for its impressive results [8, 15, 33]. NMT models adhere to an end-to-end encoder-decoder approach. The encoder maps a given source sentence into a continuous vector which is termed as context vector. This vector contains all the necessary information which can be extracted from the input sentence itself. Finally, based on the context vector, the decoder predicts its translation, word by word, in the target language. Initially, Sutskever et al. [30] proposed Recurrent Neural Networks (RNNs) for both encoding and decoding. However, one of the weaknesses of a basic RNN is that it runs into vanishing gradient due to dependencies between word pairs. In order to deal with these issues, a popular alternative is either to use LSTM [23] or GRU [14] instead of Vanilla RNN cell.

Although a lot of research has been done on NMT for European languages, few studies have used the neural approach to translate Arabic sentences [4, 11, 16, 19, 27]. Hence, one of the main purposes of this work is to combine the strengths of two widely used architectures which are: Bidirectional LSTM (BiLSTM) and Convolutional neural network (CNN) [24] and then develop a novel Deep Learning (DL) model called C-BiLSTM for Arabic MT. C-BiLSTM utilizes CNN to extract the regional features from the input sequence, and are fed into BiLSTM to deal with the long-term dependency problem and obtain the sentence representation.

Word embeddings, also called a distributed representation of words, are techniques/methods for representing an input word in such a way that it can be distinguishable from all other input words. The core idea of word embeddings is that words used in the same contexts have a high proportion of similar meaning [22]. Several methods have been developed to obtain such representations from a corpus and has achieved an important advancement in many NLP tasks [9–11].

In our work, we present the benefit obtained from the use of three different word embedding models which are: Continuous Bag of Word (CBOW) and Skip-Gram (SG) and FastText. Our scheme is based on the combination of Word2vec and FastText representations to generate EWVs for each input sentence. The resulting vectors are then used as inputs to the proposed C-BiLSTM model.

The following sections are organized as follows: The methodology is detailed in Sect. 2. The performances of the proposed method is evaluated in Sect. 3. Finally, the paper is concluded in Sect. 4.

2 Proposed Methodology

As introduced earlier, the idea of the present work is to use three word embedding models for features extraction phase. Capturing several word features from the same input sentence possibly incorporates some features those that were not captured by the other model. In this paper, after the text preprocessing step, the word representations of a given input sequence using the three embedding models are first generated. Then, the minimum (element-wise) of the resulting representations is computed using the Minimum layer to form the EWVs of the same input sequence. Finally, these word vectors are used for training and testing the proposed C-BiLSTM model. Figure 2 shows the main architecture of our work.

Hereafter, we present the Arabic preprocessing used in this study, which includes the cleaning, the normalization and the tokenization of the Arabic sentences. Word2vec and FasText models will then be introduced followed by a brief description of the different components that embody the proposed C-BiLSTM model.

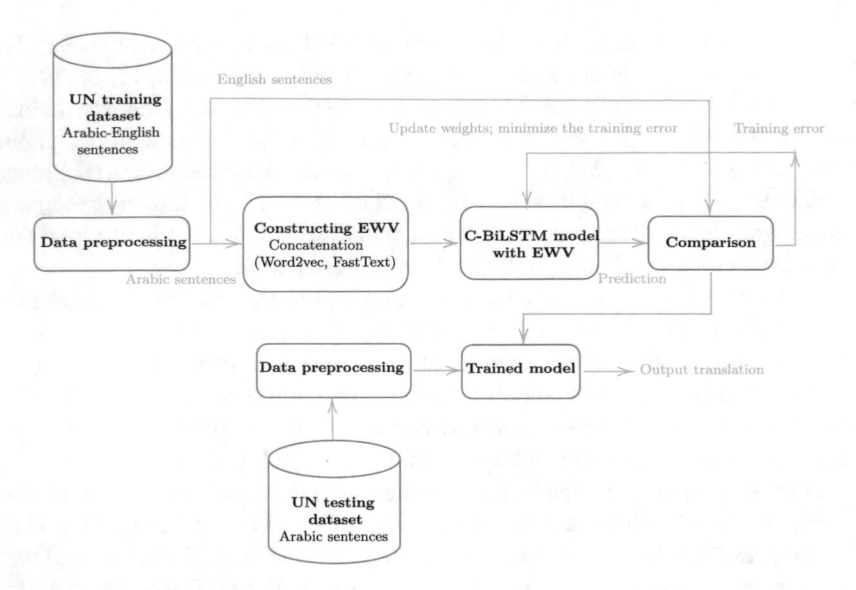

Fig. 2 The main architecture

2.1 Arabic Preprocessing

In this paper, we aimed to investigate the performance of the proposed Arabic MT with UN dataset [32]. In order to keep only the relevant words, several preprocessing operations of Arabic sentences have been applied. First, we remove the non-Arabic letters and the tashkeel symbols. We manually correct words that have missing letters and we remove all punctuation characters. Then, we normalize some variants of Alif ا, Ya ي and Ta marbota ة in all schemes, where (آ, أ, إ) are replaced with ا, the ي is replaced with Alif Maqsura ى, the ة is replaced with ه. Finally, each Arabic sentence is transformed into a list of tokens with a morphology-aware tokenization scheme. Thereby, Farasa is used to split all clitics using SVM to rank potential word segmentations [1]. It has shown to have a significant effect on NMT [4, 6, 27], particularly in the case of morphologically rich languages such as Arabic.

2.2 Word2vec

Word2vec is a prediction-based model proposed by Mikolov et al. [25] for learning dense vector representations of the unique words from a large unlabeled corpus. It is a shallow neural network model that learns the mapping of words to a point in a vector space. Word2vec offers two models namely CBOW and SG as well as advanced optimization techniques including hierarchical softmax and negative sampling. As shown in Fig. 3, CBOW learns the vector representations by predicting

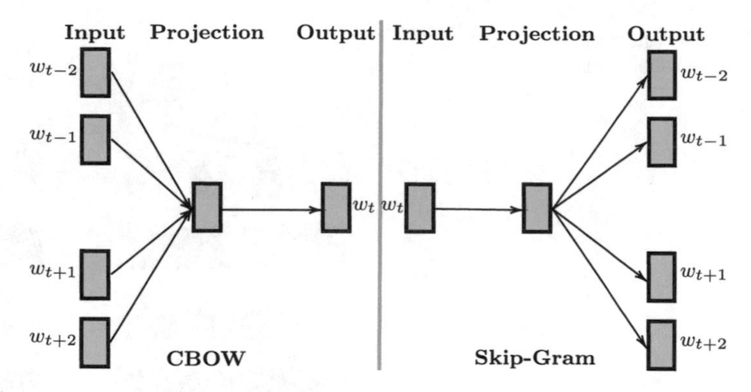

Fig. 3 CBOW and SG training model illustrations

the center word given all its context words while SG learns the vector representations by predicting each context word separately based on the center word. The two key parameters for training CBOW or SG embeddings are: the embedding dimension which is the dimension of the word vectors and the length of the context window representing the number of words that should be used before and after the center word as context for training the word vectors.

2.3 *FastText*

Embeddings models like CBOW and SG assign dense vector representations to words which are treated as atomic entities. However, they cannot handle the Out Of Vocabulary (OOV) problem since they ignore internal sub-word information (i.e., sequences of adjacent characters) which is an important consideration for languages with a large and rich vocabulary. Bojanowski et al. [13] proposed FastText in 2016, it is essentially an improvement of the word2vec. The model treats each word in the vocabulary as a bag of character n-grams, the characters embeddings generated are concatenated to generate the vector representation of the word. As a result, FastText model is able to build vectors of OOV words and can represent the morphology and lexical similarity of words. In order to further improve the quality of the words vector representations, the authors in [7] proposed another model called Probabilistic FastText. The model uses Gaussian Mixture Models (GMMs) to represent each word in the vocabulary as a GMM of K components. Moreover, it takes advantage of FastText model discussed earlier, this combination allows the model to capture different word senses, in addition to being able to generate better vector representations of OOV words.

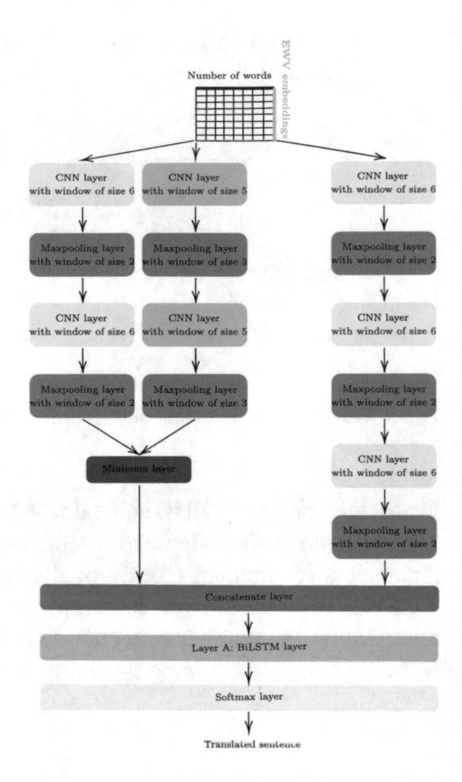

Fig. 4 The architecture of C-BiLSTM for Arabic MT

2.4 C-BiLSTM Model

The architecture of the C-BiLSTM model is illustrated in Fig. 4. It consists of two main components: CNN and BiLSTM. In the following, we describe how we apply CNN and BiLSTM to capture both regional as well as temporal features from the EWVs features.

2.4.1 CNN

The one-dimensional convolution involves a filter (kernel) passing over a sequence and extracting features at different positions. In this work, each input sentence is modeled as a matrix by concatenating its EWV embeddings as columns. The generated matrix is used as input of three multi-layer stacked CNN networks. The first one is built by stacking three convolutional layers with a window of size six. In order to reduce the amount of parameters and computation in the network, Max pooling is added after each convolution with a pooling length of two. The second one is generated by stacking two convolutional layers with a window of size five. Max pooling is added after each convolution with a pooling length of three. The last one is built by stacking two convolutional layers with a window of size six. As the other CNN networks, Max pooling is added after each convolution with a pooling length of two. The minimum (element-wise) of the outputs generated by the second as well as the third CNN networks is computed using the Minimum layer. Finally, the outputs of the Minimum layer and those of the first CNN network are concatenated and used as inputs of the BiLSTM layer.

2.4.2 BiLSTM

RNNs are a type of neural networks that are capable to process sequential data $(x_1, x_2, .., x_n)$ by propagating historical information via a chain-like neural network architecture [17]. It looks at the current input x_t as well as the previous output of hidden state noted h_{t-1} at each time step to represent the sequence data. However, baseline RNNs become unable to handle long-term dependencies when the gap between the relevant information and the point where it is needed becomes large [28]. To overcome this problem, LSTM was first introduced in [24] and has risen to prominence as a state of the art in MT [31]. Then, the GRU was proposed with low complex equations. In order to make use of the amount information seen at the previous as well as the future steps, Bidirectional RNN (BiRNN) was invented [18]. Combining BiRNN with LSTM/GRU gives Bi (LSTM/GRU) that could exploit long-range context in both input directions.

2.4.3 Softmax

The last layer of our proposed model is composed of a classification unit using Softmax activation function due to our multi-class problem. It gives the probability distribution of all the unique words in the target language. The predicted word at each time step is selected as the one with the highest probability.

3 Experimental Setup and Results

3.1 Hyperparameters and Training Setup

In this part, we investigated the impact of the hyperparameters on Arabic MT system to select the best ones to use during training. First, we adjust the learning rate dynamically using the LearningRateScheduler callback. At the beginning of every epoch, this callback gets the updated learning rate value from the schedule function. We run this for 100 epochs and measure the loss at each epoch starting from a learning rate of 10^{-6} to 1. We plot the results to try to find the optimal value; see Fig. 5.

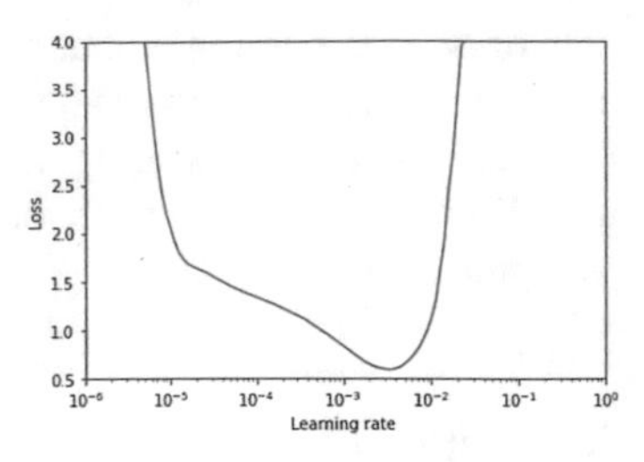

Fig. 5 Training error for different learning rate values

Table 1 The details of the proposed model parameters

Hyperparameter	Values
Loss function	Categorical crossentropy
Weight initialization	Glorot
Dropout	0.3
Optimizer	RMSprop
The maximum sentence length	20
Validation size/Test size	0.2/0.3

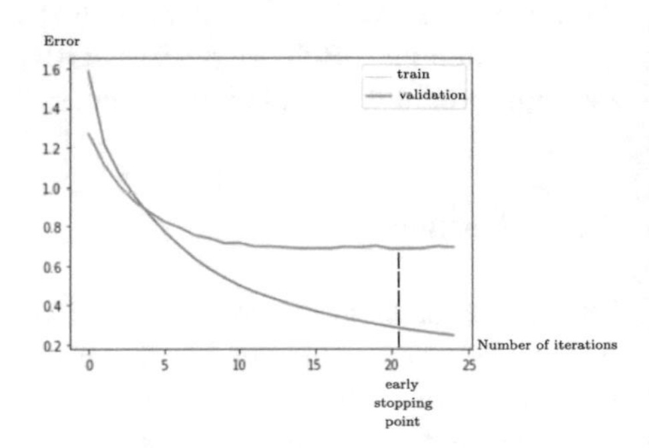

Fig. 6 Early stopping based on validation error

It can be seen that the optimal learning rate is approximately 2.5×10^{-3}. A manual tuning was used in order to optimize the values of the other hyperparameters. The details of the proposed model parameters are shown in Table 1.

Moreover, an early stopping technique based on validation error was used. As shown in Fig. 6, its main role is to stop training as soon as the validation error reaches a minimum.

3.2 Combination of Word Embedding Techniques

The Arabic language is known for its lexical sparsity which is due to the complex morphology of Arabic [2]. For example, the character ف (fa) in the word فكتب (fakataba) is a prefix, however, the same character in the word فرق (firaq) is an

Table 2 Word vector representations training parameters

Model	Size of word vectors	Size of the context window	Number of negatives sampled	Negative
Word2Vec (CBOW/SG)	200	5	1×10^{-2}	100
FastText	200	8	2×10^{-2}	300

original character. To avoid this issue, we first segment the words using Farasa into stems, prefixes and suffixes. It has demonstrated to be significantly better (in terms of accuracy and speed) than the state-of-the-art segmenters; MADAMIRA [29] and Stanford [26] on MT as well as Information Retrieval (IR) tasks.

In order to appraise the effectiveness of the combination based word embedding techniques, investigations are conducted on a subset of the online available UN dataset. First, we use Gensim tool[1] and Gensim's native implementation of FastText[2] to generate FastText embeddings and Word2Vec ones based on the SG and CBOW models. Table 2 shows the best hyperparameters found to generate 200 dimensional embeddings with FastText, CBOW and SG models. A random search strategy [12] is used in order to optimize the values of these parameters.

Once the word representations using the three models are generated, the next step is to select the best combination of the models' word vectors. Thereby, the minimum (element-wise), the maximum and the concatenation of the word representations of a given input sentence are computed using the Minimum, the Maximum and the Concatenate layers, respectively. The obtained results are reported in Table 3.

The reported results show clearly that amongst the word representations, FastText achieves better BLEU score than CBOW and SG models. This is largely due to the fact that FastText takes into consideration the internal subword information of words, which allows the model to take into account the morphology and lexical similarity of them. Further, we combined each two representations using the Concatenate layer, the Minimum layer or the Maximum one. We notice that, in the most of the cases, the translation quality is notably better than CBOW, SG and FastText models. Moreover, when combined the three word representations, we achieve the best BLEU score when the proposed DL is trained on the minimum of the three models' word vectors. With these settings, a BLEU score of 60.40 is reached compared to 57.14, 56.24 and 58.65 obtained with CBOW, SG and FastText respectively.

[1] https://radimrehurek.com/gensim/about.html.

[2] https://radimrehurek.com/gensim/models/fasttext.html.

Table 3 Arabic MT BLEU score results

		BLEU score %
UN dataset	CBOW	57.14
	SG	56.24
	FastText	58.65
	Minimum(CBOW,SG)	58.70
	Minimum(CBOW,FastText)	57.29
	Minimum(SG,FastText)	59.20
	Minimum(CBOW, SG, FastText)	60.40
	Maximum(CBOW,SG)	58.71
	Maximum(CBOW,FastText)	59.20
	Maximum(SG,FastText)	59.46
	Maximum(CBOW, SG,FastText)	58.80
	Concatenate(CBOW,SG)	59.59
	Concatenate(CBOW, FastText)	58.41
	Concatenate(SG,FastText)	59.04
	Concatenate(CBOW, SG, FastText)	58.16

Table 4 Arabic MT performance for different RNN variants

	BLEU score %
LSTM	57.58
GRU	57.11
BiGRU	58.50
BiLSTM	60.40

3.3 *The Impact of Different RNN Variants on Arabic MT*

To study the performance of the Arabic MT under different RNNs using our model, four different variants of those are used in the layer A; see Fig. 4, which are: LSTM, GRU, BiLSTM, BiGRU.

The reported results in Table 4 showed a greater gain performance of the proposed model based on BiLSTM architecture in terms of BLEU score (BLEU score = 60.40%). These findings may be explained by the fact that LSTM uses three types of gates during the training process, the input gate, the forget gate and the output one, while GRU only uses two, the update gate and the reset one. Moreover, since the

Table 5 Comparison with state-of-the-art works using UN dataset

	BLEU score %
[6]	41.14
[27]	42.38
Our approach	60.40

computational bottleneck in our model is the softmax operation we did not remark large difference in training speed between LSTM and GRU cells. Furthermore, BiLSTM is able to exploit the historical context as well as the future one and consequently achieve best results in terms of BLEU score.

3.4 *Comparison with the State-of-the-Art Works and Qualitative Evaluation*

Various works have implemented a MT system based on DL. However, research has been rarely devoted to Arabic MT using a neural approach. Among these works, the authors in [6] used an encoder using two layers of BiGRU and a decoder using unidirectional GRU with the attention mechanism. They achieved an overall BLEU score of 41.14% compared to 60.40% using our approach. The authors in [27] used an encoder-decoder based on LSTM architecture with the attention mechanism. The obtained results showed that the translation quality using our approach is notably higher than that using the [27]'s model reaching a BLEU score of 60.40% compared to 42.38% obtained by the authors in [27]. This is largely due to the proposed DL (C-BiLSTM) model, the combination of the word representations as well as the morphology-based tokenization and orthographic normalization (Table 5).

In Table 6, some examples of the source sentences (in Arabic) from the test set and their translations in English are illustrated. These qualitative observations demonstrates that the proposed approach translates the first three examples fluently. In the example 4, the model preserves the original meaning of the input sentence. However, in the example 5, the proposed model drops the source sentence after the comma and only translates the part before.

Table 6 A few examples of translations generated by our model

Source	إذ تضع في اعتبارها جميع قرارات الجمعية العامة الأخرى ذات الصلة
Our model	bearing in mind all other relevant general assembly resolutions
Truth	bearing in mind all other relevant general assembly resolutions
Source	إذ تشير إلى قراراتها السابقة بشأن التعاون بين الأمم المتحدة وجامعة الدول العربية
Our model	recalling its previous resolutions on cooperation between the united nations and the league of arab states
Truth	recalling its previous resolutions on cooperation between the united nations and the league of arab states
Source	مؤتمر الأمم المتحدة الثالث المعني بأقل البلدان نموا
Our model	third united nations conference on the least developed countries
Truth	third united nations conference on the least developed countries
Source	٢ تلاحظ مع الارتياح الدعم الذي قدمه البلد المضيف من أجل إنشاء المركز
Our model	2 notes with satisfaction the support provided for the establishment of the centre by the host country
Truth	2 notes with satisfaction the support given to the establishment of the centre by the host country
Source	الزمالات والتدريب والخدمات الاستشارية للأمم المتحدة في ميدان نزع السلاح
Our model	united nations disarmament fellowship and and
Truth	united nations disarmament fellowship, training and advisory services

4 Conclusion

In this paper, we aimed to find a robust set of features for Arabic MT. For this propose, performance analysis of a combination of three models, namely, CBOW, SG and FastText, has been discussed using UN dataset. Moreover, a DL architecture based mainly on BiLSTM and CNN has been used for the task of MT between English and Arabic texts. The obtained results were compared with those using CBOW, SG and FastText. It has been revealed that the proposed scheme exhibits a high BLEU compared to CBOW, SG, FastText as well as the sate-of-the-art works. Further studies will aim at developing a technique that allows the model to automatically search for parts of the word representations that are relevant to predicting the translation of a source sentence.

References

1. Abdelali, A., Darwish, K., Durrani, N., and Mubarak, H.: Farasa: a fast and furious segmenter for arabic. In: Proceedings of the 2016 Conference of the North American Chapter of the Association for Computational Linguistics: Demonstrations, pp. 11–16 (2016)
2. Al-Sallab, A., Baly, R., Hajj, H., Shaban, K.B., El-Hajj, W., Badaro, G.: Aroma: A recursive deep learning model for opinion mining in Arabic as a low resource language. ACM Trans. Asian Low-Resour. Lang. Inf. Process. **16**(4) (2017)
3. Alkhatib, M., Shaalan, K.: The key challenges for Arabic machine translation, vol. 01, pp. 139–156 (2018)
4. Almahairi, A., Cho, K., Habash, N., Courville, A.C.: First result on Arabic neural machine translation. CoRR abs/1606.02680 (2016)
5. Alqudsi, A., Omar, N., Shaker, K.: Arabic machine translation: a survey. Artif. Intell. Rev. 42 (2012)
6. Alrajeh, A.: A recipe for Arabic-English neural machine translation. CoRR abs/1808.06116 (2018)
7. Athiwaratkun, B., Wilson, A.G., Anandkumar, A.: Probabilistic FastText for multi-sense word embeddings. In: Proceedings of the 56th Annual Meeting of the Association for Computational Linguistics, ACL 2018, Melbourne, Australia, 15–20 July 2018, Volume 1: Long Papers, pp. 1–11 (2018)
8. Bahdanau, D., Cho, K., Bengio, Y.: Neural machine translation by jointly learning to align and translate. In: 3rd International Conference on Learning Representations, ICLR (2015)
9. Bensalah, N., Ayad, H., Adib, A., Farouk, A.I.E.: Arabic sentiment analysis based on 1-D convolutional neural network. In: International Conference on Smart City Applications, SCA20, Safranbolu, Turkey (2020)
10. dBensalah, N., Ayad, H., Adib, A., arouk, A.I.E.: Combining word and character embeddings in Arabic chatbots. In: Advanced Intelligent Systems for Sustainable Development, AI2SD 2020, Tangier, Morocco (2020)
11. Bensalah, N., Ayad, H., Adib, A., Farouk, A.I.E.: CRAN: an hybrid CNN-RNN attention-based model for Arabic machine translation. In: International Conference on Cloud Computing and Artificial Intelligence: Technologies and Applications, CloudTech'20, Marrakesh, Morocco (2020)
12. Bergstra, J., Bengio, Y.: Random search for hyper-parameter optimization. J. Mach. Learn. Res. **13**, 281–305 (2012)
13. Bojanowski, P., Grave, E., Joulin, A., Mikolov, T.: Enriching word vectors with subword information. Trans. Assoc. Comput. Linguistics **5**, 135–146 (2017)
14. Cho, K., van Merriënboer, B., Bahdanau, D., Bengio, Y.: On the properties of neural machine translation: encoder–decoder approaches. In: Proceedings of SSST-8, Eighth Workshop on Syntax, Semantics and Structure in Statistical Translation, pp. 103–111 (2014)
15. Cho, K., van Merrienboer, B., Gulcehre, A., Bahdanau, D., Bougares, F., Schwenk, H., Bengio, Y.: Learning phrase representations using RNN encoder-decoder for statistical machine translation. In: EMNLP, pp. 1724–1734. ACL (2014)
16. Durrani, N., Dalvi, F., Sajjad, H., Vogel, S.: QCRI machine translation systems for IWSLT 16. CoRR abs/1701.03924 (2017)
17. Goodfellow, I., Bengio, Y., Courville, A.: Deep Learning. The MIT Press (2016)
18. Graves, A., Mohamed, A., Hinton, G.E.: Speech recognition with deep recurrent neural networks. In: IEEE International Conference on Acoustics, Speech and Signal Processing, ICASSP, pp. 6645–6649 (2013)
19. Habash, N., Sadat, F.: Arabic preprocessing schemes for statistical machine translation. In: Proceedings of the Human Language Technology Conference of the NAACL, Companion Volume: Short Papers, pp. 49–52, June 2006
20. Hadla, L., Hailat, T., Al-Kabi, M.: Evaluating Arabic to English machine translation. Int. J. Adv. Comput. Sci. Appl. **5** (2014)

21. Hadla, L., Hailat, T., Al-Kabi, M.: Evaluating Arabic to English machine translation. Int. J. Adv. Comput. Sci. Appl. **5**, 68–73 (2014)
22. Harris, Z.: Distributional structure. Word **10**(2–3), 146–162 (1954)
23. Hochreiter, S., Schmidhuber, J.: Long short-term memory. Neural Comput. **9**, 1735–1780 (1997)
24. LeCun, Y., Bottou, L., Bengio, Y., Haffner, P.: Gradient-based learning applied to document recognition. Proc. IEEE **86**, 2278–2324 (1998)
25. Mikolov, T., Sutskever, I., Chen, K., Corrado, G.S., Dean, J. Distributed representations of words and phrases and their compositionality. In: Advances in Neural Information Processing Systems 26: 27th Annual Conference on Neural Information Processing Systems, pp. 3111–3119 (2013)
26. Monroe, W., Green, S., Manning, C.D.: Word segmentation of informal Arabic with domain adaptation. In: Proceedings of the 52nd Annual Meeting of the Association for Computational Linguistics (Volume 2: Short Papers), pp. 206–211, June 2014
27. Oudah, M., Almahairi, A., Habash, N.: The impact of preprocessing on Arabic-English statistical and neural machine translation. In: Proceedings of Machine Translation Summit XVII Volume 1: Research Track, MTSummit, pp. 214–221 (2019)
28. Pascanu, R., Mikolov, T., Bengio, Y.: On the difficulty of training recurrent neural networks. In: Proceedings of the 30th International Conference on International Conference on Machine Learning - Volume 28, ICML 2013, pp. III-1310–III-1318 (2013). JMLR.org
29. Pasha, A., Al-Badrashiny, M., Diab, M., El Kholy, A., Eskander, R., Habash, N., Pooleery, M., Rambow, O., Roth, R.: MADAMIRA: a fast, comprehensive tool for morphological analysis and disambiguation of Arabic. In: Proceedings of the Ninth International Conference on Language Resources and Evaluation (LREC 2014), pp. 1094–1101, May 2014
30. Sutskever, I., Vinyals, O., Le, Q.V.: Sequence to sequence learning with neural networks. CoRR (2014)
31. Sutskever, I., Vinyals, O., Le, Q.V.: Sequence to sequence learning with neural networks. In: Advances in Neural Information Processing Systems 27: Annual Conference on Neural Information Processing Systems 2014, 8–13 December 2014, pp. 3104–3112 (2014)
32. Tiedemann, J.: Parallel data, tools and interfaces in OPUS. In: Proceedings of the Eighth International Conference on Language Resources and Evaluation (LREC 2012), Istanbul, Turkey, pp. 2214–2218 (2012)
33. Vaswani, A., Shazeer, N., Parmar, N., Uszkoreit, J., Jones, L., Gomez, A.N., Kaiser, L.U., Polosukhin, I.: Attention is all you need. In: Advances in Neural Information Processing Systems, vol. 30, pp. 5998–6008 (2017)

Security Analysis of Nikooghadam et al.'s Authentication Protocol for Cloud-IoT

Mourade Azrour, Jamal Mabrouki, Yousef Farhaoui, and Azidine Guezzaz

Abstract The IoT has grown rapidly formerly it has changed our everyday practice. It has penetrated all domains including agriculture, industry, environment, etc. Furthermore, IoT allows to anyone to have information remotely. The question here is how we can secure the private data? In order to resolve this issue many researchers have proposed authentication protocol for controlling the access to private data in IoT environment. So, recently Nikooghadam et al. presented a lightweight authentication and session key agreement protocol for Cloud-IoT. Afterwards, they proved that their proposed protocol can resist against several attacks. Nevertheless, in this study, we confirm that Nikooghadam et al.'s protocol is vulnerable to password guessing attack.

Keywords Authentication · IoT · Cloud · Internet of Things · Security · Cryptanalysis

1 Introduction

In the last decades, the Internet of Things (IoT) and the Cloud computing have known an important development. The IoT refers to the large number of connected objects to the internet. These objects are able to collect process or transfer data. For collecting data the devices are equipped with various sensors that can measure some parameters from physical word, including temperature, humidity, water level, light, wend speed and so ones. The IoT devices are equipped with microcontrollers, which are responsible to process the sensed data. IoT can profit from the wireless networks

M. Azrour (✉) · Y. Farhaoui
Department of Computer Science, IDMS Team, Faculty of Sciences and Techniques, Moulay Ismail University, Errachidia, Morocco

J. Mabrouki
Laboratory of Spectroscopy, Molecular Modeling, Materials, Nanomaterial, Water and Environment, CERNE2D, Faculty of Science, Mohammed V University, Rabat, Morocco

A. Guezzaz
Technology High School of Essaouira, M2SC Team, Cadi Ayyad University, Marrakesh, Morocco

N. Gherabi and J. Kacprzyk (eds.), *Intelligent Systems in Big Data, Semantic Web and Machine Learning*, Advances in Intelligent Systems and Computing 1344,
https://doi.org/10.1007/978-3-030-72588-4_18

technologies to communicate the collected data. Due to its significance role and the possibility to make our life more easier, IoT have used in various daily domains, such as water management [1], healthcare [2], environment controlling [3], desaster management [4], smart house [5], smart weather [6].

On the other hands, Cloud computing refers to paradigm that allows to the client to access via network to the computing resources easily. The cloud provider preserves a group of servers that can used whenever the client need them. The reserved resource gives the impression that is unlimited to the client [7].

Thanks to the limited capability of IoT devices for processing the large number of collected data, cloud computing service can be used for resolving this issue. However, this solution must protect the private data. Hence, the information should be accessed only by the authorized parties. So, the authentication protocols are necessary in this case. In 2018, Kumari et al. [8] presented an authentication protocol based on elliptic curve cryptography (ECC) for Cloud-IoT. Then, the authors proved that their protocol offer security device anonymity and mutual authentication, as it is secured against some known attacks such as stolen-verifier attack, replay attack, man-in-the-middle attack, impersonation attack, and password-guessing attack. However, Nikooghadam et al. [9] have demonstrated that Kumari et al.'s is vulnerable against replay attack, stolen-verifier attack, denial-of-service attack and so ones. As result, Nikooghadam et al. proposed a lightweight authentication and session key agreement protocol. Then, they have showed that the proposed scheme is secured against various attacks. Nevertheless, in this paper, we demonstrated that Nikooghadam et al.'s protocol cannot withstand against server secret and user password guessing attacks.

The remainder of this paper is arranged as succeeding. The second section is reserved for presenting the related works. In the third section, we review Nikooghadam et al.'s protocol. In the fourth section, we detailed the weaknesses of Nikooghadam et al.'s protocol. Finally, the last section concludes the paper.

2 Related Works

With the rapid evolution of new technologies the security of private life, the access control and authentication are coming important day by day and are receiving much importance especially after the invention of IoT. Therefore, many authentication protocols are proposed in the last two decades [9–27].

In 2007, Wong et al. [28] proposed lightweight and hash-based user authentication protocol. Nonetheless, it has been discovered that it is vulnerable against various attacks namely replay attack, forgery attack and stolen verifier attack.

In 2009, Das [14] proposed a new password based authentication protocol. Nevertheless, this protocol does not provide nor mutual authentication nor the session key exchange. In the same year, Malasri et al. [26] suggested a two-tier scheme for medical sensor network based on Elliptic Curve Cryptography. Unfortunately, this protocol is discovered vulnerable against denial-of-service and relay attacks.

In 2010, in order to secure the communication in WSNs Yuan et al. [10] are based on biometrics technics for proposing an authentication scheme. Then, they confirmed that the proposed protocol is secured against numerous attacks including replay attack and stolen verifier attack. However, the protocol cannot resist against privileged insider attack, denial-of-service attack, and impersonation attack.

In 2012, He et al. [29] proposed a new scheme named as ReTrust. That is a lightweight authentication protocol for actual medical applications based on sensor network. Nevertheless, this protocol cannot resist against two attacks that are forgery and password guessing as will it is not able to provide forward secrecy service.

The chaotic map computation is applied by Mishra et al. [13] in 2014 to proposed an authentication and key agreement protocol for telecare medicine information systems. Nonetheless, this protocol is discovered vulnerable to password guessing attack.

In 2015, Jiang et al. [15] demonstrated that Chen et al.'s suggested protocol in 2011 [30] cannot resist against password guessing attack. So, for improving the authentication and to response to the issues of Chen et al.'s scheme Jiang et al. presented a new authentication protocol. Nevertheless, the proposed solution cannot deal with password guessing and user impersonation attack.

In 2017, Azrour et al. [31] have demonstrated that the SIP authentication protocol proposed by Farash et al. is vulnerable against certain known attacks, including denial of service attack and denning Sacco attack. In 2019, Azrour et al. [32] proved that the Ye et al.'s [11] proposed authentication scheme for IoT is weak and suffer from some security problems.

Very recently, various authentication protocols have been proposed [12, 25, 27, 33]. For example, Sharma and Kalra [12] proposed a lightweight user authentication scheme for cloud-IoT based healthcare services. Then, they confirmed that their protocol is secured against various attack and it has a low computation cost if it is compared to other related protocols.

3 Review of Nikooghadam et al.'s Protocol

In this section, we review briefly Nikooghadam et al.'s proposed authentication protocol for IoT. This protocol consists of two phases: registration and authentication. The notations used here are depicted in Table 1.

3.1 Registration Phase

In this phase, the embedded device and the cloud server achieve the following three steps:

Table 1 Notation used

Symbol	Signification
ED_i	Embedded device
ID_i	Identity of ED_i
pw_i	Password of ED_i
CS	Cloud server
x_s	Secret key of CS
T_1, T_2, T_3, T_4	The current time
$a_i, b_i, c_i, e_i, q_i, z_i$	High entropy random numbers
$\oplus$	XOR operation
$\|$	Concatenation

- Step 1. The ED_i chooses its unique identity ID_i and password pw_i. Then, it selects randomly two numbers a_i and b_i. Next, it computes three values: $Mpw_i = h(pw_i \| a_i \| b_i \| ID_i)$, $HID_i = h(ID_i \| b_i)$, and $d_i = a_i \oplus b_i$. Finally, the device sends the information HID_i, Mpw_i, d_i, a_i to the cloud server.
- Step 2. After getting the information, the cloud server calculates $v_i = h(HID_i \| Mpw_i)$. Then, it selects randomly two numbers c_i and z_i. Next, it calculates $B_i = h(HID_i \| x_s)$ and $E_i = B_i \oplus Mpw_i$. After that, it stores z_i and HID_i in secure database. Finally, it computes $A_i = E_{x_s}(c_i \| HID_i \| d_i \| a_i)$ and sends the message $A_i, E_i, z_i, v_i, c_i, b_i, a_i$ back to the embedded device.
- Step 3. When, the embedded device receive the server response, it computes $T_i = A_i \oplus Mpw_i$. Then, it stores this information secretly $T_i, E_i, z_i, v_i, c_i, b_i, a_i$.

3.2 Authentication Phase

The authentication phase of Nikooghadam et al.'s protocol consists of four steps that are:

- Step 1. The embedded device EDi enters its ID_i^* and pw_i^*. Then, it computes the following parameters: $Mpw_i^* = h(pw_i^* \| a_i \| b_i \| ID_i^*)$
  $HID_i^* = h(ID_i^* \| b_i)$
  $v_i^* = h(HID_i^* \| Mpw_i^*)$
  $B_i = E_i \oplus Mpw_i$
  $v_i^* \stackrel{?}{=} v_i$
  $d_i = a_i \oplus b_i$
  $cd_i = c_i \oplus d_i$
  Afterwards, it chooses randomly a number e_i and picks the values of T_1. Then, it computes $A_i = T_1 \oplus Mpw_i$ and encrypts $M_i = E_{B_i}(T_1 \| e_i \| A_i)$. Next, it sends the message cd_i, M_i, T_1, HID_i to the cloud server CS through a public channel.

- Step 2. After receiving device's message, the cloud server selects firstly a fresh values of T_2 and verify the validity of $T_2 - T_1 \leq \Delta T$. In the case, it is incorrect, the session is finished. If not, based on the received HID_i, the cloud server computes $B_i = h(HID_i \| x_s \| z_i^*)$. Then uses B_i, to decrypt M_i as $DEC_{B_i}(M_i) = (T_1^*, A_i^*, e_i^*)$ and $DEC_{x_s}(A_i^*) = (HID_i^*, a_i^*, c_i^*, d_i^*)$. Then calculates $cd_i^* = c_i^* \oplus d_i^*$ and checks $cd_i^* \stackrel{?}{=} cd_i$. If OK, the embedded device is authenticated. It also verifies if $T_i^* \stackrel{?}{=} T_1$ to confirm that the timestamp has not been changed during the message exchange. Next, the server selects randomly a number q_i, achieves the following calculations, and forwards the message (N_i, T_2) to the device. $Q_i = h(A_i \| B_i)$
  $s_i = q_i \oplus B_i$
  $w_i = h(cd_i^* \| e_i^*)$
  $N_i = EA * (s_i \| T_2 \| w_i)$.
- Step 3. After the reception of server's response, the embedded device picks timestamp T_3 and verifies the freshness of the message by inspection the correctness of $T_3 - T_2 \leq \Delta T$. Then, it decrypts $DEC_{A_i}(N_i) = (s_i^*, T_i^*, w_i^*)$. Next, it computes $w_i^{'} = h(cd_i \| e_i)$ and authenticates the cloud server by checking $w_i^{'} \stackrel{?}{=} w_i^*$. The embedded device then computes the following and forwards MN_i and T_3 to the cloud server: $q_i^* = s_i^* \oplus B_i$
  $Q_i = h(A_i \| B_i)$
  $sk = h(e_i \| B_i \| Q_i \| q_i \| z_i)$
  $MN_i = h(sk \| q_i \| s_i \| Q_i)$.
  Step 4. Upon getting the value of MN_i, the cloud server checks the correctness of $T_4 - T_3 \leq \Delta T$. If ok, it generates the session key $sk = h(e_i \| B_i \| Q_i \| q_i \| z_i)$ and $MN_i^* = h(sk \| q_i \| s_i \| Q_i)$. Finally, it verifies $MN_i^* = MN_i$ to authenticate the embedded device.

4 Weaknesses of Nikooghadam et al.'s Protocol

In this section, we show that the authentication protocol proposed by Nikooghadam et al. is vulnerable against Server secret and user password guessing attacks.

4.1 Server Secret Guessing Attack

Guessing of the secret is the most known attack in authentication protocols. So, in contrary of Nikooghadam that proved that their protocol is very secured against numerous attacks, we found that this protocol is vulnerable to secret guessing attack that can be executed by an insider. After the registration phase, the devices gets this value $E_i = B_i \oplus Mpw_i$, after computing $Mpw_i = h(pw_i \| a_i \| b_i \| ID_i)$, the insider can extract $B_i = E_i \oplus Mpw_i = h(HID_i \| x_s)$. Afterwards, he can perform the algorithm we give here, which can be applied for guessing the sever secret.

Algorithm 1: GuessingX

```
input : D list of secret, HID_i, B_i
output: x
begin
    for i ← 1 to sizeof(D) do
        x ← D_i;
        if B_i == h(HID_i || x) then
            return x;
        end
    end
end
```

4.2 User Password Guessing Attack

In this protocol the attacker can guess use's password. To do that, the attack has firstly to intercept the first message in the authentication phase to have HID_i,Mpw_i,d_i and a_i. Then executed the second algorithm.

Algorithm 2: GuessingPW

```
input : HID_i,Mpw, d_i ,a_i, D_Pw(listofpasswords), D_ID(listofidentities)
output: pw
begin
    b_i ← a_i ⊕ d_i;
    for i ← 1 to sizeof(D_ID) do
        Id' ← D_ID;
        if HID_i == h(Id' || b_i) then
            ID ← D_i ;
            break;
        end
    end
    for j ← 1 to sizeof(D_pw) do
        pw' ← D_pw;
        if Mpw_i == h(pw' || a_i || b_i || ID) then
            return pw';
        end
    end
end
```

5 Conclusion

In this paper, we firstly proposed the importance of the authentication services in Cloud-IoT environment. Then, we present the various related protocol proposed used for authentication in various online services. Afterwards, we reviewed briefly the authentication scheme proposed by Nikooghadam et al. Finally, we have demonstrated that Nikooghadam et al.'s authentication protocol is vulnerable to server secret and user password guessing attacks. Our future work is to propose a new efficient secure authentication scheme for Could-IoT service.

Conflict of Interest

The author declares that he has no conflict of interest.

References

1. Mabrouki, J., Azrour, M., Farhaoui, Y., El Hajjaji, S.: Intelligent system for monitoring and detecting water quality. In: Farhaoui, Y. (ed.) Big Data and Networks Technologies, vol. 81, pp. 172–182. Springer, Cham (2020)
2. Abu Bakar, N.A., Wan Ramli, W.M., Hassan, N.H.: The internet of things in healthcare: an overview, challenges and model plan for security risks management process. Indones. J. Electr. Eng. Comput. Sci. **15**(1), 414 (2019). https://doi.org/10.11591/ijeecs.v15.i1.pp414-420
3. Alam, F., Mehmood, R., Katib, I., Albogami, N.N., Albeshri, A.: Data fusion and IoT for smart ubiquitous environments: a survey. IEEE Access **5**, 9533–9554 (2017). https://doi.org/10.1109/ACCESS.2017.2697839
4. Sukmaningsih, D.W., Suparta, W., Trisetyarso, A., Abbas, B.S., Kang, C.H.: Proposing smart disaster management in urban area. In: Huk, M., Maleszka, M., Szczerbicki, E. (ed.) Intelligent Information and Database Systems: Recent Developments, vol. 830, pp. 3–16. Springer, Cham (2020)
5. Sayuti, H., et al.: Smart home and ambient assisted living based on the internet of things. Int. J. Electr. Comput. Eng. IJECE **7**(3), 1480 (2017). https://doi.org/10.11591/ijece.v7i3.pp1480-1488
6. Kulkarni, A., Mukhopadhyay, D.: Internet of things based weather forecast monitoring system. Indones. J. Electr. Eng. Comput. Sci. **9**(3), 555–557 (2018)
7. Dillon, T., Wu, C., Chang, E.: Cloud computing: issues and challenges. In: 2010 24th IEEE International Conference on Advanced Information Networking and Applications, Perth, WA, pp. 27–33 (2010)
8. Kumari, S., Karuppiah, M., Das, A.K., Li, X., Wu, F., Kumar, N.: A secure authentication scheme based on elliptic curve cryptography for IoT and cloud servers. J. Supercomput. **74**(12), 6428–6453 (2018). https://doi.org/10.1007/s11227-017-2048-0
9. Nikooghadam, M., Amintoosi, H.: Secure communication in CloudIoT through design of a lightweight authentication and session key agreement scheme. Int. J. Commun. Syst., e4332 (2020). https://doi.org/10.1002/dac.4332
10. Yuan, J., Jiang, C., Jiang, Z.: A biometric-based user authentication for wireless sensor networks. Wuhan Univ. J. Nat. Sci. **15**(3), 272–276 (2010). https://doi.org/10.1007/s11859-010-0318-2

11. Ye, N., Zhu, Y., Wang, R., Malekian, R., Lin, Q.: An efficient authentication and access control scheme for perception layer of Internet of Things (2014). https://doi.org/10.12785/amis/080416
12. Sharma, G., Kalra, S.: A lightweight user authentication scheme for cloud-IoT based healthcare services. Iran. J. Sci. Technol. Trans. Electr. Eng. **43**(S1), 619–636 (2019). https://doi.org/10.1007/s40998-018-0146-5
13. Mishra, D., Srinivas, J., Mukhopadhyay, S.: A secure and efficient chaotic map-based authenticated key agreement scheme for telecare medicine information systems. J. Med. Syst. **38**(10), 120 (2014). https://doi.org/10.1007/s10916-014-0120-3
14. Das, M.L.: Two-factor user authentication in wireless sensor networks. IEEE Trans. Wirel. Commun. **8**(3), 1086–1090 (2009). https://doi.org/10.1109/TWC.2008.080128
15. Jiang, Q., Ma, J., Li, G., Li, X.: Improvement of robust smart-card-based password authentication scheme. Int. J. Commun. Syst. **28**(2), 383–393 (2015). https://doi.org/10.1002/dac.2644
16. Saxena, N., Grijalva, S., Chaudhari, N.S.: Authentication protocol for an IoT-enabled LTE network. ACM Trans. Internet Technol. **16**(4), 1–20 (2016). https://doi.org/10.1145/2981547
17. Maurya, A., Sastry, V.N.: Fuzzy extractor and elliptic curve based efficient user authentication protocol for wireless sensor networks and internet of things. Information **8**(4), 136 (2017). https://doi.org/10.3390/info8040136
18. Hong, S.: Authentication techniques in the internet of things environment: a survey. Int. J. Netw. Secur. **21**(3), 462–470 (2019)
19. Azrour, M., Ouanan, M., Farhaoui, Y.: A new secure SIP authentication scheme based on elliptic curve cryptography. In: International Conference on Information Technology and Communication Systems, pp. 155–170 (2017)
20. Azrour, M., Farhaoui, Y., Ouanan, M.: A new secure authentication and key exchange protocol for session initiation protocol using smart card. Int. J. Netw. Secur. **19**(6), 870–879 (2017). https://doi.org/10.6633/IJNS.201711.19(6).02
21. Azrour, M., Ouanan, M., Farhaoui, Y.: A new enhanced and secured password authentication protocol based on smart card. Int. J. Tomogr. Simul. TM **31**(1), 14–26 (2018)
22. Azrour, M., Ouanan, M., Farhaoui, Y.: SIP authentication protocols based on elliptic curve cryptography: survey and comparison. Indones. J. Electr. Eng. Comput. Sci. **4**(1), 231 (2016). https://doi.org/10.11591/ijeecs.v4.i1.pp231-239
23. Shah, T., Venkatesan, S.: Authentication of IoT device and IoT server using secure vaults. In: 2018 17th IEEE International Conference on Trust, Security and Privacy in Computing and Communications/12th IEEE International Conference on Big Data Science and Engineering (TrustCom/BigDataSE), New York, NY, pp. 819–824 (2018). https://doi.org/10.1109/TrustCom/BigDataSE.2018.00117
24. Yang, S.-K., Shiue, Y.-M., Su, Z.-Y., Liu, I.-H., Liu, C.-G.: An authentication information exchange scheme in WSN for IoT applications. IEEE Access **8**, 9728–9738 (2020). https://doi.org/10.1109/ACCESS.2020.2964815
25. Sharma, G., Kalra, S.: Advanced lightweight multi-factor remote user authentication scheme for cloud-IoT applications. J. Ambient. Intell. Human. Comput. **11**, 1771–1794 (2020). https://doi.org/10.1007/s12652-019-01225-1
26. Malasri, K., Wang, L.: Design and implementation of a securewireless mote-based medical sensor network. Sensors **9**(8), 6273–6297 (2009). https://doi.org/10.3390/s90806273
27. Lee, J., Kim, M., Yu, S., Park, K., Park, Y.: A secure multi-factor remote user authentication scheme for cloud-IoT applications. In: 2019 28th International Conference on Computer Communication and Networks (ICCCN), Valencia, Spain, pp. 1–2 (2019). https://doi.org/10.1109/ICCCN.2019.8847031
28. Hu, F., Jiang, M., Wagner, M., Dong, D.-C.: Privacy-preserving telecardiology sensor networks: toward a low-cost portable wireless hardware/software codesign. IEEE Trans. Inf. Technol. Biomed. **11**(6), 619–627 (2007). https://doi.org/10.1109/TITB.2007.894818
29. He, D., Chen, C., Chan, S., Bu, J., Vasilakos, A.V.: ReTrust: attack-resistant and lightweight trust management for medical sensor networks. IEEE Trans. Inf. Technol. Biomed. **16**(4), 623–632 (2012). https://doi.org/10.1109/TITB.2012.2194788

30. Cheng, Z.-Y., Liu, Y., Chang, C.-C., Chang, S.-C.: An Improved Protocol for Password Authentication Using Smart Cards, vol. 22, no. 4, p. 10 (2012)
31. Azrour, M., Farhaoui, Y., Ouanan, M.: Cryptanalysis of Farash et al.'s SIP authentication protocol. Int. J. Dyn. Syst. Differ. Equ. **8**(1/2), 77–94 (2018)
32. Azrour, M., Ouanan, M., Farhaoui, Y., Guezzaz, A.: Security Analysis of Ye et al. Authentication protocol for internet of things. In: Farhaoui, Y., Moussaid, L. (ed.) Big Data and Smart Digital Environment, vol. 53, pp. 67–74. Springer, Cham (2019)
33. Kumar, M., Verma, H.K., Sikka, G.: A secure lightweight signature based authentication for Cloud IoT crowdsensing environments. Trans. Emerg. Telecommun. Technol. **30**(4), e3292 (2019). https://doi.org/10.1002/ett.3292

Internet of Things for Monitoring and Detection of Agricultural Production

Jamal Mabrouki, Maria Benbouzid, Driss Dhiba, and Souad El Hajjaji

Abstract The creation of information based advancements, otherwise called savvy culture, recognizes and bolsters the decrease of money related, regular and cultural issues. As the total populace develops exponentially, it is fundamental to audit current cultivating practices to satisfy the needs of nourishment security. Insightful sensor frameworks give more data on water needs and harvests. This data can be utilized to computerize the water supply framework and get ready ranchers to advance their water system arrange. The data procured in the initial step is moved to the cloud. Unclassified Exceptional Readiness for Distributed Power Generation, which is a significant data measure that the producer of a cell phone application can utilize. This paper presents and assesses the idea of remote detecting gadget.

Keywords Internet of things · Monitoring · Detection · Soil · Sustainable · Irrigation · Agricultural

1 Introduction

Proximal and remote detecting innovations are key segments of accuracy agribusiness, supplementing research facility investigation of soil and plant tests by giving point by point information that describe the spatial heterogeneity of harvest developing conditions [1]. By utilizing sensors combined with variable rate hardware, it is accepted that an ideal application rate can be characterized for every area in a field. Be that as it may, this isn't valid. Equipment and programming can deal with enormous measures of information at the ranch or field level, however this doesn't ensure that the sources of info applied will give a financially savvy reaction. By and large, when

J. Mabrouki (✉) · M. Benbouzid · S. El Hajjaji
Laboratory of Spectroscopy, Molecular Modeling, Materials, Nanomaterials, Water and Environment, CERNE2D, Faculty of Science, Mohammed V University in Rabat, AV Ibn Battouta, BP 1014, Agdal, Rabat, Morocco
e-mail: jamal.mabrouki@um5r.ac.ma

D. Dhiba
International Water Research Institute IWRI, University Mohammed VI Polytechnic (UM6P), Benguerir, Morocco

N. Gherabi and J. Kacprzyk (eds.), *Intelligent Systems in Big Data, Semantic Web and Machine Learning*, Advances in Intelligent Systems and Computing 1344,
https://doi.org/10.1007/978-3-030-72588-4_19

the scope of financially ideal application rates is built up, it is legitimate to expect that the likelihood of expanding benefits is higher. In this manner, before actualizing variable rate innovation, it is important to guarantee that current developing conditions require the executives proper to the field or homestead [2]. Regularly, two separate areas with various ideal application rates are required. On the off chance that these conditions are met, at that point it gets sensible to redistribute ranch inputs dependent on proposal maps or by utilizing constant sensors. This compares to the second degree of control. The primary degree of control should begin with the detailing of the objective enhancement work. The most widely recognized supposition that will be that the target to be improved is gainfulness. Creation forms are typically measured as far as the effect on yield of various seeding or preparation rates [3]. Key parameters for such a creation work incorporate geography, soil conditions, atmosphere and the historical backdrop of plot the board in earlier years. What's more, productivity relies upon the cost of yields and the expense of homestead inputs [4]. It is critical to allude to past understanding, seed and manure preliminary information, and a fundamental comprehension of soil forms, plant physiology and meteorology [5]. This introduction talks about the idea of two degrees of control for the usage of adjusted administration on a field or ranch utilizing on-board choice help apparatuses and keen tractor innovations, two specialized territories as of now being considered by the Precision Agriculture and Sensor Systems look into gathering.

2 Literature Review

Horticulture is the fundamental critical area of the Moroccan economy. Right now, the rancher can utilize innovation to apply authority over the adjustment of yield the executives and water use elements. Ranchers have acquired new advances and apparatuses to expand their benefits on account of the introduction of cloud innovations, which have expanded the quantity of educated shoppers and arrived at exceptional temperature esteems as of late [6]. Tragically, numerous ranchers despite everything utilize conventional cultivating techniques, bringing about low harvest and organic product yields. Be that as it may, any place there has been robotization and individuals have been supplanted via programmed machines [7–10]. A large portion of the archives include the utilization of sensors that gather information from a few sorts of sensors and afterward utilize a wifi system to send it to distributed storage. The information gathered gives extra data on explicit ecological conditions, which would then be able to be utilized to screen the framework [11]. The observing of ecological conditions isn't satisfactory and exhaustive to improve farming profitability. There are numerous different components that majorly affect profitability. These incorporate bug and nuisance assaults, which can be checked by splashing crops with the fundamental bug sprays and pesticides. Estimating soil N (nitrogen), P (phosphorus) and K (potassium) content is important to decide how a lot of extra supplements should be added to the dirt to expand crop fruitfulness. Soil fruitfulness is recognized utilizing NPK sensors. Nitrogen, phosphorus and potassium are significant segments of soil manure. Knowing their focus in the dirt can prompt supplement lack or bounty in soils used to help crop creation. Dampness sensors are very basic gadgets for estimating dampness in nature. In fact, this is the gadget used to gauge barometrical mugginess. A hygrometer identifies the perceptions and breaks down the temperature of the mugginess and air. The proportion of the stickiness of the air at an offered temperature to the most noteworthy mugginess is called relative dampness. Relative moistness turns into a significant factor in the quest for well-being [12, 13]. This application at required interims of controlled measures of plant watering. It is utilized to develop farming harvests, keep up scenes and replant upset soils in dry zones and in times of underneath normal precipitation. It has certain necessities in horticultural creation, including ice assurance, weed development in oat fields and shirking of soil procurement [14]. Temperature assumes a significant job in farming creation. It profoundly affects crop development, improvement and yields, the frequency of bugs and maladies, and water and manure necessities. Climate factors add to ideal harvest development, advancement and yield. Application at required interims of controlled measures of plant watering. It is utilized to develop farming harvests, keep up scenes and replant upset soils in dry zones and in times of underneath normal precipitation. It has certain necessities in horticultural creation, including ice assurance, weed development in oat fields and shirking of soil procurement. Temperature assumes a significant job in farming creation. It profoundly affects crop development, improvement and yields, the frequency of bugs and maladies, and water and manure necessities. Climate factors add to ideal harvest development, advancement and yield [15–18].

3 Materials Techniques and System Structure

3.1 System Structure

This segment depicts the writing study, an examination of soil quality and a productive water system framework utilizing agro-sensors. recommends that the framework ought to develop the fitting yield that suits the dirt. The dirt is tried by different sensors, for example, a hydrogen particle fixation gadget, a temperature gadget and a dampness gadget. The gathered qualities are sent by means of a Wi-Fi switch to the home director and the harvest affirmation is framed through the portable application. When the dirt temperature rises, the programmed water system framework is appointed. The picture of the harvest is caught and sent to the circle chief to lawfully discuss pesticides. suggests that "Customized Irrigation System with temperature confirmation" should make an automated water framework instrument on the recognizable proof of the dirt dampness substance. It likewise permits control of the steady temperature of the homestead, which is an extraordinarily earnest factor for creation, as demonstrated by the yield. The utilization of suitable techniques for the water framework is basic for development [20]. It plays a little scope ATmega16 controller

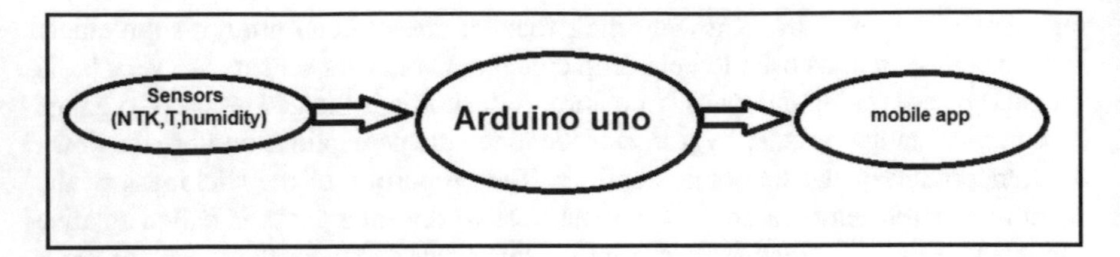

Fig. 1 Proposed system flow diagram

on the AVR card that is altered to gather data on factor soil dampness conditions utilizing a dampness acknowledgment system and enacts the water direct at any area where the ranch needs water. Prescribed a remote temperature and soil molecule sensor framework introduced in the plant's subsoil. To get to it, an entry unit controls the information from the sensors, initiates actuators and guides the information to a web application. A gauge was created with limit expectations of soil temperature and dampness adjusted to control the measure of water in a path dependent on a micro-controller. suggested a system that will have a compact device that gives a gauge of pH and assesses nitrogen (N), phosphorus (P) and potassium (K) from the pH of the dirt. They utilized the gathering count to envision sensible harvests dependent on quality and furthermore it gives the proper fertilizers required for this appearance [21, 22] (Fig. 1).

3.2 Materials

a. *Arduino UNO R3*
Arduino Uno is a microcontroller board dependent on the ATmega 328P (information sheet). It has 14 advanced information/yield pins (6 of which can be utilized as PWM yields), 6 simple data sources, a 16 MHz quartz precious stone, a USB association, a force supply attachment, an ICSP header and a reset button. It contains all that you have to help the microcontroller; simply associate it to a PC with a USB link or force it with an AC/DC connector or battery to fire it up. You can tinker with your UNO without stressing a lot over accomplishing something incorrectly. In the most pessimistic scenario, you can trade the chip for a couple of dollars and start without any preparation. "Uno" signifies one in Italian and was picked to check the arrival of Arduino Software rendition 1.0 (IDE). The Uno card and form 1.0 of Arduino Software (IDE) were the reference adaptations of Arduino, which has now advanced into later forms. The Uno card is the first in a progression of Arduino USB cards, and is the reference model for the Arduino stage; for a total rundown of current, past and obsolete cards, see the Arduino Card Index [19, 23] (Fig. 2).

Fig. 2 Arduino cards UNO R3

b. *NPK Sensor*
The proximity of soil gases (N), phosphorus (P) and metal components (K) is distinguished by the use of an optical-electrical sensor. This sensor should make it possible to select the extent of additional indoor components to be added to the soil to expand the crop yields for these supplemental area units. This can improve soil levels and reduce the inconvenient use of pesticides to increase the value of the land. The NPK estimate of the area of the example unit is controlled by the assimilation of each supplement by daylight. The electro-optical gadget is a control gadget comprising 3 LEDs as light source and a photodiode for light detection. The recurrence of the LEDs is chosen by the optical event of each supplement [24]. The supplement enters the sun through the LED and in this way the photodiode converts the residual light reflected from the reflector into current. The framework integrates Arduino's relationship to acquire the competence of these things, so that the efficiency of the inductive load is expanded - again in an alphanumeric presentation by browsing. The electrical-optical system will examine calculations of the NPK content of dirt by performing tests on different soils [25].

c. *Temperature and Humidity Sensor*
The temperature and stickiness sensor is aligned with the advanced performance of the sign. This sensor includes a resistive-type adhesiveness estimation segment and an NTC temperature estimation portion. The alignment coefficients are stored as drafts in the WBS memory, which are used by the sensor's internal sign recognition process. The single-cable sequential interface makes mixing the frames quick and easy. Its small size, low force utilization and signal transmission up to 20 m make it the ideal choice for various situations, including the most demanding. The segment is a 4-pin single line pin bundle. It is useful to combine it and extraordinary bundles can be given on customer request [26] (Fig. 3).

Various kinds of sensors can be utilized for the measurement of soil moistness. In this framework, hygrometer [27, 28] serves as a dirt moisture sensor. The hygrometer

Fig. 3 Temperature and humidity sensor

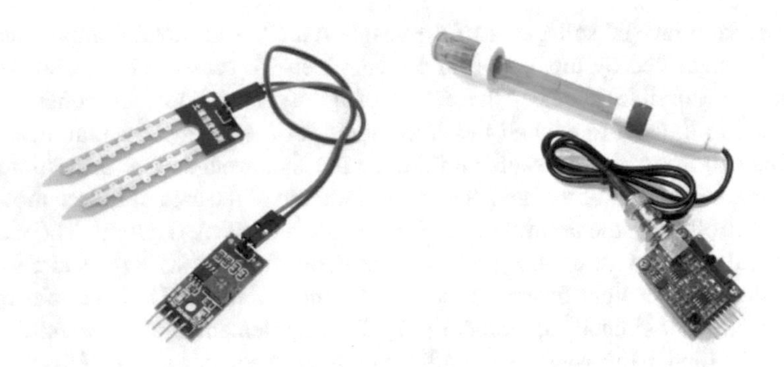

Fig. 4 Soil Humidity Sensor Hygromete and pH Sensor

has both advanced and simple outputs. The sensor output is provided as a contribution to the Arduino Uno.

MW soil moisture sensor module Soil moisture sensor compatible with Arduino This is a simple water sensor that can be used to distinguish soil moisture. The output of the module is significant when the soil moisture deficiency or yield rate is low. It can be used in the module's plant watering gadget, and the plants in your nursery do not need to be monitored by people. Possibility to customize the power by changing the advanced potentiometer (blue) Operating voltage: 3.3 V–5 V With power indicator (red) and computerized yield marker (green) With LM393 comparator stable [29, 30] (Fig. 4).

d. *pH Sensor*

pH of soil is utilized to know the sort of soil whether it is acidic kind of soil or fundamental kind of soil. This can be known by the ph estimation of the dirt. The ph estimation of soil shifts from 0 to 14. On the off chance that the ph estimation of soil is in the middle of 0 to 7 it is of acidic kind. In the event that the ph estimation of soil is 7, at that point it is impartial sort and on the off chance that it is in the middle of 7 to 14, at that point it is essential kind of soil. In view of the ph esteem we can propose the rancher the kind of yield to be utilized [31, 32].

e. *Ethernet Module*
The Ethernet module (Fig. 5) uses a new ENC28J60 microchip to manage the network protocol requirements. The module is connected to most standard microcontrollers via the SPI interface. It provides + 3.3 V electrical sustenance and baud rates up to 20 MHz (ENC28J60 Ethernet) [33].

f. *WIFI MODULE*
Espressif Systems' Smart Connectivity Platform (ESCP) is a lot of elite, high integration wireless SOCs, intended for space and force compelled versatile stage fashioners. It provides unsurpassed capacity to implant WiFi abilities inside different frameworks, or to work as a standalone application, with the most minimal expense, and negligible space requirement. ESP8266EX offers a total and independent WiFi organizing arrangement; it tends to be utilized to have the application or to offload WiFi organizing capacities from another application processor.[34] ESP8266 is a Wifi Serial Transceiver Module an integrated chip component designed for the needs of today's connected world. This chip offers a complete and integrated Wi-Fi networking solution, which can be used as an application provider or to separate all Wi-Fi networking functions from other application processors [35] (Fig. 6)

ESP8266 has on-board processing and storage capabilities that allow the chip to be integrated with sensors or with certain device applications via input output pins with only short programming. ESP8266 has on-board processing and storage capabilities that allow the chip to be integrated with sensors or with certain device applications via input output pins with only short programming [36].

Fig. 5 Arduino ENC28J60 Ethernet Module

Fig. 6 ESP8266 Wifi Module

3.3 Machine Learning Technology

Neural system strategy is an AI procedure which has a firm capacity to learn and can include the non straight connection between the sources of info and yields of a framework. A portion of the neural system explicit applications for water system and water asset the board incorporate soil dampness forecast, crop yield expectation, water system water request forecast [11, 37]. A NN technique is utilized here to foresee dampness and temperature angles terms of their capacity to deliver solid strategies reproducing complex procedures.

Feed Forward Neural Network is a NN strategy which can't recall past qualities. This paper depends on a dynamic planning try whereby the fitting option is, introduced by the Recurrent Neural Network (RNN). A RNN has internal identity circled cells, empowering information to be maintained from past strides in time. The Long Short-Term Memory Network (LSTM), a level of RNNs, is picked for its proper application in overseeing non direct unique frameworks [12]. The LSTM requires significant pre-preparing of info information and is fit for keeping up helpful information over various time stages. Time arrangement information are clung to the model as contributions on dissipation, precipitation and temperature. Solid water table profundity expectation for LSTM models outlines their capability to hold and gain from information from long haul time succession. This capacity is especially worthwhile in soil dampness water system whereby the current soil dampness content is dependent upon past dampness, soil nutrient(NPK), precipitation and temperature information [38].

4 Results and Discussion

The current framework helps in acquiring the data about nature of the dirt to discover right harvest for their yield with the help of Raspberry pi board alongside not many sensors like pH sensor, dampness sensor and temperature sensor to gather the information from those sensor and sends to distributed storage to store the estimations of separate sensor. It gives android application office to the end-client to get their sensor esteems. It very well may be gotten to through entering the IP address of their gadget. Be that as it may, this framework can't anticipate the climate condition in earlier and won't recommend the best harvest to yield. It neglects to tell with respect to the water system at the necessary time. To examine the dirt quality, the proposed framework utilizes mugginess sensor, temperature sensor and NPK sensor to get insights regarding the dampness content in the dirt, temperature of the encompassing and Nitrogen, phosphorous, potassium substance in the dirt. These sensors are associated with the Arduino UNO board. A committed wi-fi module is associated with the board that assists with sending information to the cloud to process the information assembled by various sensors. The assembled information are Temperature, Humidity and NPK values. Those information from the sensors are put away in Influx DB cloud which is a high availability open source distributed storage. Information are refreshed to Influx DB much of the time. Grafana dashboard which is joined with Influx DB shows the graphical portrayal of the accumulated information which are put away in Influx DB. The Grafana Dashboard is appeared in Fig. 7.

So as to envision the future estimation of the information, it is handled utilizing an AI system. The procedure utilized right now the RNN-LSTM strategy, which is utilized to foresee the future incentive with long haul conditions. In light of the anticipated estimations of temperature, dampness and NPK, the framework can distinguish the suitable yield and the kind of water system to be favored for the harvest will be recommended to the client, for example the ranchers. This is finished utilizing a portable application. The application is made utilizing the android studio programming. This application will show soil temperature, moistness and supplements. It will likewise give a caution if the temperature drops. Dampness zone to guarantee water system. It utilizes the temperature and supplement estimates to examine the

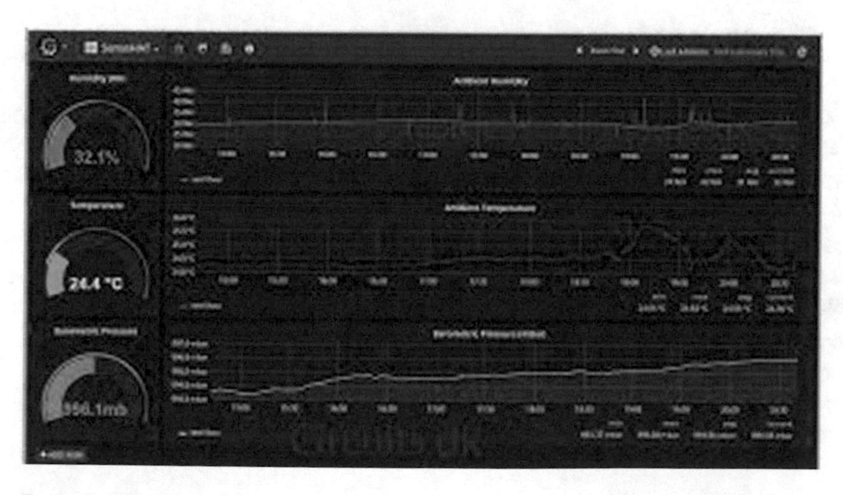

Fig. 7 Grafana Dashboard

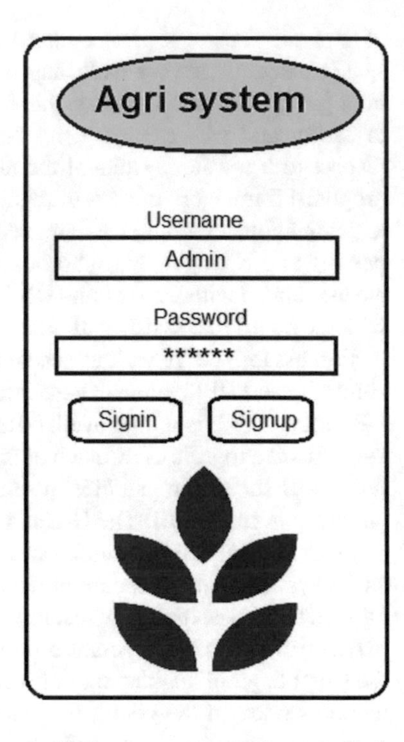

Fig. 8 App Layout

best harvest. The rancher can sign in as appeared in Fig. 8 to see subtleties and get proposals for good harvest yield as appeared in Fig. 8.

5 Conclusion

In this study, the project for monitoring the humidity, temperature and nutrients of citrus fruits based on the Internet platform of things was proposed. The project concerns early warning and decision making. as basic objectives, and provide a reference solution for the large-scale cultivation of citrus fruits.

Right now, gives the insight concerning supplement substance, dampness and temperature of the dirt with the assistance of various sensors associated with Arduino UNO board which is cost productive. From NPK sensor, supplement substance of the dirt are gotten. RNN-LSTM strategy gives the anticipated qualities that causes rancher to think about water system required planning likewise it recommend the yield to plant for the specific timeframe. So the rancher can get the great yield of development.

References

1. Gondechawar, N., Kawitkar, R.S.: IOT abased smart agriculture. Int. J. Adv. Res. Comput. Commun. Eng. **5**, 838–842 (2016)
2. Yadav, P., Swanson, C., Penning, T., Wallace, J.: Study of moisture sensors' response to drying cycles of soil. In: Conference: Irrigation Show Technical Program, At Las Vegas, NV (2019)
3. Na, A., Issac, W., Vashney, S., Khan, E.: An IOT based system for remote monitoring of soil characteristics. In: International Conference on Information Technology (InCITe)-The Next Generation IT Summit (2016)
4. Masrie, M., Rosman, M.S.A., Sam, R., Janin, Z.: Detection of nitrogen, phosphorous and potassium (NPK) nutrients of soil using optical fiber Transducer. In: 4th IEEE Conference on Smart Instrumentation, Measurement of Applications (ICSIMA) (2017)
5. Regalado, R.G., Dela Cruz, J.C.: Soil PH and nutrient (Nitrogen, phosphorous and potassium) analyzer using colorimetry (2016)
6. Ananthi, N.: IOT based smart soil monitoring system for agricultural production. In: IEEE International Conference on Technological Innovation in ICT for Agricultural and Rural Development (2017)
7. Bhardwaj, S., Goel, S., Sangam, V., Bhasker, Y.: Automatic irrigation system with temperature monitoring. Int. Res. J. Eng. Technol. (IRJET)
8. Gutiérrez, J., Villa-Medina, J.F., Nieto-Garibay, A., Porta-Gándara, M.Á.: Automated irrigation system using a wireless sensor network and GPRS module. IEEE Trans. Instrum. Meas. **63**(1), 166–176 (2014)
9. Gaikwad, S.V., Galande, S.G.: Measurement of NPK, temperature, moisture, humidity using WSN. Int. J. Eng. Res. Appl. **5**, 84–89 (2015)
10. Mhaiskar, S., Patil, C., Wadhai, P., Patil, A., Deshmukh, V.: A survey on predicting suitable crops for cultivation using IoT. Int. J. Innov. Res. Comput. Commun. Eng. **5**(1), 318–323 (2017)
11. Adeyemi, O., Grove, I., Peets, S., Domun, Y., Norton, T.: Dynamic neural network modelling of soil moisture content for predictive irrigation scheduling. Sensors **18**, 3408 (2018)
12. Ashish. InfluxDB To Grafana: Visualizing Time Series Data in Real Time, 06 March 2017. https://www.codementor.io/ashish1dev/influxdb-to-grafana-visualizing-time-series-data-in-real-time-5hxhaq0uj
13. Zhang, J., Hu, S., Long, Z., Kou, Q.: The wireless data transmission system based on GPRS and its discussion for application. J. Electron. Meas. Instrum. **23**, S1 (2009)
14. Wang, N., Zhang, N., Wang, M.: Wireless sensors in agriculture and food industry—recent development and future perspective. Comput. Electron. Agric. **50**, 1–14 (2006)
15. Braginsky, D., Estrin, D.: Rumor routing algorithm for sensor networks. In: Proceedings of the First ACM International Workshop on Wireless Sensor Networks & Applications, Atlanta, GA, USA, 28 September 2002, pp. 22–31 (2002)
16. Luo, J., Eugster, P.T., Hubaux, J.P.: Route driven gossip: probabilistic reliable multicast in ad hoc networks. In: Proceedings of the Joint Conference of the CiteSeer IEEE Computer and Communications, San Francisco, CA, USA, 30 March–3 April 2003, vol. 3, pp. 2229–2239 (2003)
17. Tian, B., Zhao, X.L., Yao, Q.M., Zha, L.: Design and implementation of a wireless video sensor network. In: Proceedings of the 2012 9th IEEE International Conference on Networking, Sensing and Control (ICNSC), Beijing, China, 11–14 April 2012, pp. 411–416 (2012)
18. Andrea, G., Lucchi, M., Tavelli, E., Barla, M., Gigli, G., Casagli, N., Chiani, M., Dardari, D.: A robust wireless sensor network for landslide risk analysis: system design, deployment, and field testing. IEEE Sens. J. **16**(16), 6374–6386 (2016)
19. Lee, H.C., Ke, K.H., Fang, Y.M., Lee, B.J., Chan, T.C.: Open-source wireless sensor system for long-term monitoring of slope movement. IEEE Trans. Instrum. Meas. **66**(4), 767–776 (2017)
20. Wu, J., Kong, Q., Li, W., Song, G.: Interlayer slide detection using piezoceramic smart aggregates based on active sensing approach. IEEE Sens. J. **17**(19), 6160–6166 (2017)
21. Ramesh, M.V., Kumar, S., Venkat Rangan, P.: Wireless sensor network for landslide detection. In: ICWN, pp. 89–95 (2009)

22. Fei, Y.J., Xu, Z.J., Feng, L.: The research of internet of things in agricultural production and management. In: Proceedings of the Fifteenth Session of the Annual Meeting of the Association of China, the Tenth Venue: Conference on Information Technology and Agricultural Modernization, Guiyang, China, 25–27 May 2013. (in Chinese)
23. Li, D.L.: Internet of things and wisdom of agriculture. Agric. Eng. **2**, S126 (2012)
24. Nandha Kumar, G., Nishanth, G., Praveen Kumar, E.S., Archana, B.: Arduino based automatic plant watering system with internet of things. Int. J. Adv. Res. Electr. Electron. Instrum. Eng. **6**(3) (2017)
25. Kumar, M.S., Ritesh Chandra, T., Pradeep Kumar, D., Sabarimalai Manikandan, M.: Monitoring moisture of soil with low cost home made soil moisture sensor and Arduino Uno. In: Advanced Computing and Communication Systems, 10th October 2016. IEEE (2016)
26. Kamalaskar, H.N., Zope, P.H.: International Journal of Engineering Sciences & Technology (IJESRT) survey of smart irrigation research system. ISSN 2277-9655
27. Otazú, V.: Manual on Quality Seed Potato Production Using Aeroponics, vol. 44. International Potato Center (CIP), Lima, Peru (2010). https://cippotato.org.research/publication/manaul-on-quality-seed-potato-production-usingaeroponics
28. Soil Hygrometer sensor (smartprototyping.comlSoil-HygrometerDetection-Module-Soil-Moisture-Sensor-For-Arduino.html
29. Wang, C., Zhang, A., Karimi, H.R.: Development of La3+ doped CeO2 thick film humidity sensors. Abstr. Appl. Anal. **2014**, 6 (2014). Article ID 297632
30. Chen, Z., Lu, C.: Humidity sensors: a review of materials and mechanisms. Sens. Lett. **3**(4), 274–295 (2005)
31. Zor, S.D., Cankurtaran, H.: Impedimetric humidity sensor based on nanohybrid composite of conducting poly (diphenylamine sulfonic acid). J. Sens. **2016**, 9 (2016). Article ID 5479092
32. Asao, T.: Hydroponics - A Standard Methodology for Plant Biological Researches, 1st edn. Intech, Rijeka (2012)
33. Borgognone, D., Colla, G., Rouphael, Y., Cardarelli, M., Rea, E., Schwarz, D.: Effect of nitrogen form and nutrient solution pH on growth and mineral composition of self-grafted and grafted tomatoes. Sci. Hortic. **149**, 61–69 (2013)
34. Patil, V.C., Al-Gaadi, K.A., Biradar, D.P., Rangaswamy, M.: Internet of things (Iot) and cloud computing for agriculture: an overview. In: Proceedings of Agro-Informatics and Precision Agriculture (AIPA 2012), India, pp. 292–296 (2012)
35. Assam Agricultural University, Jorhat, Agro-Climatic Planning for Agricultural Development in the State of Assam: Draft Outline for the Eighth Plan Period (1994)
36. Whitmore, A., Agarwal, A., Da Xu, L.: The Internet of Things-A survey of topics and trends. Inf. Syst. Front. **17**(2), 261–274 (2014)
37. Pascual, V.J., Wang, Y.-M.: Impact of water management on rice varieties, yield, and water productivity under the system of rice intensification in Southern Taiwan. Water (2017)
38. Thakare, S., Bhagat, P.H.: Arduino-based smart irrigation using sensors and ESP8266 WiFi module. In: Second International Conference on Intelligent Computing and Control Systems (ICICCS 2018) (2018)

Plant Diseases Detection and Classification Based on Image Processing and Machine Learning

Assia Ennouni, My Abdelouahed Sabri, and Abdellah Aarab

Abstract To further the progress of sustainable development and solve real-life problems we have seen many processes are applied in our life, like artificial intelligence and decision-making. Morocco is one of the countries that rely heavily on agriculture and food production. So, food production is considered the basic needs of a human being for that we have seen fast advancements in agriculture productivity to meet the projected demand. However, with the time passing by, all species of plants are subjected to various types of diseases that cause huge damage. Although the observation of variation in the infected part of the leaf plant is very important but not enough because the perception of the human eye is not so much stronger. The identification of plant diseases is a very important task in the agriculture area. So, the best identification means there is a huge gain on agricultural productivity, quality, and quantity. To detect plant diseases in an earlier stage we require efficient and precise techniques to assist farmers in decision-making. This article presents, first, an overview of plant diseases from leaves images and different disease classification approaches that can be used for plant leaf disease detection.

Keywords Smart agriculture · Computer vision · Image processing · Plant leaf diseases detection · Classification

1 Introduction

Morocco is one of the countries that rely heavily on agriculture. So global production must increase to meet the projected demand. In this field, we have seen fast advancements in agriculture productivity especially in both food production quality and quantity Therefore, the quantity and quality of plant products get reduced by diseases. A plant disease is, by definition, anything that stop a plant from behaving naturally and not yielding enough [1]. Thus, a plant is considered diseased when it is infected by an actor causing abnormal physiological or biochemical behavior such as growth and/or functions which are abnormal [16]. So, plant disease detection

A. Ennouni · M. A. Sabri (✉) · A. Aarab
LISAC Laboratory, Faculty of Sciences Dhar-Mahraz, University Sidi Mohamed Ben Abdellah, Fez, Morocco

N. Gherabi and J. Kacprzyk (eds.), *Intelligent Systems in Big Data, Semantic Web and Machine Learning*, Advances in Intelligent Systems and Computing 1344,
https://doi.org/10.1007/978-3-030-72588-4_20

plays an important role because the beginner's farmers cannot identify in the early stage the majority of plant diseases In addition, their visual identification is a more difficult task and may not be effective or precise. Early detection could be fruitful and beneficial. So, the application of computer vision and machine learning algorithms are recently used to help and assist decision makers in medicine or supervision [18, 19]. In the same vision to help farmer, image-processing algorithms can certainly be used to assist in detecting and classifying plant diseases by analyzing leaf images [20]. This paper presents a recent review of plant leaves diseases from images. So this study has been divided into six sections as follows: The first one gives a brief introduction. The second one examines the various types of plant diseases. In the third section, the plant disease detection process is presented and described in detail including the main steps. Then we propose various approaches that are discussed in the literature. Table 1 summarizes research that has been reported to solve this issue. Finally, our conclusions are drawn in this study. Plant diseases detection process using leaf images (Fig. 1).

The disease identification process using leaf images incorporates in four main steps as follows:

1. Acquisition,
2. Preprocessing and segmentation,
3. Features extraction,
4. Classification and diagnosis results.

Initially after image acquisition, we need to pick out the plant, which is influenced by the disease, and then gather the leaf of this plant and take a snapshot of leaf and load the leaf image into our detection system. The next step is preprocessing; during this step, and distortion removal improves images. The most used preprocessing techniques are color space transition, enhancement, smoothing, and cropping. All

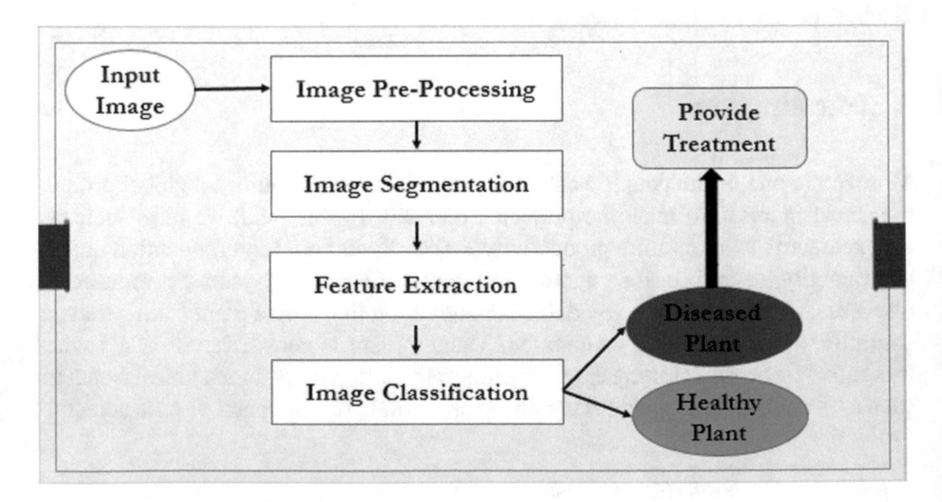

Fig. 1 Plant diseases detection process

these techniques enhance image features for the analysis step. Then the segmentation step is applied for getting the meaningful representation of the image and the useful correlated segments along with the OI (Objects of Interest). In this step, we can use K-means, edge detection or any other algorithms, etc. Features are extracted from the leaf or/and the plant afterward. Texture, color or shape features are the most used in such application. Several approaches exist to extract features from an image. Color features can be extracted using Color Histogram and/or Color Coherence Vector. Texture features can be obtained from Local Binary Pattern (LBP) or Complete Local Binary Pattern (CLBP) [2]. Finally, Extracted features are then used to classify and identify the healthy or infected leaf parts [3]. Artificial Neural Network (ANN) and Support Vector Machine (SVM) are widely used as classification algorithm.

2 Plant Diseases

Plant disease is damage of the normal state of a plant that disrupts or changes its vital functions [16]. All species of plants are exposed to diseases. Plant diseases emergence and prevalence change from season to season, depending on environmental conditions, crops and cultivated varieties. Some plants are particularly subject to some kind of diseases while others are more resistant to them.

We can classify plant diseases either to the nature of the causal agent or to the infectious or noninfectious. In this section, we learn about plant pathology and studies plant disease including disease reasons and how manage healthy plants. We mention various kind of diseases that can affect the leaves and their symptoms that can help us to make the crop healthy and fruitful. Therefore, the various types of leaf diseases are Bacterial—Fungal—Viral as shown below in Fig. 2.

A. **Bacterial plant diseases**

Symptoms of plants bacterial infection are much like fungal plant disease symptoms. The infecting bacteria are microorganism that can only be seen with a microscope. The infection is identified by the presence of leaf spots, wilts, blights, cankers, scab, soft rots, and overgrowth. The infection can appear on leaves, fruits, stems or/and blossoms. There are Millions of referred to be beneficial or healthy bacteria that live everywhere, on skin and in the intestines for instance [4]. Beneficial bacteria are involved in diverse processes such as digestion and nitrogen fixation in the roots for example. And there are also pathogenic bacteria that cause serious diseases that can be fatal in plants (Fig. 3).

B. **Fungal plant diseases**

Fungus, plural fungi, any of about 144,000 known species, which covers the yeasts, rusts, smuts, mildews, molds, and mushrooms. Fungi that caused fungal, the most popular in the environment, are not very risky, but can be hurtful. There are also many funguses like organisms, including slime molds and omycetes (water molds).

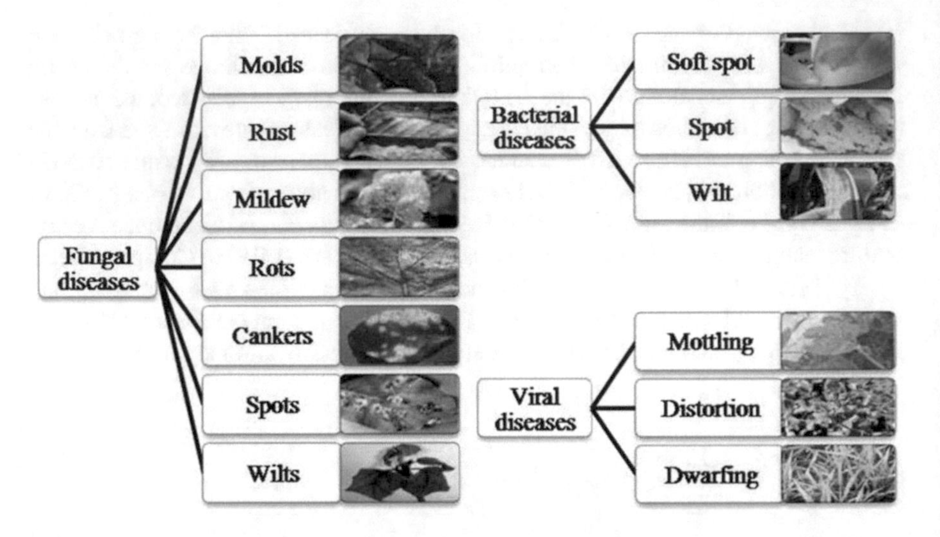

Fig. 2 Plant disease types [19]

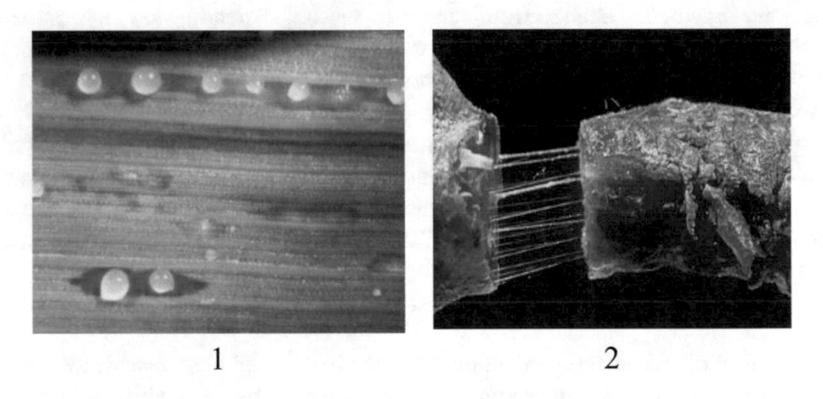

Fig. 3 (1) Bacterial leaf blight on wheat. (2) Bacterial strand test on cut stems

Fungi are considered among the most common plants on earth and which are of great importance and useful in medicine and for the environment. Fungus live freely in soil or water; others establish parasitic or symbiotic relationships with plants or animals [16] (Fig. 4).

C. **Viral plant diseases**

Viruses are intracellular nucleic acid particles with a protein layer that infect living organisms and can replicate. Viroids are viral-like particles without a protein layer. Viroids and viruses are mainly transmitted by insects, nematodes and fungi by introducing them during feeding [18]. Symptoms of this disease type vary rely on the particular type of virus causing contagion or infection. Viral diseases can affect any

Fig. 4 Soybean infected with Sclerotinia

Fig. 5 Necrotic spot virus on pepper leaves

zone of the plant. Viruses are microscopic particles that cannot be seen by eyes but with a microscope. Plant viruses are mostly isometric (polyhedra) or rod-shaped [17] (Fig. 5).

3 Literature Review

This section presents various researches for detecting the disease in plant leaf using an image processing technique and provides review on existing proposed approaches that are used in plant disease detection.

Paper [5] presents a new approach for plant leaf disease classification in which Hue Saturation Intensity (HIS) is used. Initial image is transformed first and then the image will be segmented using the FCM algorithm. Features used are spot color, size, and shape. Finally, classification is based on NN (Neural Networks).

Authors in paper [6] propose an approach to detect citrus leaf disease. K-means clustering algorithm is used to segment the leaf and to define the diseased areas.

The Gray-Level Co-Occurrence matrix (GLCM) is used to extract textural features and classification is done using SVM algorithm.

In [7] segmentation is first used to identify the diseased region. An optimal threshold based on weighted Parzen-window is adopted. The quotient of the infected spot surface and the leaf areas is used to estimate plant diseases.

In the paper [8], the authors propose to use the K-means clustering algorithm to extract the infected object. Textural features are extracted using color co-occurrence technique. The classification is based also on NN (Neural Networks).

Authors in paper [9] proposed two steps to identify the affected part of the high-resolution multispectral Sugar beet leaves disease. Initially, segmentation based on the K-means algorithm is performed on the initial image. Secondly, the diseases classification is done using the NN classifier (Neural Networks).

Paper [10] proposes a two phases an approach for Cotton diseases detection and classification. First, to identify the affected region, edge detection based segmentation is used, and then disease classification is performed using the Homogeneous Pixel Counting algorithm.

The idea proposed in [11] suggested three main steps to detect and classify leaf diseases. The RGB color space is used and thresholding is applied to the green component to segment the image. Textural features are extracted based on the Color Co-Occurrence method. Features are then used and passed to the SVM classifier to classify the leaf into different disease categories.

Diseases Detection and classification of cotton Leaf Spot has been treated in [12]. Skew divergence approach is used to detect the Edge and to segment the leaf. Features extraction was based on Color, GA and texture followed by PSO features selection to improve the classification accuracy. Classification is done using the SVM and Back Propagation Neural Network (BPNN) algorithms.

A multiscale segmentation approach is proposed in [13] where the multilevel wavelet decomposition is used for detecting the plant disease with Real-time weed discrimination. As a preprocessing step, the histogram equalization is used and features are obtained from Wavelet decomposition. Disease classification is obtained using the Euclidean distance method.

To classify the grapefruit peel diseases like a canker, greasy spot, copper burn, and wind scar, authors in [14] used color texture features obtained from the Spatial Gray-level Dependence Matrices (SGDM). Finally, the classification is based on the squared distance method.

Paper [15] presents popular methods based on machine learning algorithms such as Kmeans, NN, CCM to detection and classify agricultural products disease.

A summary of the reviewed papers with algorithm names and classification accuracy are presented in Table 1 and Fig. 6.

Table 1 List Summary of reviewed articles with algorithm names and classification accuracies

Papers	Techniques	Accuracy
[5]	GLCM, Kmeans, SVM	94,5%
[6]	HIS, FCM, NN	96,5%
[7]	Threesholding, Kmeans Clustering	94%
[8]	Threefold, Kmeans, color cooccurrence, NN,	96%
[9]	Kmeans, NN	88,9%
[10]	Edge detection, Color cooccurance	80%
[11]	Edge detection, Transformation, SVM	87%
[12]	SkewDivergence, CYMK feature, SVM, BNP, Fuzzy	86%
[13]	Multilevel wavelet Detection, Histogram Euclean distance	85%
[14]	SGDM, Squared distance method	96%
[15]	Kmeans, NN, CCM	95%

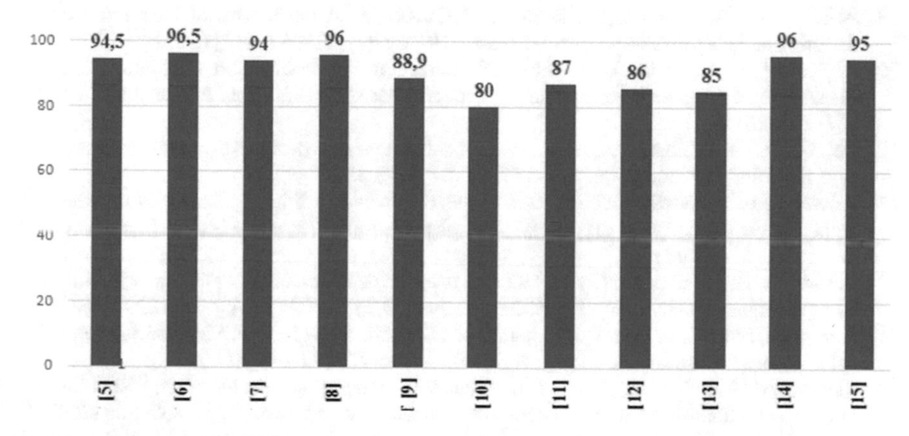

Fig. 6 Accuracy values of reviewed articles

4 Conclusion

Our research underlined the importance of image processing in the agriculture area by giving a recent review of different leaf disease detection and classification techniques. Thus, three types of plant diseases and their symptoms were examined and explained namely bacterial, fungal and viral. Then the general process has been presented and explained. Then various approaches for detection were given in detail and discussed in the literature section. Finally, to conclude the rate of classification which was based on the study of previous approaches in the detection of plant diseases yields a low precision in the suggested process. In future, many other techniques can be used to improve the rate of classification. So, we will propose to incorporate another step to plant disease detection process to improve the accuracy of classification into our future system.

References

1. https://cropwatch.unl.edu/soybean-management/plant-disease
2. Dubey, S.R., Jalal, A.S.: Detection and classification of apple fruit diseases using complete local binary patterns. In: IEEE Computer and Communication Technology (ICCCT), 2012 Third International Conference on Computer and Communication Technology, Allahabad, pp. 346–351 (2012)
3. Omrani, E., Khoshnevisan, B., Shamshirband, S., Saboohi, H., Anuar, N.B., Nasir, M.H.N.M.: Potential of radial basis function-based support vector regression for apple disease detection. Measurement **55**, 512–519 (2014)
4. Varshney, S., Dalal, T.: A novel approach for the detection of plant diseases. Int. J. Comput. Sci. Mob. Comput. **5**(7), 44–54 (2016). ISSN 2320-088X
5. Phadikar, S., Sil, J., Das, A.K.: Classification of rice leaf diseases based on morphological changes. Int. J. Inf. Electron. Eng. **2**(3), 460–463 (2012)
6. Detection of leaf diseases and classification using digital image processing 2017 International Conference on Innovations in Information, Embedded and Communication Systems (ICIIECS)
7. Jun, W., Wang, S.: Image thresholding using weighted Parzen Window estimation. J. Appl. Sci. **8**(5), 772–779 (2008)
8. Al-Hiary, H., Bani-Ahmad, S., Reyalat, M., Braik, M., Alrahamneh, Z.: Fast and accurate detection and classification of plant diseases. IJCA **17**(1), 31–38 (2011) (2010)
9. Al-Bashish, D., Braik, M., Bani-Ahmad, S.: Detection and classification of leaf diseases using K-means-based segmentation and neural networks based classification. Inf. Technol. J. **10**(2), 267–275 (2011)
10. Revathi, P., Hemalatha, M.: Classification of cotton leaf spot diseases using image processing edge detection techniques, pp. 169–173. IEEE (2012). ISBN
11. Arivazhagan, S., Newlin Shebiah, R., Ananthi, S., Vishnu Varthini, S.: Detection of unhealthy region of plant leaves and classification of plant leaf diseases using texture features. CIGR **15**(1), 211–217 (2013)
12. Revathi, P., Hemalatha, M.: Cotton leaf spot diseases detection utilizing feature selection with Skew divergence method. Int. J. Sci. Eng. Technol. **3**(1), 22–30 (2014). ISSN 2277-1581
13. Kim, D.G., Burks, T.F., Qin, J., Bulanon, D.M.: Classification of grapefruit peels diseases using color texture feature analysis. Int. J. Agric. Biol. Eng. **2**(3), 41–50 (2009)
14. Siddiqil, M.H., Sulaiman, S., Faye, I., Ahmad, I.: A real time specific weed discrimination system using multi-level wavelet decomposition. Int. J. Agric. Biol. **11**(5), 559–565 (2009)
15. Tripathi, M.K., Maktedar, D.D.: Recent machine learning based approaches for disease detection and classification of agricultural products. In: International Conference on Electrical, Electronics and Optimization Techniques (ICEEOT) (2016)
16. https://www.britannica.com/science/plant-disease/Epiphytotics. Accessed 28 Dec 2019
17. Isleib, Michigan State University Extension, 19 December 2012. https://ohioline.osu.edu/factsheet/plpath-gen-5Jim
18. Ennouni, A., Filali, Y., Sabri, M.A., Aarab, A.: A review on image mining. In: 2017 Intelligent Systems and Computer Vision (ISCV), Fez, Morocco, 17–19 April 2017, pp. 1–7. Print on Demand (PoD) ISBN: 978-1-5090-4063-6. Electronic ISBN: 978-1-5090-4062-9. https://doi.org/10.1109/ISACV.2017.8054916
19. Sabri, M.A., Filali, Y., Ennouni, A., Yahyaouy, A., Aarab, A.: An overview of skin lesion segmentation, features engineering, and classification. In: Intelligent Decision Support Systems: Applications in Signal Processing, pp. 31–52. De Gruyter, Berlin. https://doi.org/10.1515/9783110621105-002. ISBN 9783110621105
20. Saleem, M.H., Potgieter, J., Arif, K.M.: Plant disease detection and classification by deep learning. Plants **8**(11), 468 (2019). https://doi.org/10.3390/plants8110468

Textual Matching Framework for Measuring Similarity Between Profiles in E-recruitment

Islam A. Heggo and Nashwa Abdelbaki

Abstract Finding the most relevant jobs for job-seeker is one of the ongoing challenges in the area of e-recruitment. Text mining, information retrieval and natural language processing are some of the key concepts for matching a number of profiles and find the most textually similar job profiles given the job-seeker profile. The mission of finding the relevant jobs that are textually similar to the job-seeker profile depends mainly on the quality of three mains phases. The first early phase is basically constructing the profiles that should be involved in the matching phase through gathering data from different sources and representing this data into a form of important keywords. The second phase is the matching phase which works on mining and analyzing the text of these constructed profiles to find the similar profiles. The final phase is ranking these produced similar profiles by relevance. We introduce our intelligent similarity computing engine which works on predicting the personalized jobs for each job-seeker. It aims to empower the recommendation effectiveness in the area of e-recruitment by applying the three aforementioned phases on the problem of job recommendation. Its purpose is to recommend the textually relevant jobs based on computing the textual similarity distance between jobs and job-seekers' profiles.

Keywords Information retrieval · NLP · RS · Relevancy scoring · Search engine · Ranking · Personalization · TF-IDF · BM-25 · Textual matching

1 Introduction

For finding the top textually similar jobs to the job-seeker profile, we need to utilize more than one concept such as information retrieval, text mining, natural language processing and information filtering. This process of recommendation relies mainly on three phases. Our first phase is to collect all the needed text and data to build our

I. A. Heggo (✉) · N. Abdelbaki (✉)
School of Communication and Information Technology, Nile University, Giza, Egypt
e-mail: i.heggo@nu.edu.eg

N. Abdelbaki
e-mail: nabdelbaki@nu.edu.eg

N. Gherabi and J. Kacprzyk (eds.), *Intelligent Systems in Big Data, Semantic Web and Machine Learning*, Advances in Intelligent Systems and Computing 1344,
https://doi.org/10.1007/978-3-030-72588-4_21

similarity model on. This data can be extracted from the content-based information which is manually pre-filled by the user to describe explicitly the preferences of user or the behavioral-based information which is dynamically collected from the users' activities that describes the implicit preferences of each user [1]. Both of content-based data and behavioral-based data can structured or unstructured. The structured data is found in profiles of job-seekers and jobs in the form of information that has predefined finite set such as gender, educational degree, industry…, etc. The unstructured data is found in profiles of job-seekers and jobs in in the form of information that does not belong to a definite set of options such as job description, job-seeker occupations, job-seeker skills…, etc. The second phase is analyzing all of these types of data, content-based data and behavioral data whether structured or unstructured. The text mining of this information is a mandatory phase to filter out the most influential words and phrases to emphasize on. The final phase is retrieving the most relevant jobs in a sorted order by relevancy score where the top jobs would have the minimum distance in similarity space to the job-seeker profile.

All these phases are enveloped by the knowledge of recommendation systems because text mining is not the only vital mission to achieve our objective. Our ultimate purpose is to recommend the most interesting jobs for each job-seeker by obtaining the textual similarity between the two entities, jobs and job-seeker. We built this engine to eventually recommend the jobs with higher similarity score to improve the recommender systems techniques in the field of e-recruitment and decrease the rate of unemployment worldwide.

We discuss the numerous phases of a textual-based recommendation engine by finding the similarity between the content of job-seekers and jobs. The paper begins with the various conceptual information retrieval and filtering models to retrieve the top matching jobs in both of sub-sections called Boolean independence model and vector space model. Another sub-section called textual processing and matching layers illustrate our developed text mining and matching layers. The textual similarity model which aims to extract the influential terms for matching from job-seeker profile and job post. Before evaluation, we discuss in Sect. 3.5 the different relevancy scoring formulas for ranking results based on their textual relevancy score. Finally, we define the most suitable metrics for evaluating such frameworks and decide the most accurate approaches for efficiency measuring. Our framework evaluation is found in the last section called results and deductions.

2 Literature Survey

In spite of the significance of large-scale textual matching layers and textual similarity systems on this challenging field of e-recruitment, very few numbers of academic researches have been conducted on the specifics of these engines. It is uncommon to find researches addressing a practical technique of analyzing the text, data and information of job posts and job-seekers profiles. It is rare as well to find researches addressing the approaches of jobs ranking according to their relevancy and matching level to the job-seeker profile. However, we have utilized the principles of text mining, information retrieval and NLP from similar industries to address the topic of establishing a powerful textual matching framework which can accurately compute the similarity between job-seeker profiles and job posts. We investigated the leading job search engines LinkedIn and Glassdoor. They use users' historical actions such as search, view job, and apply to job, in addition to the explicit filled preferences. These data are manipulated to fetch similar job posts. LinkedIn and Glassdoor rely mainly on keyword-based search to query their pool of jobs. Despite they are great systems, sometimes they recommend irrelevant jobs that could be biased to user's behavior rather than user's profile content (e.g. skills, experiences). For instance, users' profiles contain experiences such as "Senior Software Engineer", "Lead Software Engineer" and "Product Manager". Nevertheless, LinkedIn keeps recommending jobs like database administrator and project coordinator. The same problem exists in Glassdoor where it keeps recommending jobs like "English Language Teacher" which is a completely irrelevant job for a job-seeker who filled the experiences and preferences sections in the personal profile by another different job title like "Senior Software Engineer". The Glassdoor platform did not only disrespect the profile information like experiences and preferences but also it did not consider the behavioral information properly where this job-seeker kept searching for "Software Engineer" jobs at Glassdoor frequently. However, all these search queries were not enough to make Glassdoor platform understand the preferences to recommended efficiently proper matching jobs.

3 Hybrid Textual Matching Methodology

We propose the methodology of analyzing both of job-seekers profiles and job posts. Our methodology starts by deciding the influential type of data for an efficient recommendation before discussing the needed information retrieval, processing and ranking models.

3.1 Behavioral-Based and Content-Based Extracted Text

It is influential to gather more text from numerous data sources to be able to recommend precise results. Out first phase is deciding the text that should be considered for the recommendation and matching phase. On the job side, considering most of the text in the job post would be useful. On the job-seeker side, the prefilled content of job-seeker's profile is not enough but also considering the behavioral-based information will lead to higher matching and recommendation quality. There is a lot of worthy information can be collected from the frequent interactions of users [2–4]. For example, the tracking of the jobs which the user had applied to, then extract only the most frequent job-titles and keywords from these jobs to finally construct the dynamic

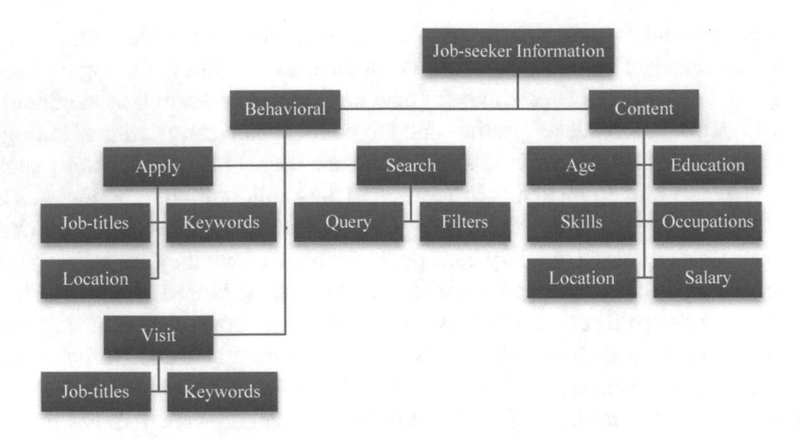

Fig. 1 Behavioral-based and content-cased collected text

Table 1 Representation of presence and absence of terms in documents

	Doc #1	Doc #2	Doc #3	Doc #4	Doc #5	Doc #6	Doc #7
Software	1	0	1	0	1	0	0
Java	**0**	0	0	**1**	0	1	0
NLP	**1**	1	1	**0**	1	0	1
Search	**1**	1	0	**1**	1	0	1
Engine	**0**	0	1	**1**	0	1	0

job-seeker profile and use it for matching with the jobs' profiles. Another behavioral-based information source is the most recurrent search queries and filters which job-seekers keep using and search by. They are also highly valuable to understand what job-seekers are looking for when these queries are frequently and periodically used. Getting the continuously visited jobs by job-seeker and how long each visit takes is another good factor to consider, but it is less important than the searched queries and previous applications. Therefore, using some behaviorally collected keywords will be definitely useful. Figure 1.

3.2 *Boolean Independence Model of Information Retrieval (BIM)*

BIM is a very simple model based on sets theory and Boolean Algebra. It considers documents as a set of terms and the queries are Boolean expressions over terms. Query could be a simple query as a one term query, or it could be more complex as multi-terms query which contains multiple of terms. If the document meets the query expression then it is verified as a relevant document [5, 6]. As shown in Table 1,

Table 2 The corresponding applied boolean algebra to BIM example

	(Search ∧ Engine) ∨ Java ∨ NLP					(Search ∧ Engine) ∨ Java ∨ NLP			
Doc #1	1	0	0	1	Doc #4	1	1	1	0
	0		1			1		1	
	1					**1**			

it contains record for each document whether it contains each term out of all the different terms or not. Each cell becomes 1 if the corresponding term is found in the corresponding doc and 0 if term is not found.

- Consider the following query
  (Search AND Engine) OR Java OR NLP
- The corresponding boolean algebra expression is as follow
  (Search ∧ Engine) ∨ Java ∨ NLP
- That expression will return two documents: - Doc#1, Doc #4 (Table 2)

3.3 *Vector Space Model of Information Retrieval (VSM)*

A vector space model or term vector model is a more advanced model where two vectors are defined. The first vector is the query vector. The second vector is the document vector. Each term in the query is considered as vector dimension in the query vector. Each term in the document is considered as vector dimension in the document vector. The query in our case is the gathered and extracted influential keywords from the job-seeker"s profile and the document in our case is the extracted most influential keywords from the jobs" profiles. The relevance of the document to the query is computed through calculating the scalar product of those two vectors [5, 6]. It manipulates document like a bag of words. A document is a t-dimensional vector $d = [d_1, d_2 \ldots, d_t]$. The ith entry of the vector is the boosting weight or importance of term d_i in the document (the job's profile). A query is a t-dimensional vector as well where q vector equals to $[q_1, q_2 \ldots, q_t]$. The ith entry of the vector is the boosting weight or importance of term q_i in the query (the job-seeker's profile). Then the dot product of above defined vectors is calculated as illustrated in Eq. 1. Each job (document) is scored according to its matching level to the job-seeker's profile (query). The dot product produces a non-negative real number. The document matches the query if the computed score is greater than zero. Only documents with non-zero score are returned.

$$q.d = \sum_{1}^{t} q_i d_i = q_1 d_1 + q_2 d_2 + \ldots + q_t d_t \quad (1)$$

Simply we can set each dimension in the document vector to one if the document includes the term d_i and zero if the document does not include it. Instead, we use more accurate and efficient boosting weights can be set, like the term frequency in the document to differentiate between jobs that include the term once and other documents which use the term more frequently. So, if a job includes specific term 5 times then it will have 5 as a term weight (d_i) in the document vector. We used multiple of boosting weights that we describe later in the textual processing and matching layers section. Example of these used boosting weights is the term frequency, the inverse term frequency which refers to the novelty of that term even if it has a higher term frequency but it is not so valuable in our matching and the text length which we used to normalize the above dot product results. Text length attribute normalizes the term frequency. So, if a job includes specific term 5 times and the whole job description consists of 100 other terms then it will have 5/100 as a normalized term weight (d_i) in the document vector. However, if a job includes the same specific term 2 times but the whole job description consists of 20 other terms then it will have 2/20 as a normalized term weight (d_i) in this document vector. That means that the second document will have a better relevancy score when matching with this specific term even it its term frequency higher in the first document. Because the normalize term frequency in the first document is 0.05 and the normalized term frequency in the second document is 0.1.

3.4 *Textual Processing and Matching Layers*

We need to prepare the text and emphasize on the crucial factors that can significantly produce better recommendations. We adopted many concepts to enhance the textual relevancy matching between job post and job-seeker profile.

Term Identification. Number of matched query terms is proportionally correlated to the document relevancy. In some cases, the query terms can be ORed terms not ANDed terms, then the most relative documents are the ones which have more query terms.

Typo and Misspelling Tolerance (Fuzzy Search). Damerau–Levenshtein distance algorithm is useful to achieve a typo tolerance similarity algorithm. Damerau–Levenshtein distance algorithm is the advanced algorithm of Levenshtein distance. Levenshtein algorithm is built to measure the similarity between two strings by counting minimum number of single-character operations required to transform one word into the other. These operations include character deletions, character insertions and character substitutions. Damerau–Levenshtein added the transposition of two adjacent characters operation. It is important for retrieving relevant results even if the query was mistyped. For example, if the query is *Jave developer*. Damerau– Levenshtein will figure out that this query needs one-character substitution to match *Java developer* by substituting the "e" to "a" in *Java*. In English language, we can set a generic rule to handle misspellings and overcome the typos challenge like allowing 1 distance tolerance for short words or 2 distance tolerance for longs words. But in Arabic language, it is kind of complex challenge to set a generic rule like English.

Table 3 Various misspelled variations of Arabic job title

5 characters	6 characters	7 characters	8 characters
دلفرى	ديلفرى	دليفيرى	ديليفيرى
دلفري	ديلفري	دليفيري	ديليفيري
	دليفرى	ديليفرى	
	دليفري	ديليفري	
	دلفيري	ديلفيرى	
	دلفيرى	ديلفيري	

Table 3 shows an example of a job title which is practically posted by users in too many variances. This job title is written in too many forms, some forms were a word of 5 characters, 6 characters, 7 characters and even 8 characters. The same word is posted in 16 different variations with a maximum distance 4 to transform one variation to another variation. Therefore Damerau–Levenshtein distance approach will be helpful in many cases, but there are some other complex situations where Damerau–Levenshtein approach will not help and synonyms approach will help better.

Synonyms. Using synonyms is the approach of query augmentation by including the similar words or synonyms [4, 8]. It is helpful to append the similar words to achieve wide range of accurate results. This tells our matching engine to behave similarly with the similar words as *HR*, *human resources* and *personnel specialist* or *senior* and *Sr.* There are many synonym sets, but it is dangerous to use a generic synonym set [9]. The ultimate solution is to build your own domain-based set to be accurate. For example, in Arabic "**شيخ**" could be a synonym for "**عجوز**" but "**شرم الشيخ**" is a city that cannot be a synonym for "**شرم العجوز**". Another English example is *substitute* which is a synonym for *alternative*, but we cannot augment *jquery-substitute* with *jquery-alternative*. The context is very important. Polysemy problems which means that a word can have more than one meaning depending on the context like *fair* could mean *exhibition* or the adjective of *fairness* which means *unbiased*, another Arabic example is "**طيار**" could mean *pilot* or *delivery boy* in some Arabic slang languages Table 4 shows an example of related words and synonyms in the e-recruitment domain. The left column are some English synonyms for HR jobs, the right column is come Arabic synonyms for HR jobs.

We propose some dynamic methodologies to partially automate the correlated keywords in the e-recruitment domain. Basically, by using crowd behavior on searching and applying, we can understand the relations between words to some extent. One model of these models is using the collaborative filtering model that

Table 4 Synonyms sample for HR jobs

HR Jobs Synonyms in English	HR Jobs Synonyms in Arabic
Human Resources Officer	مسئول موارد بشرية
HR Specialist	شئون موظفين
Personnel Officer	شئون عاملين
Human Capital Specialist	اخصائى موارد بشرية
HR Clerk	موارد بشرية
Personnel Specialist	موظف شئون عاملين

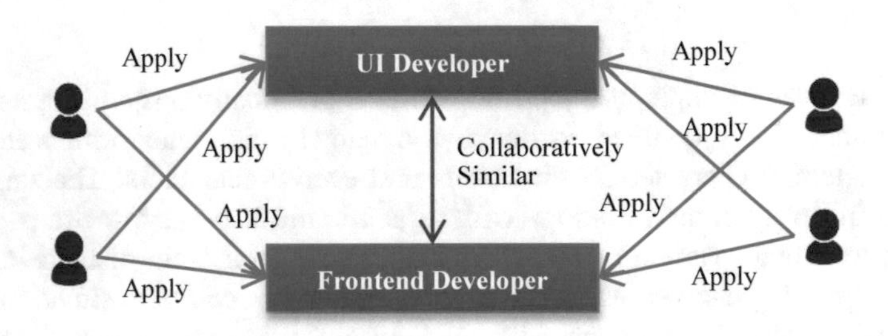

Fig. 2 Collaboratively correlated keywords

monitors what crowd concurrently applies to. Our assumption is when a lot of job-seekers apply to jobs titled *X* and the same job-seekers apply as well to jobs titled *Y* for a long timeframe, it means that X and Y are somewhat correlated. As Fig. 2 shows, if many job-seekers applies to jobs titled *UI Developer* and *Frontend Developer* jobs frequently, then it means that the collaborative co-occurrence of *UI Developer* and *Frontend Developer* is high and these job-titles are sort of correlated.

Similarly, that model can be duplicated and assisted by the crowd searching behavior. This model depends on the collaborative filtering model that monitors what crowd concurrently searching for. Our assumption is when a lot of job-seekers search for keyword *X* and the same job-seekers search also for keyword *Y* frequently for a long timeframe, it means that X and Y are somehow correlated. As Fig. 3 shows, if many job-seekers concurrently search for both of keywords **اخصائى تعيينات** and **اخصائى توظيف** frequently, then it means again that the collaborative co-occurrence of **اخصائى تعيينات** and **اخصائى توظيف** is high and these keywords are somehow correlated.

There are the cluster-based behavioral models that cluster job-seekers and track their applying behavior and searching behavior. The first model of them is the cluster-based behavioral model that monitors the applying behavior of each job-seekers cluster. The assumption of this model that when a lot of job-seekers belong to some cluster *X* keep applying to job titled *Y* frequently for a long timeframe, then we can

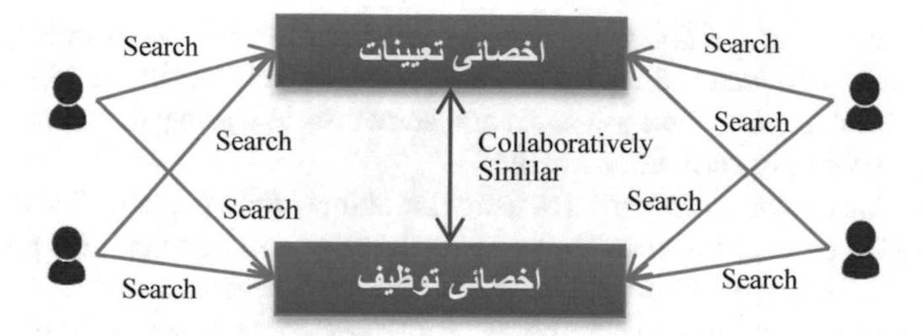

Fig. 3 Example of collaboratively correlated keywords

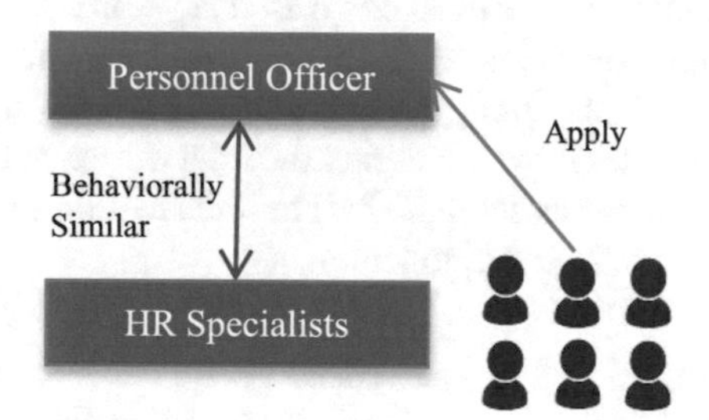

Fig. 4 Cluster-based behaviorally correlated keywords

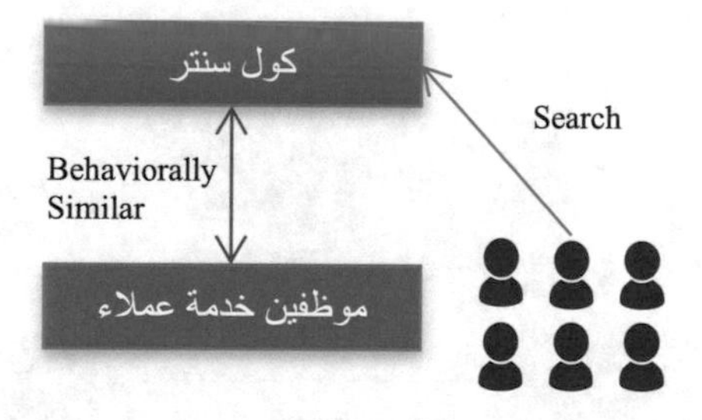

Fig. 5 Example of cluster-based behaviorally correlated keywords

conclude that X and Y are somewhat correlated. As Fig. 4 shows, if many job-seekers who worked as *HR Specialist* apply to jobs titled *Personnel Officer* frequently, then it means that the behavioral co-occurrence of *HR Specialist* and *Personnel Officer* is high, and these keywords are sort of correlated.

The second model of the cluster-based behavioral models is the model that monitors the searching behavior of each job-seekers cluster. The assumption of this model that when a lot of job-seekers belong to some cluster *X* keep searching for the keyword *Y* frequently for a long timeframe, then we can conclude that X and Y are somehow correlated. As Fig. 5 illustrates, if many job-seekers who worked as خدمة عملاء search for keyword كول سنتر frequently, then it means that the behavioral co-occurrence of خدمة عملاء and كول سنتر is high and these terms are rather correlated.

Combining all of the above mentioned collaborative and behavioral synonyms partial detector will lead to an overall relations graph similar to the one illustrated in Fig. 6. It will map all of these job-titles and keywords to a group of correlated terms that should be finally verified manually.

Content-based synonyms partial detector is another efficient model that is based on calculating the co-occurrence of keywords in the corpus. It computes the appearance frequency of the keywords together. The higher co-occurrence of the keywords X and Y, the more correlation score between the keywords X and Y. Table 5 shows one of our experiments on the keywords analysis to find some similar or correlated keywords based on the co-occurrence of these keywords together in the same job post. For example, it finds that the most correlated keywords to PHP are Information Technology (IT), Computer Science, Software Development, HTML and CSS. It computes the correlation by counting the times both of X and Y appeared together. As Eq. 2 illustrates, it calculates it by processing the intersection between the documents includes X and the documents include Y. The intersection produces the count of documents contain both of X and Y together.

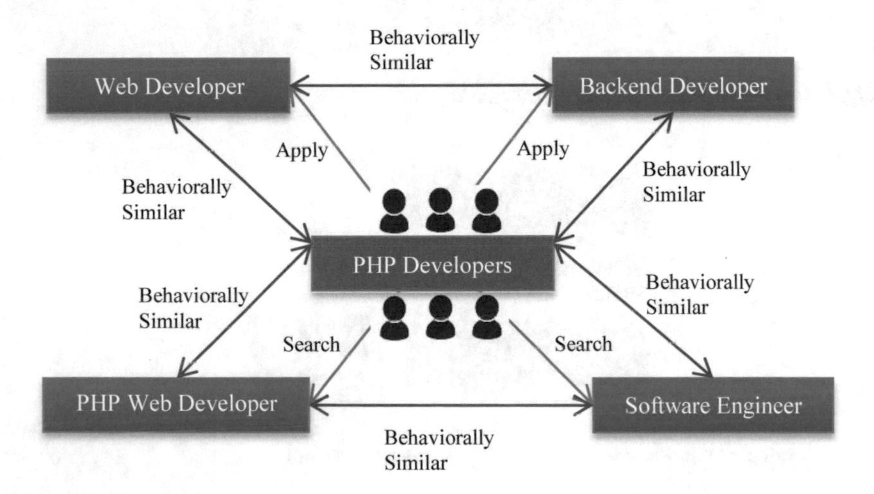

Fig. 6 The combination of all synonyms' partial detection models

Table 5 Results of content-based synonyms partial detector

Keyword	Correlated keywords	Keyword	Correlated keywords
PHP	Information technology (IT)	JSP	Information technology
	Computer science		Java
	Software development		Computer science
	HTML		Software development
	CSS		JavaScript

$$Correlation_{x,y} = D_x \cap D_y \tag{2}$$

$$Correlation_{x,y} = J_{x,y} \tag{3}$$

But this technique is not the most accurate technique to predict the best correlated keyword or term most correlated. Because the most correlated keywords are not simply the most popular keywords in a set. They are the terms that have appeared with the current processed keyword than any other keyword. It is not enough to consider only the times both X and Y appeared together but statistically it is more efficient to consider how many times each X or Y appeared in corpus. It is clear in the first experiment that Information Technology (IT) was the most correlated keyword to PHP and JSP as well because the Information Technology (IT) is a widely popular used keyword that is generic enough to be used in this most of this industry's jobs. Our second experiment followed the Eqs. 4 and 5. The results of our second experiment were far better than the results from our first experiment as it is shown in Table 6. We can see how the results were better by generating Laravel as the most reasonable correlated keyword to PHP and JSF was the most correlated keyword to JSP.

$$Correlation_{x,y} = \frac{D_x \cap D_y}{D_x \cup D_y} \tag{4}$$

$$Correlation_{x,y} = \frac{J_{x,y}}{J_x \mid J_y} \tag{5}$$

Another emerging concept is the word2vec. It is the work of Mikolov at Google on efficient vector representations of words which is firstly published at 2013 [10]. Simply it represents each word in a vector of weights. Traditionally word can be presented in one-hot encoding (1-to-N encoding) where every element in the vector is associated with a word in the vocabulary. The word is simply encoded by producing a vector in which the corresponding element is set to one, and all other elements are zero. For example, if our vocabulary consists of 4 words: - software, engineer, senior and development respectively. Therefore, the vector of word "engineer" would be "0100". The vector is visualized in Table 7.

By using such encodings, there will not be meaningful comparisons that we can conduct between word vectors. Word2vec concept uses a distributed representation of

Table 6 Enhanced results of optimized content-based synonyms partial detector

Keyword	Correlated keywords	Keyword	Correlated keywords
PHP	Laravel	JSP	JSF
	MySQL		Servlets
	CSS		Junit
	HTML		Java EE
	jQuery		J2EE

Table 7 Example of one-hot encoding vector

Vocabulary	Software	Engineer	Senior	Development
	0	1	0	1

Table 8 Example of word2vec encoding vector

	Software	Engineer	Senior	Development
Encoding of 'Business'	0.27	0.32	0.65	0.77
Encoding of 'Backend'	0.63	0.67	0.58	0.85

a word. It builds a multi-dimensional vector, a vector with several hundred dimensions (e.g. 300). Each word is represented by a distribution of weights across all of the 300 dimensions/elements. Now instead of a one-to-one mapping (one-hot) between an element in the vector and a word, the representation of a word becomes spread across all of the elements in the vector, and each element among the vector contributes to the definition of various words (Table 8).

These vectors are constructed based on the contextual co-occurrence of words in the corpus similarly to the previous discussed technique content-based synonyms partial detection. Such vector comes to represent in some abstract way the latent semantic meaning of a word. By examining a large corpus, it is possible to construct word vectors that can capture the relationships between words in an astonishingly meaningful way.

Terms Proximity. If the query has two terms or more, proximity role is about identifying how physically near are those terms in the relevant matched documents. Example if query is *software engineer*, then the engine will retrieve documents which have *software engineer* where its proximity can be considered as zero because it is the same like search query, but it is reasonable to retrieve documents which contains *software development engineer* where its proximity can be considered as one because there is one word "*development*" which is laid between the query terms "*software*" and "*engineer*".

Attribute Importance and Field Weights. The attributes importance criterion identifies the most important matching attribute(s) of the document. Basically, each stored document has many attributes as job titles, job description... etc. So, if user searches for *"operation manager"*, it is meaningful to get jobs have job titles similar to *"operation manager"* in the top list, preceding other jobs that have *"operation manager"* in the job description.

Tokenization. It is the process of splitting corpus based on our defined rules and producing single tokens and terms to achieve an efficient textual matching between job-seeker and job profiles [7]. Our rules are based on tokenizing on whitespaces, character-numeric transition and some special characters. An example of character-numeric transition is "PHP5" and "CSS3" will be tokenized to "PHP" "5" and "CSS"

Table 9 Tokenization example

C/C++		NLP Engineer		PHP5		asp.net		Human Resources (hr)		
C	C++	NLP	Engineer	PHP	5	asp	net	Human	Resources	hr

Table 10 Text normalization of some Arabic characters

Arabic Unnormalized Characters	آ	ا	أ	إ	مُ	ة	ه	ى	ي	ســـــ
Arabic Normalized Characters	ا	ا	ا	ا	م	ه	ه	ى	ى	س

"3", these tokens will match any other profile contains "PHP5", "CSS3", "PHP" and "CSS" without versioning. An example of special characters tokenization is "asp.net" and will be tokenized to (asp*)* (net*)*, these tokens will match any other profile contains "asp.net" "asp" and ".net" (Table 9).

Term Frequency (TF). It means that the most relevant document which has the highest term frequency, term frequency refers to the number of times that term appeared in the document [5–7]. Example like searching for *accountant*, the first retrieved document should be the one which contains the maximum count of word *accountant*.

Term Novelty, Inverse Document Frequency (IDF). It tends to obtain the term's importance. It needs to define a weight for each term according to their importance, so if a term is so common in all documents then it should get a low score than the other terms which more meaningful [5–7]. Example like searching for "Software Engineer at Microsoft Egypt", the term "Egypt" here is meaningless if all jobs are already in Egypt, it will get bad results if the engine emphasized on that term to return the documents which could contain dozens of Egypt. Instead it should return the documents which contain highest occurrences of "Software", "Engineer" and "Vodafone".

Text Normalization. It is used to transform all characters variations to only one variation to simply retrieve the same term in case of writing it in different variation [5–7]. This layer is important for some languages like Arabic. In Arabic, we need to normalize all forms of "Alef" to only one form (آ , ا , أ , إ), the same is applied for all "teh" (ه, ة) and dotless "yeh" (ى, ي). For Arabic, removing any type of diacritics and character stretching is required. Example of diacritics is adding "damma", "shadda", "fatha" or "kasra" like (مُمَثِّل). An example of stretching characters in Arabic as (تطويـــــل) (Table 10).

Stop Words. For gaining meaningful matching results, we had to eliminate the most common useless words that contain no valuable meaning in our context such as the, in, at, it, for, on, is…, etc. Table 11 shows a sample of English and Arabic stop words.

Table 11 Stop words samples in English and Arabic

English Stopwords	Arabic Stopwords
in	فى
from	من
to	الى
on	على
and	و

Capitalization and Case Folding. It is often convenient to normalize and lower case every word's characters for better accurate matching. That will lead to make 'Java', match 'java' and 'JAVA'. Counterexamples include 'IT' vs. 'it' and 'US' vs. 'us' where uppercases words means different latent semantic than the lowercases words. So it is influential to handle that carefully [11].

HTML/Tags Cleansing. Removing unwanted tags and characters is another important cleaning layer [11]. Such layer will empower our engine to eliminate tags such as <hr> html tags to avoid retrieving irrelevant results when searching for *HR* which refers to *Human Resources*. Usually HTML tags are not an important factor in our case-study to keep them but these tags can be important in other applications like considering bold words and the underlined keywords as an important terms or considering hyperlinks to perform page scraping and add the data of this external page.

Stemming. It is the process of transforming the term into its origin form. Aggressive English stemmer (Snowball, Porter) will stem *development*, *developers* into *develop*. Counterexamples, it stems *international* to *intern* and *accountant* to account. K-stemmer, is a light English stemmer that stems *developers* to *developer*.

As it is shown in Table 12 for stemming some words by the two most popular stemmers Porter stemmer (similar to Snowball stemmers) and K-stemmer, we can find that aggressive stemmers like Porter or Snowball success in stemming many e- recruitment terms such as 'development', 'developers', 'developing', 'developer', 'designing', 'designers' and 'designer'. All these terms are stemmed to their origin to match the different variations of the same word. However, these aggressive stemmers aggressively stem some other words unsuccessfully to be the exact stem of other different words not only the different variations of the same word. For example, we find that 'hospitality' from the tourism is stemmed to 'hospit' and 'hospital' from the healthcare industry is also stemmed to 'hospit' term. They are totally different keywords with completely different meaning from significantly different industries. Another failed stemming example is 'engineering' term that is stemmed to 'engin' to match other irrelevant stems like 'engin' the stem of 'engine' term. There is the light English stemmer which is called K-stem. It did not aggressively stem words

Table 12 Stemming some words by Porter stemmer and K-stemmer

English term	Aggressive stemmer	Light stemmer
Development	Develop	Development
Developer	Develop	Developer
Developing	Develop	Develop
Developers	Develop	Developer
Manager	Manag	Manager
Management	Manag	Management
Managing	Manag	Manage
Design	Design	Design
Designing	Design	Designing
Designers	Design	Designer
Country	Countri	Country
Countries	Countri	Country
International	Intern	International
Intern	Intern	Intern
Accountant	Account	Accountant
Accounting	Account	Account
Hospitality	Hospit	Hospitality
Hospital	Hospit	Hospital

that could be possibly matched with irrelevant other words like ‘engineering’ and ‘engine’, ‘international’ and ‘intern’. Nevertheless, light stemmer will not be able to understand that ‘managing’, ‘manager’ and ‘management’ are all referring to the same exact word. K-stem will not be able to stem all these variations to the same exact stem. It means that it is sort of a trade-off between the two kinds of stemmers.

There is no one stemmer that is ultimately better. It depends on the case-study and the platform own data. For the e-recruitment domain, we can either use some kind of aggressive stemmers but protect some keywords from stemming like hospitality or use one of the light stemmers and add group some keywords as synonyms like ‘managing’, ‘manager’ and ‘management’ or customize their stem based on our domain-based knowledge.

There is also a fine light Arabic stemmer that is developed by L. Larkey, L. Ballesteros and M. Connell [12]. It removes many affixes whether prefixes or suffixes. The prefixes like (ال, بال, وال, كال..., etc.) and suffixes like (ات, ة, ين, ون..., etc.). There is another Arabic Porter stemmer that is pretty similar to the Larkey Arabic light stemmer. Both stemmers are different in some minor cases. One of the cases is the words with suffixes ات which refers to plural females in some words like مندوب refers to singular male and مندوبة refers to a singular female while مندوبات refers to plural females. Another example is سكرتير refers to a singular male and سكرتيرة refers to a singular female while سكرتيرات refers to plural females. Larkey Arabic light stemmer stems all of these three variations to one same stem, as it is illustrated in

Table 13 Example of stemming some Arabic words by light stemmer and Porter stemmer

Arabic Term	Light Stemmer	Porter Stemmer
مبرمجين	مبرمج	مبرمج
سواقون	سواق	سواق
سواقين	سواق	سواق
سواقيين	سواق	سواقى
مندوبات	مندوب	مندوبا
مندوبين	مندوب	مندوب
مندوبة	مندوب	مندوب
بالقاهرة	قاهر	قاهر
الامارات	امار	امارا
عامل	عامل	عامل
عمال	عمال	عمال
كاليفورنيا	يفورنيا	يفورنيا

Table 13. Porter Arabic stemmer will handle the plural female differently from the other variations. Porter will not remove the 'ات' like what Arabic light stemmer did.

Text Length Normalization. We previously mentioned that we consider the term frequency in the document (job post) and the term frequency in the corpus (all job posts) to compute the term frequency parameter and term novelty parameter. That makes it significant to consider the content length as well. There are lengthy job posts that could contain the term as many as its appearance in shorty job posts [13, 14]. The assumption here is that shorty job descriptions which have the same term frequency of other lengthy job descriptions focus narrowly more on this term more than the lengthy jobs that focus loosely on this term and other terms

3.5 Relevancy Ranking Models

There are many developed ranking formulas to score the relevant documents. There are Okapi BM25 and different customizations of TF-IDF which are widely involved in search engines. They are ranking functions to rank matching documents according to their relevancy to a given search query. We utilized these equations to score each job and rank them based on their matching level to the job-seeker profile [15–17].

Naive TF-IDF. It is the simplest variation of tf-idf which is defined as the multiplication of term frequency by inverse document frequency [16].

$$tf\text{-}idf = tf(t, d) \cdot idf(t) \tag{6}$$

Where tf(t, d) is the raw frequency of a term in a document, it is the number of times that term *t* occurs in document *d*.

$$tf(t, d) = frequency \tag{7}$$

idf(t, d) is the inverse document frequency, which is the logarithm of dividing total number of documents (n) by the number of documents (n_t) contain the term *t*

$$idf(t) = \log\left(\frac{n}{n_t}\right) \tag{8}$$

Adjusted TF-IDF Scoring Function. More sophisticated tf-idf is clarified below

$$score(q, d) = coord(q, d) \cdot queryNorm(q) \cdot \sum_{t \in q} (tf(t, d) \cdot idf(t)^2 \cdot t.getBoost() \cdot norm(t, d)) \tag{9}$$

coord(q, d) is a score factor based on how many of the query *q* terms are found in the document *d*. The more query terms in a document the higher is the score [17].

queryNorm(q) is just a factor for normalizing, it is used to make scores between queries comparable, but it does not affect individual document score as it is applied to all documents. All ranked documents are multiplied by the same factor

$$queryNorm(q) = \frac{1}{\sqrt{q.getBoost()^2 + \sum_{t \in q} (idf(t) \cdot t.getBoost())^2}} \tag{10}$$

t.getBoost() is a search time boost of term *t* in the query *q*. *tf(t, d)* stands for term *frequency*, it is the number of times the term *t* appears in the currently scored document *d*. The more term occurrences in a document the higher is the score

$$tf(t, d) = \sqrt{frequency} \tag{11}$$

idf(t) correlates inversely to n_t (the number of documents where the term *t* appears). *n* is the number of all stored documents. This means rarer terms will highly influence the total score. This IDF formula is modified in a minor way to handle a term appearing in no documents (hence avoiding a zero denominator)

$$idf(t) = 1 + \log\left(\frac{n}{n_t + 1}\right) \tag{12}$$

norm(t, d) is another normalization function. It is responsible of document boosting, field boosting and document length normalization. *lengthNorm* is computed in accordance with the number of tokens or terms of this field in the document, so shorter fields have higher influence on the score.

$$norm(t, d) = doc.getBoost() \cdot lengthNorm \cdot \prod_{f \in d} f.getBoost() \qquad (13)$$

BM25 Scoring Function. Okapi BM25 is a probabilistic retrieval and ranking model

$$score(q, d) = \sum\nolimits_{t \in q} \left(idf(t) \cdot \frac{tf(t, d).(k_1 + 1)}{tf(t, d) + k_1.\left(1 - b + b.\frac{|d|}{avgdl}\right)} \right) \qquad (14)$$

k_1 is a variable to control non-linear term frequency normalization (saturation), default value is 1.2. In classic Lucene, TF is constantly increasing and never reaches a saturation point, so this modification is adjusted to suppress the impact of term frequency. $|d|$ is the length of the current matching document and *avgdl* is the average stored documents length. These last two variables compute how long a document is relative to the average document length, so relatively longer document than the average length will get lower score than shorter documents. b is another variable to control to what degree document length normalizes TF values, by default it equals to 0.75, it is important to finely tune how much document length influence the scoring function. The inverse document frequency of BM-25 is defined as follow.

$$idf(t) = \log \frac{n - n_t + 0.5}{n_t + 0.5} \qquad (15)$$

TF-IDF vs. BM-25. TF-IDF is a vector space model but BM-25 is defined as one of probabilistic models. However, their concepts are not very different. They both consider some kind of weights for each term as the result of calculating the product of some IDF formula variation and some TF formula variation to produce that term weight as a relevancy score of the current processed document to the given query. However, one of the most important differences between TF-IDF and BM-25 is the saturation when term appears more frequently. They both agree on giving high score for documents contain higher terms occurrences. But the impact of term frequency is always rising, therefore BM25 typically approaches a boundary for high term frequencies. Other classic TF constantly increases and does not reach a reasonable boundary. With exclusion of the document length, BM-25 term frequency formula is defined in Eq. 16. The variable k is often equal to 1.2. It is a variable that can be configured from 1.2 to 2.

$$\frac{tf(t, d) \cdot (k + 1)}{tf(t, d) + k} \qquad (16)$$

Adjusting k is useful to alter the influence of TF, it controls the saturation boundary. Higher k values leads to further saturation reach. Through stretching out the point

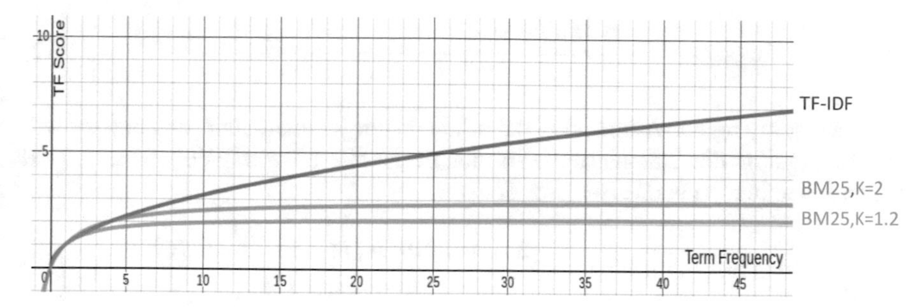

Fig. 7 Term frequency saturation of tf-idf and bm-25

of saturation, it stretches out the relevance score difference between documents with higher TF and other documents with lower TF, as shown in Fig. 7.

4 Results and Deductions

We developed this textual matching framework to measure similarity between job-seeker profile and job post among complete hybrid job recommender system. There are numerous approaches for evaluating recommender systems whether offline metrics such as recall, precision, F1 measure and normalized discounted cumulative gain (nDCG) or online real-world metrics such as click-through rate (CTR), user conversion rate, time to first click and first click rank.

4.1 Data Sample

Our data sample is collected from a real online robust e-recruitment platform. It contains 150,000 jobseekers, in addition to 10,000 unique jobs and 650,000 job applications. We initially evaluated the whole hybrid recommender system via click rank metric. The first recommended job of our developed system got the highest click rate then the second job then the third one.

4.2 Offline Evaluation

Precision. It calculates the fraction of the relevant retrieved results among all the retrieved results. The general rule of precision in information retrieval follows Eq. 17.

$$\text{Precision} = \frac{\{Relevant\ Results\} \cap \{Retrieved\ Results\}}{\{Retrieved\ Results\}} \tag{17}$$

When applying above precision definition on our job matching problem, we will use the equation that calculates the fraction of the relevant retrieved jobs among all the retrieved jobs. It is illustrated in Eq. 18

$$\text{Job Matching Precision} = \frac{\{Relevant\ Jobs\} \cap \{Retrieved\ Jobs\}}{\{Retrieved\ Jobs\}} \tag{18}$$

In statistical hypothesis and confirmatory data analysis, precision refers to the positive predictive value (PPV) which is defined as the count of true positives per the count of true positives and false positives. True positive is the event which the model makes a positive prediction and the actual result is already positive. False positive is the event which the model makes a positive prediction, but the actual result is negative

$$\text{Positive Predictive Value} = \frac{no.of\ true\ positives}{no.of\ true\ positives + no.of\ false\ positives} \tag{19}$$

We will apply this PPV on our practical job matching problem. We call it job matching practical precision value (JMPPV). It follows the standard positive predictive value definition by dividing the true positives (relevant and retrieved jobs) by true positives (relevant and retrieved jobs) and false positives (irrelevant and retrieved jobs).

$$\text{JMPPV} = \frac{Job_{relevant\,\&\,retrieved}}{Job_{relevant\,\&\,retrieved} + Job_{irrelevant\,\&\,retrieved}} \tag{20}$$

Recall. It calculates the fraction of the relevant retrieved results among all the total relevant results. The general rule of recall in information retrieval is defined in Eq. 21.

$$\text{Recall} = \frac{\{Relevant\ Results\} \cap \{Retrieved\ Results\}}{\{Relevant\ Results\}} \tag{21}$$

When applying the above recall definition on our job matching problem, it will lead to the equation that calculates the fraction of the relevant retrieved jobs among all the relevant jobs. It is illustrated in Eq. 22

$$\text{Job Matching Recall} = \frac{\{Relevant\ Jobs\} \cap \{Retrieved\ Jobs\}}{\{Relevant\ Jobs\}} \tag{22}$$

In statistical hypothesis and binary classification, recall refers to the sensitivity which is defined as the count of true positives per the count of true positives and false negatives. True positive is the event which the model makes a positive prediction and the actual result is already positive. False negative is the event which the model makes a false prediction, but the actual result is positive

$$\text{Sensitivity} = \frac{no.of\ true\ positives}{no.of\ true\ positives + no.of\ false\, negatives} \tag{23}$$

We will apply this sensitivity on our practical job matching problem. We call it job matching sensitivity value (JMSV). It is defined as the division of the true positives (relevant and retrieved jobs) by true positives (relevant and retrieved jobs) and false negatives (relevant and unretrieved jobs).

$$\text{JMSV} = \frac{Job_{relevant\&retrieved}}{Job_{relevant\&retrieved} + Job_{relevant\&unretrieved}} \tag{24}$$

F-Measure. It is a formula that combines both of precision and recall to get the benefits of the two approaches and reach a total final score. That f-score refers to the overall model accuracy. F-score is considered as a harmonic mean of precision and recall. Equation 25 shows the general f-measure equation

$$F_{\alpha} = \left(1 + \alpha^2\right) \cdot \frac{precision \cdot recall}{\left(\alpha^2 \cdot precision\right) + recall} \tag{25}$$

One of the most popular variations of f-measure is the F_1 score which weights precision the same as recall by setting α to 1. Equation 26 shows how F_1-score looks like for evaluating job matching problem

$$F_1 = (2) \cdot \frac{\text{Job Matching } precision \cdot \text{Job Matching Recall}}{\text{Job Matching } precision + \text{Job Matching Recall}} \tag{26}$$

There are some other variations like F_2 which weighs recall higher than precision and $F_{0.5}$ which weighs precision higher than recall. Formula of F_2 is illustrated in Eq. 27

$$F_2 = (5) \cdot \frac{precision \cdot recall}{4 \cdot precision + recall} \tag{27}$$

We have other novel metrics that measure the results precision and ranking accuracy such as Precision@K also called P@K. Precision@K tends to measure the precision of the top K results only. It is computed by dividing the relevant results among the top K results by K. For example, if there are 6 relevant jobs in the top retrieved 10 jobs, we can compute the precision of the top 10 recommended results (P@10) by dividing 6 by 10 (6/10). The result of P@10 will equal to 60%. To generalize the idea of P@K, we can compute at different thresholds K and calculate the average of results then compute it for many job-seekers and calculate the mean again over all the recommended jobs for each job-seeker. This generalization methodology for P@K is called Mean Average Precision (MAP). For example, based on our case-study that can be achieved by looping on each job-seeker J and calculate the P@2, P@4, P@8, P@10… and P@20, then compute the MAP for each job-seeker MAP_J by calculating (P@2 + P@4 + P@8… + P@20)/10. Now we get the mean average precision of the recommended jobs per each job-seeker MAP_J. Finally, we need to compute the mean average precision of the recommended jobs over all the job-seekers. If we have 2000 job-seekers, then we can compute the mean average precision of the recommended jobs over all the job-seekers by calculating (MAP_{J1} + MAP_{J2} + MAP_{J3} … + MAP_{J2000})/2000.

All these metrics are powerful and efficient in many applications. However, these evaluation metrics need some sort of binary rating applications to work well and scale smoothly where users have the ability to like and dislike the recommended result. Certainly, there are some e-recruitment platforms adopt this type of rating. In our case, we do not have that type of binary classification in our platform. It is even more like a unary rating platform. The only data we have about users" interest is the applying behavior. When job-seeker applies to job, it means the job- seeker finds this job relevant. This data is not enough to evaluate by precision, recall and f-measure accurately. The other solution to use these metrics is to run the similarity engine on a bunch of job-seekers and manually annotate the results by relevant or irrelevant. This solution does not scale, hard to achieve and inefficient.

Offline evaluations typically evaluate the accuracy of a recommendation model based on a ground-truth, not only the accuracy but also the novelty or serendipity of recommendations. Offline evaluations were primarily meant to recognize number of potential recommendation techniques. Then these techniques should be evaluated particularly by a user study or online evaluation to define the most efficient techniques. However, recently we find many critiques towards the assumption that offline evaluation can predict the approach's effectiveness in the real-world online evaluations and user studies [18]. There are many researchers researched this topic and showed that results from offline evaluations do not necessarily correlate with results from online evaluations [19, 20]. They stated that approaches which are efficient in offline evaluations are not necessarily efficient in real-world recommender engines.

4.3 Online Evaluation

Because of the lack of accuracy of offline evaluations to evaluate the users' satisfaction based on the above reviews and discussion, we tended to adopt the online and real-world evaluations that mostly represent a better methods to evaluate such platforms and correlate well with the users' satisfaction.

There are many online and real-world business metrics such user conversion rate, click-through rate (CTR), time to first click, first click rank. One of the important

Fig. 8 Click rank on recommended job by real online job-seekers

Event Label	Total Events
	6,610 % of Total: 2.24% (294,861)
1. 1	1,736 (26.26%)
2. 2	994 (15.04%)
3. 3	771 (11.66%)
4. 4	580 (8.77%)
5. 5	476 (7.20%)
6. 10	445 (6.73%)
7. 6	436 (6.60%)
8. 7	397 (6.01%)
9. 8	393 (5.95%)
10. 9	382 (5.78%)

evaluations for such recommendation systems is measuring the percentage of apply actions on recommended jobs relatively to the total apply actions. But you have to define a baseline for such metric to decide when the engine is working well and when it is not. We evaluated the whole recommendation engine via first click rank and click rank metric which refer to the rank of the clicked jobs. We evaluated it by using Google Analytics to count the clicks on each job in the list. The result is that the first recommended job got the highest click rate then the second then the third job, the full list of clicks rank on the top 10 results is shown in Fig. 8.

4.4 Experimental Evaluation

Our experimental results to this standalone textual matching layer and ranking technique lead to the following desired deductions: - jobs matching more query terms of the job-seeker's keywords are getting higher score than jobs matching less query terms. Jobs having more occurrences of a query term (job-seeker's keywords) are ranked higher than jobs with fewer occurrences of terms. Jobs containing more novel terms have higher score than jobs containing common terms, rare terms have more significance than the common terms. Because of using the documents' length normalization, shorter job posts having the same occurrences of query terms are getting higher score than lengthy job posts, it means this job concentrates more on these terms (skills). Jobs including query terms in the job title have better score than other jobs including terms in less significant fields such as job description (terms existence in more important fields is more significant).

5 Conclusion

We presented the textual similarity engine that computes the similarity between two dynamically constructed profiles whether the text is gathered content-based or behavioral-based. We optimized this engine particularly for the e-recruitment field. It is utilized to recommend the most relevant jobs for job-seekers by matching job-seeker profiles to job posts. By developing such engine, we can reach a powerful intelligent e-recruitment platform that understands job-seekers' experiences, capabilities and preferences to recommend the most suitable and interesting jobs. The objective of this project is to achieve the higher level of users' satisfactions. We worked on serving our purpose by efficiently understanding the conceptual methodologies for matching query with a set of documents and how can we apply these abstract methodologies on our recommendation problem. We built the text analysis and mining layers which perform a lot of text mining tasks. For example, these layers emphasize more on the novel terms in profiles, ignore useless words, remove stop words, normalize some unnormalized texts especially in Arabic language, remove diacritics and find the important synonyms and correlated keywords. After preparing the vital keywords from job-seeker profile and job post, we utilized the studied conceptual methodologies for matching query with a set of documents which is the vector space model. We considered the job-seeker's keywords as the query vector and manipulated job posts as our desired set of documents that are represented in vectors as well. Wc analyzed the various well- known similarity equations such as the basic TF-IDF, customized variations of TF- IDF and BM-25. They are important equations for ranking the recommended results based on their relevancy score to the job-seeker profile. Eventually we illustrated the needed metrics to evaluate such engine and how abstract well-known evaluation criteria can be utilized to serve the evaluation of similar engine in the e-recruitment industry. The evaluation of this textual matching framework among complete hybrid job engine proved its efficiency by the click-through rank of real online users who used this functionality extensively.

References

1. Liu, J., Dolan, P., Pedersen, E.R.: Personalized news recommendation based on click behavior. In: Proceedings of the 15th International Conference on Intelligent User Interfaces, Hong Kong, China, pp. 31–40 (2009)
2. Lu, Y., El Helou, S., Gillet, D.: A recommender system for job seeking and recruiting website. In: Proceedings of the 22nd International Conference on World Wide Web Companion, Rio de Janeiro, Brazil, pp. 963–966 (2013)
3. Rafter, R., Bradley, K., Smyth, B.: Automated collaborative filtering applications for online recruitment services. In: Proceeding of the International Conference on Adaptive Hypermedia and Adaptive Web-Based Systems, pp. 363–368 (2000)
4. AlJadda, K., Korayem, M., Ortiz, C., Russell, C., Bernal, D., Payson, L., Brown, S., Grainger, T.: Augmenting recommendation systems using a model of semantically-related terms extracted from user behavior. In: Proceeding of the Second CrowdRec Workshop. ACM RecSys, Austria (2014)

5. Büttcher, S., Clarke, C.L.A., Cormack, G.V.: Information retrieval: implementing and evaluating search engines (2016)
6. Krishnamurthy, S., Akila, V.: Information Retrieval Models: Trends and Techniques, Web Semantics for Textual and Visual Information Retrieval, Chapter 2, pp. 17–42 (2017)
7. Smiley, D., Pugh, E., Parisa, K., Mitchell, M.: Apache Solr Enterprise Search Server, 3rd edn. Packt publishing, Birmingham (2015)
8. AlJadda, K., Korayem, M., Grainger, T., Russell, C.: Crowdsourced query augmentation through semantic discovery of domain-specific Jargon. In: Proceeding of IEEE International Conference on Big Data (Big Data), USA (2014)
9. Miller, G.A.: WordNet: a lexical database for English (1995)
10. Mikolov, T., Chen, K., Corrado, G., Dean, J.: Efficient estimation of word representations in vector space. arXiv: 1301.3781 (2013)
11. Jayalakshmi, T., Chethana, C.: A semantic search engine for indexing and retrieval of relevant text documents. Int. J. Adv. Res. Comput. Sci. Manage. Stud. (IJARCSMS) **4**
12. Larkey, L.S., Ballesteros, L., Connell, M.E.: Light stemming for arabic information retrieval. In: Arabic Computational Morphology, Part of the Text, Speech and Language Technology Book Series, vol. 38, pp 221–243 (2007)
13. Singhal, A., Buckley, C., Mitra, M.: Pivoted document length normalization. ACM SIGIR Forum - SIGIR Test-of-Time Awardees, vol. 51, no. 2, pp. 176–184 (2017)
14. Singhal, A., Salton, G., Mitra, M., Buckley, C.: Document length normalization. Inf. Process. Manage. **32**(5), 619–633 (1996)
15. Ramos, J.: Using TF-IDF to Determine Word Relevance in Document Queries (2003)
16. Nayrolles, M.: Relevancy and scoring mechanisms. In: Mastering Apache Solr: A Practical Guide to Get to Grips with Apache Solr, Chapter 4, pp. 60–62 (2014)
17. Shahi, D.: Solr scoring. In: Apache Solr: A Practical Approach to Enterprise Search, Chapter 8, pp. 189–207 (2015)
18. Beel, J., Langer, S.: A comparison of offline evaluations, online evaluations, and user studies in the context of research-paper recommender systems. In: Proceeding of the 19th International Conference on Theory and Practice of Digital Libraries, Poland (2015)
19. McNee, S.M., Albert, I., Cosley, D., Gopalkrishnan, P., Lam, S.K., Rashid, A.M., Konstan, J.A., Riedl, J.: On the recommending of citations for research papers. In: Proceedings of the ACM Conference on Computer Supported Cooperative Work, pp. 116– 125 (2002)
20. Turpin, H., Hersh, W.: Why batch and user evaluations do not give the same results. In: Proceedings of the 24th Annual International ACM SIGIR Conference on Research and Development in Information Retrieval, pp. 225–231 (2001)

Printed in the United States
by Baker & Taylor Publisher Services